INTERPRETING MACH

This volume presents new essays on the work and thought of physicist, psychologist, and philosopher Ernst Mach. Moving away from previous estimations of Mach as a pre-logical positivist, the essays reflect his rehabilitation as a thinker of direct relevance to debates in the contemporary philosophies of natural science, psychology, metaphysics, and mind. Topics covered include Mach's work on acoustical psychophysics and physics; his ideas on analogy and the principle of conservation of energy; the correct interpretation of his scheme of 'elements' and its relationship to his 'historical-critical' method; the relationship of his thought to movements such as American pragmatism, realism, and neutral monism, as well as to contemporary figures such as Friedrich Nietzsche; and the reception and influence of his works in Germany and Austria, particularly by the Vienna Circle.

JOHN PRESTON is Professor of Philosophy at the University of Reading. He is the author of *Feyerabend: Philosophy, Science and Society* (1997) and *Kuhn's The Structure of Scientific Revolutions: A Reader's Guide* (2008). He edited the third volume of Feyerabend's *Philosophical Papers* (1999) and co-edited *The Worst Enemy of Science?: Essays in Memory of Paul Feyerabend* (with Gonzalo Munévar and David Lamb, 2000).

INTERPRETING MACH

Critical Essays

Edited by

JOHN PRESTON
University of Reading

CAMBRIDGE
UNIVERSITY PRESS

CAMBRIDGE
UNIVERSITY PRESS

University Printing House, Cambridge CB2 8BS, United Kingdom

One Liberty Plaza, 20th Floor, New York, NY 10006, USA

477 Williamstown Road, Port Melbourne, VIC 3207, Australia

314–321, 3rd Floor, Plot 3, Splendor Forum, Jasola District Centre, New Delhi – 110025, India

79 Anson Road, #06-04/06, Singapore 079906

Cambridge University Press is part of the University of Cambridge.

It furthers the University's mission by disseminating knowledge in the pursuit of education, learning, and research at the highest international levels of excellence.

www.cambridge.org
Information on this title: www.cambridge.org/9781108474016
DOI: 10.1017/9781108564311

First published 2021

A catalogue record for this publication is available from the British Library.

Library of Congress Cataloging-in-Publication Data
Names: Preston, John, 1957– editor.
Title: Interpreting Mach : critical essays / edited by John Preston, University of Reading.
Description: New York : Cambridge University Press, 2021. | Includes bibliographical references and index.
Identifiers: LCCN 2020041974 (print) | LCCN 2020041975 (ebook) | ISBN 9781108474016 (hardback) | ISBN 9781108463287 (paperback) | ISBN 9781108564311 (epub)
Subjects: LCSH: Mach, Ernst, 1838-1916.
Classification: LCC B3303 .I58 2021 (print) | LCC B3303 (ebook) | DDC 193–dc23
LC record available at https://lccn.loc.gov/2020041974
LC ebook record available at https://lccn.loc.gov/2020041975

ISBN 978-1-108-47401-6 Hardback

For Erik C. Banks
(1970–2017)

CONTENTS

FIGURES

TABLES

CONTRIBUTORS

†Erik C. Banks was a professor of philosophy at Wright State University, Ohio, the United States.

Pietro Gori is a researcher and invited professor in the Institute of Philosophy / Faculty of Social and Human Sciences at the New University of Lisbon, Portugal.

Luca Guzzardi is an assistant professor of philosophy of science in the Department of Philosophy at the University of Milan, Italy.

Alexandra Hui is an associate professor in the Department of History at Mississippi State University, the United States.

Alexander Klein is Canada Research Chair and associate professor of philosophy at McMaster University, Canada.

Lydia Patton is a professor in the Department of Philosophy at Virginia Tech, the United States.

John Preston is a professor in the Department of Philosophy at the University of Reading, the United Kingdom.

Friedrich Stadler is a professor emeritus of history and philosophy of science at the University of Vienna, Austria.

Richard Staley is a reader in the Department of History and Philosophy of Science at the University of Cambridge, the United Kingdom.

S. G. Sterrett is the Curtis D. Gridley Distinguished Professor of History and Philosophy of Science in the Department of Philosophy at Wichita State University, Kansas, the United States.

Michael Stöltzner is a professor in the Department of Philosophy at the University of South Carolina, the United States.

Thomas Uebel is Professor Emeritus in the Department of Philosophy at the University of Manchester, United Kingdom.

Daan Wegener is a lecturer in history of science at Utrecht University, the Netherlands.

~

Introduction

A New Mach for a New Millennium

JOHN PRESTON

Mach's Career and Principal Publications

Ernst Mach is, in my view, one of the greatest of a very significant and original group of thinkers: the philosophising scientists of the late nineteenth and early twentieth centuries. Although Mach did not think of himself as a philosopher, it is the more philosophical parts of his published works that continue to attract the most attention. Mach's ideas, their reception, and the trajectory of Mach scholarship since the late nineteenth century unfolded as follows.

Mach studied physics at the University of Vienna from 1855, becoming skilled in the techniques of the physical laboratory there. He received his doctorate in 1860, taking up his first professional appointment in physics in 1861. His first published book was a compendium of physics for medical professionals, but because Mach had by then developed an interest in physiology and psychology, it was followed in the same year (1863) by a series of lectures on psychophysics. His experimental investigations covered a variety of phenomena, including the Doppler effect, the measurement of blood pressure, and acoustics.

In 1864, Mach moved to take up a position at the University of Graz, where he again taught a variety of courses, including psychophysics, having been directly influenced by the founder of that discipline, Gustav Theodor Fechner. He gave many popular-scientific lectures, at least some of which were to girls and women at Graz's recently established *Mädchenlyzeum* ('Lyceum for Girls'). Over the next few years, he published a series of papers setting out and investigating the phenomena we now call 'Mach bands' (the apparently paradoxical exaggeration of the contrast between the edges of patches of different shades of grey that are next to one another). And as a result of his investigations into the physiology of sound sensation, especially in musical appreciation, he published *Einleitung in die Helmholtz'sche Musiktheorie: Populär für Musiker dargestellt* (*Introduction to Helmholtz's Music Theory, Popularly Represented for Musicians*) in 1866.

By 1867, though, he had moved on to Prague, as a professor of experimental physics. His groundbreaking studies of motion perception were published as

1

Grundlinien der Lehre von den Bewegungsempfindungen (*Fundamentals of the Theory of Movement Perception*) in 1875.

While still in Prague, Mach published a short book, *Die Geschichte und die Wurzel des Satzes von der Erhaltung der Arbeit* (1872, translated as *History and Root of the Principle of the Conservation of Energy*). Amply philosophical, this served to present many of his most important ideas in embryo. It can be thought of as the first in his series of distinctive 'historical-critical' studies, which aim to exhibit the historical contingency of the state of science in Mach's time, as well as the presumptuousness of the 'mechanical world view' which Mach, by that time, had come to be so sceptical about.

Mach's background in ballistic experimentation made him well-placed to conduct a series of experiments on shock waves during the second half of the 1870s and much of the 1880s. From these experiments we ultimately derive several items of physical terminology, including our way of referring to the speed of sound as 'Mach 1'. His remarkable study of the history of mechanics, *Die Mechanik in ihrer Entwickelung historisch-kritish dargestellt* (translated as *The Science of Mechanics: A Critical and Historical Account of Its Development*), the first of his books to include a subtitle referring to his 'historical-critical' method of exposition and critique, was published in 1883. It featured Mach's full critique of Isaac Newton's conceptions of space and time.

Although Mach had shelved his own manuscript *Beiträge zur Analyse der Empfindungen* (*Contributions to the Analysis of the Sensations*) for twenty years after Fechner had reacted badly to it, it was published in 1886, and later fleshed out to become a much larger volume, *The Analysis of Sensations and the Relation of the Physical to the Psychical*. Addressed principally to biologists and aiming at showing how physics and the physiological sciences could learn from one another, it has become Mach's best-known book, although it must be said that this has somewhat distorted his reputation in the process.

Problems in Prague, together with the suicide of one of his sons, dictated a move for Mach, and in 1895 he returned to Vienna, where he took up a new chair as Professor of the History and Philosophy of the Inductive Sciences. His *Popular Scientific Lectures*, testifying to the vast range of his interests and to his great pedagogical abilities, were published in 1895.

From the 1860s onwards, the tireless work of Ludwig Boltzmann had vastly increased the reputation of statistical mechanics, the kinetic theory of gases, and the atomic hypothesis. Mach's opposition to the mechanical world view and to the atomism associated with it engaged him in controversy with Boltzmann and leading figures in physics at his own alma mater, the University of Vienna. His 1896 book *Die Prinzipien der Wärmelehre, historisch-kritisch entwickelt* (translated as *Principles of the Theory of Heat: Historically and Critically Elucidated*, but not until 1986), which Mach's biographer John Blackmore memorably called 'probably the only book ever written on thermodynamics with over a dozen chapters overtly on philosophy', was largely directed against Boltzmann.

Mach's career as an experimenter and lecturer was cut short in 1898 when he suffered a stroke which partially paralyzed him. But he continued to work, revising his published books for new editions and incorporating some of his lectures on philosophy of science into his 1905 book *Erkenntnis und Irrtum: Skizzen zur Psychologie der Forschung* (translated as *Knowledge and Error: Sketches on the Psychology of Enquiry*), which he thought of as the mature statement of his epistemology.

Although Boltzmann's suicide in 1906 meant the loss of Mach's most illustrious opponent, the Berlin physicist Max Planck initiated a fierce debate with him towards the end of the twentieth century's first decade. Mach found himself forced to defend his widely based epistemological stance against Planck's rival conception of an acceptable scientific world view, a conception derived much more exclusively from contemporary physics.

Mach died in 1916, but his book *The Principles of Physical Optics* was seen through to publication by his son, Ludwig, in 1921.

In recent years, Mach's main books have also been republished in German (by Xenomoi in Berlin: www.xenomoi.de/philosophie/mach-ernst), with new introductions, in the series *Ernst Mach Studienausgabe*, edited by Friedrich Stadler, together with Michael Heidelberger, Dieter Hoffmann, Elisabeth Nemeth, Wolfgang Reiter, Jürgen Renn, and Gereon Wolters. The volumes are:

> *Die Analyse der Empfindungen und das Verhältnis des Physischen zum Psychischen* (1886), edited by Gereon Wolters (2008).
> *Erkenntnis und Irrtum* (1905), edited by Elisabeth Nemeth and Friedrich Stadler (2011).
> *Die Mechanik in ihrer Entwickelung. Historisch-kritisch dargestellt* (1883), edited by Gereon Wolters and Giora Hon (2012).
> *Populärwissenschaftliche Vorlesungen* (1896), edited by Elisabeth Nemeth and Friedrich Stadler (2014).
> *Die Prinzipien der Wärmelehre* (1896), edited by Michael Heidelberger and Wolfgang Reiter (2016).
> *Die Prinzipien der physikalischen Optik* (1921), edited by Dieter Hoffmann and Josef Pircher (2020).

Mach's Influence

Mach was already enormously influential across scholarly fields and cultural fields during his own lifetime. His works evoked reactions from fellow scientist–philosophers such as Heinrich Hertz, Wilhelm Ostwald, Georg Helm, Ludwig Boltzmann, William James, Sigmund Freud, Charles Peirce,

Oswald Külpe, Jacques Loeb, Pierre Duhem, František Wald, Karl Pearson, W. K. Clifford, W. S. Jevons, Bronisław Malinowski, Ewald Hering, Carl Stumpf, Max Planck, and Albert Einstein. The strength of Mach's influence among Russian thinkers was testified to by the fact that V. I. Lenin felt it necessary to critique it in his 1909 book *Materialism and Empirio-Criticism* in order to ensure that Russian Marxists should not be tempted by a Machian perspective which, as a thoroughly modern and scientifically informed perspective, represented a serious competitor to 'dialectical materialism'. At the same time, in more purely academic philosophical and cultural circles, Mach was influential on figures such as Edmund Husserl, Ernst Cassirer, Paul Carus, Hans Kleinpeter, Richard Avenarius, Wilhelm Jerusalem, Wilhelm Schuppe, Heinrich Gomperz, Friedrich Adler, Joseph Petzoldt, Hans Cornelius, Philip Jourdain, and Robert Musil.

In the twentieth century's second decade, following Mach's death, his philosophical work influenced two new groups of thinkers. The first group featured two of the founders of analytical philosophy: Bertrand Russell and Ludwig Wittgenstein. The second group, working in the second half of that decade, included the figures who would go on to form the Vienna Circle, perhaps most notably Moritz Schlick and Philipp Frank. It was thinkers from that Circle who, taking Mach as one of their central inspirations, cemented his reputation in the history of philosophy and established what we might think of as the 'received view' of his philosophical works. Because of the new ways in which logic and philosophy were themselves being conceived, along with an antipathy to 'psychologism' and an increasing tendency to think of science in formal terms, this received view portrayed Mach as something like a paradigm case of a *pre*-logical positivist. Even outside the more narrow confines of the Vienna Circle, though, Mach also had other admirers and defenders during this same era, such as P. W. Bridgman, B. F. Skinner, Robert Bouvier, Hugo Dingler, C. B. Weinberg, and Richard von Mises.

It was positivist readings of Mach which went on to dominate what we might think of as a third phase in the reception of his ideas, a phase beginning in the 1950s and featuring thinkers who reacted strongly against positivism, such as Karl Popper (and his followers), Gerald Holton, Francis Seaman, Peter Alexander, and John Blackmore. These readers thought of Mach as a paradigm phenomenalist, sensationalist, foundationalist, and instrumentalist, and for them such views betokened the untenability of his philosophy. In this era (1950s–1970s), Erwin Hiebert, Stephen G. Brush, Wolfram Swoboda, Floyd Ratliff, Otto Blüh, and Larry Laudan are notable examples of those who were beginning to develop a more historically informed and sympathetic take on Mach.

The complexion of English-language Mach studies began to change more rapidly in the 1970s. Paul Feyerabend, having rediscovered in the mid-1970s the Mach he had read significantly earlier, began resuscitating his reputation.

In the full flow of his own renunciation of Popperian views, Feyerabend took particular aim at accounts of Mach produced by followers of Popper and Imre Lakatos. His earliest published paper on this subject also includes a nod to work on Mach by Laudan, a contribution of whose in the mid-1970s might be grouped with Feyerabend's in this respect.

Since then, there has been something of a flowering of Mach scholarship, with Feyerabend's increasing encouragements lying alongside continuing Mach-related publications in the 1980s from Blackmore, Gerald Holton, and Erwin Hiebert, but being joined by important work from new figures such as Gereon Wolters, Rudolf Haller, Klaus Hentschel, Henk Visser, Michael Matthews, John Norton, Aldo Gargani, and Brian McGuinness, who, having brought Mach's *Erkenntnis und Irrtum* to publication for an English-speaking audience in 1976, did the same for his *Wärmelehre* a decade later. In the 1990s, the new voices included Friedrich Stadler, Andy Hamilton, S. G. Sterrett, Julian Barbour, Ursula Baatz, Michael Stöltzner, and, in the first decade of the twenty-first century, Dario Antiseri, Karl Hayo Siemsen and his son Hayo Siemsen, Paul Pojman, Michael Heidelberger, Jaakko Hintikka, Robert DiSalle, and Gary Hatfield.

The most recent 'turn' in Mach scholarship, though, can be thought of as largely due to the work of Erik C. Banks, to whose memory this volume is dedicated. Erik received his undergraduate degree from Bennington College, Vermont, and his PhD from the City University of New York in 2000 for his thesis 'Ernst Mach's World Elements', with Arnold Koslow as his dissertation advisor. This was based on his study of Mach's *Nachlass* at the Deutsches Museum, Munich, during the summer of 1999. It was subsequently published as his first monograph, under that same title (Banks 2003). In 2006, Erik joined the faculty at Wright State University, Ohio. He published several very important articles on Mach, and in 2014 his second monograph, *The Realistic Empiricism of Mach, James, and Russell: Neutral Monism Reconceived*, was published by Cambridge University Press.

Erik's work as a whole situates Mach firmly in the history of the sciences to which Mach contributed (notably physics, physiology, and psychology), as well in the history of philosophy. It also displays great potential for showing how Mach's work might contribute to contemporary debates in, for example, metaphilosophy and naturalistic metaphysics, philosophy of mind, and philosophy of science (integrating history and philosophy of science).

Erik died unexpectedly in August 2017. He was originally my co-editor on this volume, and I am deeply grateful to him for his inspiration and for all of the work he put into it. A small indicator of his great dedication to Mach and to the production of this volume is the fact that he had already finished his chapter for the volume, more than three years before it goes to press.

Erik's work undoubtedly contributed to the recent renaissance of interest in Mach. The Ernst Mach Centenary Conference of June 2016, organised by the

Vienna Circle Institute, University of Vienna, and the Austrian Academy of Sciences, featured a host of scholars of many aspects of Mach's work. Its very substantial proceedings were published in 2019 in two volumes edited by Friedrich Stadler.

The Volume's Papers

In this volume, Alexandra Hui concentrates on Mach's work in psychophysics, surveying the many ways in which, using a great variety of resources, including multimodal descriptions, he described psychophysical experiences. She compares the ways in which he discussed psychophysics in his popular lectures, scientific publications, and personal correspondence, and she argues that Mach's celebrated techniques of presentation and argumentation went far towards cultivating a psychophysical imaginarium, a space devoted to the cultivation of the imagination.

Richard Staley, too, focuses on Mach's work in psychophysics, showing how that dovetailed with his activity in physics. He takes up Albert Einstein's remark that the influence of Mach's writings on the evolution of the sciences was such that even his opponents were unaware how much of Mach's thinking they had absorbed 'with their mother's milk'. Staley argues that Mach's studies of sense perception sought to recover basic perceptual experience sensitive to the relations between different sensations of space and time, helping to initiate Mach's work on physical space and time. He shows how Mach's studies of mass, action and reaction, and inertia were not only conceptually linked, but also derived and emerged from perceptual studies and bodily experience, exhibiting the way in which Mach's ambition was to find an epistemology which could examine *all* human experience.

Daan Wegener's chapter addresses various aspects of the nineteenth-century history of energy conservation and emphasises the centrality of the law of conservation of energy within Mach's work, but also connects that law with more general concerns in Mach's philosophy of science via his distinctive take on the nature of concepts and their meaning.

Mach's *Knowledge and Error* includes much material on the idea of analogy, prefigured by an important article of 1902. These discussions of analogy in ordinary life and in natural science are the subject of the chapter by S. G. Sterrett. She shows how rich and subtle was Mach's account, which distinguished analogy from similarity in a way that carried over to the natural sciences, and also in a way which brings to prominence the *value* of analogy, which is now being rediscovered in disciplines such as archaeology. For Mach, analogy in natural science is a relation between systems of concepts, and the use of this method can be powerful in extending knowledge, since science is more variegated than just scientific enquiry into the unknown. Where opportunities to use analogy are wasted, the scientist can be left clinging to an inadequate theory.

How Mach stood with respect to both the American Pragmatist tradition and, more widely, the very idea of a pragmatist philosophy has been a subject of renewed interest in recent years. In this volume, Thomas Uebel's contribution argues that Mach's historicist naturalism can be considered an original form of pragmatism. For Uebel, Mach's principle of scientific significance and its background conception of knowledge deeply exemplify the rejection of scepticism and acceptance of fallibilism which characterise pragmatism.

Focusing more specifically on the relationship between Mach and one of the leading pragmatist thinkers, William James, Alexander Klein shows that this influence ran both ways. In experimental matters concerning volition and the feeling of muscular effort, it was James who changed Mach's mind. Mach, for his part, might have exerted a modest influence on James's philosophical outlook but, surprisingly, philosophical issues only motivated a small fraction of their intellectual exchanges.

The relationship between Mach's thought and that of an apparently more intellectually distant near-contemporary, Friedrich Nietzsche, has also come under scrutiny in recent years. Pietro Gori here provides a thorough account of this association, arguing that the consistency of their views is substantial. Despite their interests being different, both Mach and Nietzsche were concerned with the same issues about our intellectual relationship with the external world, dealing with the same questions and pursuing a common aim of eliminating worn-out philosophical conceptions. Gori shows that not only did they converge on what we now know as the problem of realism versus anti-realism in the philosophy of science, but also they both rejected 'representational' (realist) conceptions of science in favour of a certain sort of pragmatic anti-realism, whose focus was on the role science plays as a means of orientation.

Pragmatism also figures here in the contribution of Lydia Patton, who presses for a far fuller and more robust understanding of Mach's notion of the 'economy of science' than the one associated with the received view. She argues that Machian 'economy' appeals not only to the continuity between scientific experiences and concepts, but also to the increasing complexity of scientific concepts, emphasising both continuities between experiences on the one hand and areas of divergence that promote the branching of scientific concepts and methods on the other. She examines the roles of abstraction, pragmatism, and history in Mach's economy of science, arguing that his overarching concern, in accounting for the role of the scientist in the economy of science, is with what she calls the pragmatic history of the experiencing and creative knower, rather than with exclusive, reductive phenomenological or biological explanations.

In a related vein, Luca Guzzardi here analyses the relation that two important doctrines of Mach's epistemology – his scheme of 'elements' and his idea of the economy of thought – bear to his 'historical-critical' approach. He argues, not only against the received view, but also going further than scholars such as Erwin Hiebert, that there is a more profound, structural relationship between Mach's conception of history and the anti-metaphysical remarks which open *The Analysis of Sensations* and introduce his scheme of elements. Guzzardi also proposes that recognising this can afford novel insights into Mach's doctrine of the economy of thought.

The ways in which Mach's work was understood and used by the Vienna Circle are examined in this volume first by Friedrich Stadler. Stadler begins by detailing the reception of Mach in what Rudolf Haller called the 'first Vienna Circle' (1907–1912), then moving on to its reception by the Vienna Circle proper (1924–1936). With a wealth of material, he shows how Mach influenced nearly all the members of the Circle in his capacity as an empiricist, a critic of metaphysics, and an advocate of the unity of science. But Stadler also argues that Mach's influence is somewhat bifurcated, with certain logical empiricists admiring him for overcoming an old-fashioned aprioristic philosophy and for reformulating an anti-Kantian empiricism, while others were more appreciative of his epistemological and methodological incentives and his conception of the unity of science.

Michael Stöltzner then compares the reception of Mach's work in Austria with its reception in Germany, arguing that the stricter, German interpretation of Mach finally prevailed. He points to a discrepancy between Mach's own conceptions of causality and natural laws and the less generous ones then popular among German physicist–philosophers, who also tended to think of Mach's scheme of elements as phenomenalism. In the light of the new physics of relativity and quantum theory, both based upon abstract principles, Mach's attempt to supply a physiological foundation of physical quantities appeared unattractive, and his epistemological principles (such as economy) were stripped of their biological and physiological bases. The Vienna Circle's eventual choice of the name 'Logical Empiricism' over 'Logical Positivism' suggested that positivism (associated with phenomenalism) had been superseded by the broader framework of empiricism, albeit a framework which still allowed the figures in question to pay tribute to Mach in a general way.

The first chapter of Mach's book *The Analysis of Sensations* (originally published in 1886, but then again in new and much-expanded editions up until 1906) undoubtedly formed much of the basis of the 'received view' and its association of Mach with phenomenalism, sensationalism, and positivism, and it has been something of a flypaper for Mach's critics. In his paper here, John Preston examines in particular the opening chapters of that book and tries to show that there is scope to read it in a non-phenomenalist way. The project of Mach's book, he argues, was to supply something which would allow the unification of physics, physiology, psychophysics, and psychology. The resulting 'scheme of elements', inspired by Mach's work in psychophysics, was meant to identify 'elements' which stand in relations that both the physical sciences *and* psychological sciences could examine. Mach's monism, on this account, is a proposal from within science that offers a non-metaphysical ontology, an ontology which allows very different kinds of science to investigate epistemically accessible features of reality.

Erik C. Banks did a great deal to revive the way of thinking about Mach's work that Bertrand Russell and William James shared – the idea of 'neutral monism' – and his chapter for this volume builds on that proposal, showing along the way the inferiority of the received view of Mach as a crude positivist. Erik argues that this received view is a myth, and that Mach should rather be counted as an ancestor of the kinds of realism developed by the American 'New Realists' in the early twentieth century, which then developed in the work of Wilfrid Sellars, Grover Maxwell, and Herbert Feigl. Erik's chapter ends by ranging a non-panpsychist version of this 'Russellian monist' perspective against the work of contemporary figures such as Saul Kripke, David Chalmers, and Galen Strawson, and envisaging a revival of Mach's ideas that would parallel the way in which Russell's ideas in the philosophy of mind have recently been revived.

Reference

Banks, Erik C. 2003. *Ernst Mach's World Elements: A Study in Natural Philosophy.* Kluwer Academic Publishers.

Ernst Mach's Piano and the Making of a Psychophysical Imaginarium

ALEXANDRA HUI

... we have to complete observed facts by analogy.

—Mach (1886, p. 13)

In the summer of 2018, I revisited Professor Mach. It had been over a decade since I'd made my way up the wide stairs and down the halls of the Deutsches Museum in Munich to the light-filled reading room of the museum's archives. This time, I passed Hermann von Helmholtz's Steinway piano on the stair landing. The archives had also acquired more of Ernst Mach's unpublished writings. I was eager to examine these, and I found re-engaging with the materials I had studied so carefully before to be something like visiting an old friend. Here were Mach's careful drawings of the inner ear bones. There was his quickly jotted recipe for risotto Milanese.

In the ten years since I had completed that project on nineteenth-century psychoacoustics and music, the history of science scholarship has crystallised its engagement with how things, phenomena, and concepts become objects of enquiry. Scholars frame this as how scientific objects come into being (Rheinberger and Fruton 1997; Daston 2000; Landecker 2007). In my own research, I'm interested in the phenomena that occur prior to the scientist's engagement with objects of enquiry. That is, I'm curious about the sensory perceptual processes that in turn frame the scientist's approach to and eventual understanding of scientific objects. The investigative object's coming into being is the culmination of an earlier crystallisation of the investigator's individual sensory perceptual framework.

I begin with the assumption that hearing is historical. That is, not only have sounds changed over time, but how individuals have heard them, what elements they found to be meaningful, and so on, have also changed over time. From there, we can begin to think about how the scientific ideas about hearing are both a clue to their developers' – the scientists' – own hearing and how they also altered their ways of hearing. I think we can presume – or at least I do – that, for the scientists, the process of studying sound and the sensory perception of it not only created new knowledge, but also altered the scientists' very perceptual frameworks. They altered their own bodies. Then, as they conveyed their scientific ideas, assumptions, and practices, the scientists made similar sensory perceptual frameworks possible in those around them and after them. Indeed, it was Mach that initially drew me to psychoacoustics. It was Mach's writings that first and firmly cemented my understanding of the profound implications of such a psychophysical world view for the practice of science, both as an individual and as a species. In this chapter, I document how Mach solidified his early interest in psychophysics, expanded its principles both as an experimental instrument and as a philosophical world view, and mobilised a very specific understanding of psychophysical parallelism and analogies to develop new ideas and articulate them to others.

A Psychophysical Imaginarium

Most generally, we understand an imaginarium to be a (not necessarily physical) space dedicated to the cultivation of ideas. Mach's psychophysical imaginarium was such a space in a layered, threefold way. It was, from the 1860s onward, an experimental programme. It was also, by the 1880s, a monistic world view that framed his thinking about scientific questions (to be potentially answered through psychophysical study). And finally, Mach's psychophysical imaginarium was a stage in a longer, evolutionary arc of humanity's understanding of the world. Mach toggled between these uses of his psychophysical imaginarium as needed to creative and fruitful ends. He also developed techniques to facilitate parallel imaginaria in others, such as readers and lecture audiences.

Mach began his 1863 *Vorträge über Psychophysik* with a swan dive into the ongoing preoccupation of the scientific community with experimental and representational precision (Holmes and Olesko 1995). The concern, among the sciences outside physics especially, was articulating the messiness of life processes as general, mathematical laws. Mach suggested, however, that fussing over which sciences were inherently exact and which were not was wrongheaded. The distinction could only be made during developmental stages, implying that all sciences worked towards exactitude and universality. Indeed, he claimed, psychology (the science of psychical phenomena) as well as psychophysics (the science of the interrelation of physical and psychical phenomena) were well on their way to becoming fully exact doctrines. To illustrate as much, Mach devoted the three lectures of *Vorträge über Psychophysik* to the history and current research of psychophysics, framed most fundamentally as a tension between individual sensory experiences and universal laws (Mach 1863, pp. 4–5).[1]

---

[1] Tellingly, Mach made the provocative claim here that physics was nothing more than applied mechanics and linked the observations of astronomers to Adolphe Quetelet's statistical distribution of individual difference.

Let's return to Mach's understanding of psychophysics. In the third and longest lecture in his 1863 *Vorträge über Psychophysik*, Mach defined the task of psychophysics to be the determination of the exact relationship between physical stimulus and psychical sensation by means of experience, observation, and experiment (Mach 1863, p. 12). The measurement of the stimulus was straightforwardly mechanical. The measurement of the sensation was, Mach continued, more difficult. But it had been done by others in the form of just-noticeable-difference measurements, most notably by his friends and colleagues Ernst Heinrich Weber and Gustav Fechner. In the 1830s and 1840s, Weber had performed a series of experiments on the sensation of touch, incrementally changing the difference between weights placed, while visually obscured by a piece of cardboard, in an experimental subject's hands, documenting whether the subject correctly noticed the difference (Weber 1834, 1846). He performed a similar set of experiments with temperature difference, as well as measurements of the variation in touch sensitivity across the body.

Fechner had struggled through these decades to reconcile his extreme empiricism with his phenomenalism (Heidelberger 2004, pp. 73–74). He was critical of strictly materialist science as too reductive, limited to an individual's own consciousness (Fechner 1851, pp. 1–14, 289–293). He instead proposed a system in which individual consciousness was connected to an immortal, all-knowing consciousness. What he termed 'day view' (*die Tagesansicht*) science consisted of a direct realism in which physical appearances existed objectively but were also interconnected with a higher consciousness (Fechner 1879/1994). In this world, physical and psychical experiences were merely two different perspectives of the same event, both of which were real. In October of 1850, Fechner was struck with what he described as an epiphany that Weber's experiments on touch sensitivity demonstrated what he had suspected was a correspondence between an arithmetic series of psychical intensities and a geometrical series of physical intensities. In his two-part 1860 publication *Elemente der Psychophysik*, Fechner presented this psychophysical monism in mathematical form, now called the Fechner–Weber law. The stimulus x was related to the sensation y as follows: $y = a \log (x/b)$, where a and b were constants (Fechner 1860b, p. 13). This was not necessarily a resolution to the mind–body problem, so much as a treatment of the functional relationship between the psychical and the physical.

After deriving it, Mach explained that Fechner's mathematical expression could be applied to measure a variety of sensory experiences (Mach 1863, p. 17). Mach briefly described several examples, ranging from light intensity to the sensation of tone pitch to the sense of time, as well as some subsidiary principles of Fechner's law. He then devoted the remainder of the discussion to exploring the psychophysical differences between the eye and ear. Initially, Mach believed these would be similar, parallel sensory organs. Turning to the recent work of Fechner, Helmholtz, and Wilhelm Wundt, it became clear that while both adhered to the Fechner–Weber law, the organs functioned quite differently to perceive light, colour, space, pitch, and volume. Mach appears to have especially admired that Fechner was able to carry out all of his psychophysical examinations without making any assumptions about the nature of the psychical nor the actual processes that connected stimulus and sensation. He was able to, Mach implied, stick to the 'facts of experience' (*Thatsachen der Erfahrung*), and so precision and therefore progress in science carried on (Mach 1863, p. 39).

Mach then asked, if the physical and psychical were so closely related, how should one think about their connection? Fechner understood physical stimulation and psychical sensation to be two different points of view regarding the same experience. Perhaps, Mach continued, this parallelism explains how the observer of a brain sees an event as an electric current, but for the owner of the brain, an event is the colour green (Mach 1863, p. 40). As Fechner described, the view of a sphere changes whether one is on the convex outside or concave inside. Mach found this to be a tidy way to cope with conflicting appearances/ observations, such as wave versus particle explanations for the dispersal of light. From here, Mach pivoted to attack the conception of atoms generally. He pointed out that physicists found it quite difficult to imagine atoms – as they were then understood – as centres of force. What, other than nothing, could be at the centre? And what did it mean for one centre of force to be acting on another? 'Let us confess it!' Mach continued, 'nothing of the externality of the atom' could be reasonably extracted (Mach 1863, p. 41).[2] This is Mach's earliest articulated phenomenalist, anti-atomist position.[3] It was a direct consequence of his initial engagement with psychophysics.

For Mach, the psychophysical framework was the only way for the sciences of physics, physiology, and psychology to progress, for they were inextricably connected. Furthermore, psychophysics meant that exact research did not need to be abandoned as one ventured beyond the 'realm of the plainly sensible' (*das Gebiet des Handgreiflichen*). This must in part explain his immediate enthusiasm for Fechner's work.[4] Mach later described this period

---

[2] '*Auch Physiker haben schon die Schwierigkeit gefühlt, sich die Atome materiell vorzustellen, und einige betrachten daher die Atome als blosse Kraftcentra. Doch ist ein Kraftcentrum für sich eigentlich nichts. Und was heisst es wohl, wenn man sagt, ein Kraftcentrum wirke auf ein anderes? Gestehen wir es kurz! Wir können dem Atom vernünftiger Weise keinerlei Aussenseite abgewinnen, sollen wir aber überhaupt etwas denken, so müssen wir demselben eine Innenseite beilegen, eine Innerlichkeit einigermassen analog unserer eigene Seele*' (Mach 1863, p. 41).

[3] Other scholars have traced Mach's engagement with atomism to his 1896 *Die Prinzipien der Wärmelehre historische-kritisch entwickelt* (Brush 1968).

[4] Fechner completed the first volume of *Elemente der Psychophysik* at the end of 1859 and the second volume during the following summer. Mach's lectures embracing and elaborating on the philosophical implications of a psychophysical world view dated to 1863.

as one during which his views were quite unstable. He described how, as a teenager, he was struck by the superfluity of the Kantian 'thing in itself', and soon after, on a bright summer day, he became aware of the world, his own ego/soul included, appearing as a single shimmering, coherent mass of sensations. For Mach's own research in the 1860s, 'alternating study of the physics and physiology of the senses, and through historico-physical studies', a psychophysical approach was the only way forward (Mach 1886, p. 21).[5] Two decades later, in his *Beiträge zum Analyse der Empfindungen*, Mach defended his past approach, explaining that he sought only to adopt a point of view in physics that was consistent across other domains of science.[6]

He had by then, however, developed his own, more specialised guiding principle for the study of the sensations: the *principle of the complete parallelism of the psychical and physical* (Mach 1886, p. 28). Between 1860 and 1880, Mach moved past Fechner's conception that the psychical and physical were two sides of the same reality – the inside and outside of a sphere. For Mach, this shared reality was predicated on a duality, whereas his view was one of unity (Mach 1891). He jettisoned the distinction 'between things and sensations, between outer and inner, between the material and spiritual world' (Mach 1886, p. 12). The world could instead be understood, Mach explained, to be made up entirely of elements, such as colours, sounds, and pressures. These elements could be described functionally as sensations or as physical properties, but scientific study had to break out of habitual stereotyped conceptions of a dualistic world. Mach offered his readers an illustration of the psychophysical experience/observation of himself observing across sensory modalities from different points of view and circumstance (Figure 1.1).

Admittedly, moving across the false borders dividing the psychical from the physical became more difficult when observing individuals other than oneself or non-humans or the influence of one's own body on one's sensations. Here, Mach explained that observed facts 'must be supplemented by analogy' (*eine beobachtete Thatsache durch Analogie ergänzen*) (Mach 1886, p. 13).[7] If psychophysics was the experimental programme best suited to exploring Mach's monistic understanding of the world, analogy was one of the critical tools for the practice of psychophysics. Despite their status as not observed

[5] He had just completed an examination of the controversy between Christian Doppler and Joseph Petzval and was able to demonstrate the effect of changed colour or tone as the observer changes position in relation to the source of the wave (Mach 1861).

[6] Atomism did not, he emphasised, meet this requirement – not in 1863, nor in 1886 (Mach 1886, p. 21).

[7] Mach had several different understandings of analogy. Some of these are addressed elsewhere in this volume. Here, I address only this narrow conception of analogies as non-facts that contribute to the practice of science.

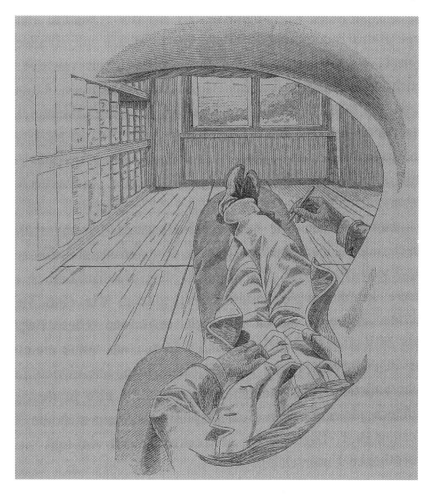

Figure 1.1   Mach reclined on his sofa with his right eye closed, observing himself observing (Mach 1886, p. 14)

facts, analogies held epistemological power. To sum up, the psychophysical imaginarium as experimental programme made epistemic use of a principle of psychical and physical parallelism and analogy as supplement.

Mach's Vienna

The monistic position Mach asserted in 1886 followed two decades of psychophysical experimentation. I deliberately employ the word 'followed' here in the strictly chronological sense. It is the historian's task to document the swirl of intellectual and cultural resources that Mach drew upon in the development of his own ideas, but it is also possible for the historian to overreach and posit direct, causal connections. Perhaps we can line up some analogies though? Mach himself described his intellectual position to be unstable in the 1860s, so let's begin there. What was he up to then?

Mach was in Vienna in 1863, working with Josef Popper-Lynkeus of the Vienna Physical Institute on a series of experiments on sound sensation. Mach also met Eduard Kulke that summer, both having been drawn into a lively discussion on the nature of musical tones at the Café Griensteidl (Mach 1906, pp. x–xi). The two men remained close friends, meeting up almost daily until Mach moved to Graz and then corresponding regularly until Kulke's death in 1897. Kulke had trained in mathematics and physics, but he built a career writing short stories, plays, and music criticism. Much of his fiction was devoted to portraying the folk life of Moravian–Jewish villages – ghetto tales of a sort. Mach noted that Kulke's Judaism had limited his academic advancement, and he spoke admiringly of his friend's courage to defy the 'raw anti-Semitism' he frequently encountered in the performing arts world of *fin-de-siècle* Vienna (Mach 1906, pp. x–xiii).

The deeply entwined nature of music and politics (and raw anti-Semitism) are best exemplified by Richard Wagner's essay 'Das Judenthum in der Musik'. Wagner revised this in 1869 to attack the music critic Eduard Hanslick, who had criticised the Viennese performance of *Lohengrin*, and he claimed that Hanslick's concealed Judaism rendered his writings anti-German. Several music scholars have argued that Wagner intended the character of Sixtus Beckmesser, who embodied several contemporary Jewish stereotypes, in his 1868 *Die Meistersinger von Nürnberg* opera to be a caricature of Hanslick.[8] These cruel episodes must be situated in several larger trends in the European music world, including a growing engagement with non-Western music, an interest in the music of 'internal others' and folk music, as well as the so-called War of the Romantics (Bohlman 2000; Born and Hesmondhalgh 2000; Radano et al. 2000). The allegiances in this war fell along stylistic and philosophical lines. The Leipzig-based composers Johannes Brahms and Robert and Clara Schumann, the violinist Joseph Joachim, and the critic Hanslick repeatedly clashed with the Wagner-led *Neudetusche Schule*.

Kulke loved Wagnerian opera, and he surrounded himself with such Wagnerians as Franz Liszt, Hans von Bülow, Peter Cornelius, and Anton Bruckner. Kulke described his passion for Wagner's music as an 'aesthetic heresy' (*ästhetische Ketzerei*) prompted by the opening chords of the *Tannhäuser* opera. When others mocked him for finding Wagner's music

---

[8] There was an ongoing debate on this matter among German studies and music scholars through the 1990s, but it seems to have been definitively resolved by Thomas Grey (Grey 2002, 2008, 2009).

beautiful – even pointing out that it was not acoustically possible for his harmonies to be beautiful – Kulke maintained that it was beautiful to him (Kulke 1906, pp. vii–viii). His individual, subjective experiences of music were valid.

Mach had at this time just completed an effort to physically then physiologically explain the role of attention in individuals' subjective experiences of sound. This phenomenon was and continues to be termed 'accommodation in hearing'. An example: the opening chords of the overture of Wagner's *Tannhäuser* began with the woodwind section. As the instruments wound through their mysterious melody, a listener could focus on, say, the clarinets and hear their melody distinct from the other instruments. Focusing instead on the bassoons in the same passage, the listener would experience something else. The listener could deliberately toggle their attention from instrument to instrument or take in the ensemble as a whole, thereby altering their individual aural experience of the performance. Applying Gustav Kirchhoff's theory of equal absorption and transmission to the eardrum, Mach placed a rubber tube in an experimental subject's ear. He then sang softly while moving the other end of the tube back and forth past his own ear. While Mach believed he had demonstrated a kymographic model of the hearing organ, the observations did not offer insight into the mechanism of accommodation (Hui 2013).

In the summer of 1863, with Popper-Lynkeus and his students, Mach attempted a new experimental approach to study accommodation in hearing, this time with a vibrating tuning fork held in his teeth, along with one end of a rubber tube in one of his ears and the other in an assistant's ear (Mach 1865). Mach would alter his attention, focusing first on the tuning fork's fundamental tone, then the overtones. For Mach, the focus of his attention was strong and distinct. The assistant, however, heard no change as Mach redirected his attention. Over the next decade, Mach mapped out the topography of the middle ear to better understand the senses of balance and acceleration. In 1871, he began a series of experiments with Johann Kessel on cadaver ears (Mach 1872b; Mach and Kessel 1872, 1874; Kessel 1874). The elaborate experimental set-up included the middle ear of a cadaver, weights, pulleys, a Lissajous vibration microscope, and a tuning fork. Mach and Kessel were able to demonstrate that changed tension on the tensor tympani muscle altered the transmission of sound through the ossicles (ear bones), but this was only the mechanical/physiological component of the accommodation phenomenon. An attempt to perform a similar series of experiments on a living person with the assistance of an 'ear mirror' did not replicate their previous results. Accommodation in hearing was a psychophysical phenomenon; it could not be understood in the 'realm of the plainly sensible' only. Mach would spend the next decade mobilising both parallelism and analogy to investigate it.

Psychophysical Parallelism and Epistemic Analogies

As I turn to Mach's unpublished writings now – his laboratory notebooks, lecture notes, and correspondence – I want to run a little fast and loose with Mach's understanding of the role of parallelism and analogy in experimental science. Briefly, in what follows, I will present the more formal entries of experimental programmes, recorded observations, and lecture notes against the less formal marginalia and loose-leaf inserts. These latter often took the form of brief lists of seemingly unrelated objects, concepts, and names. Christoph Hoffmann has shown how Mach's research notebooks cannot simply be read as 'passive reflections of experimental operations or cognitive processes' (Hoffman 2003, p. 183). Rather, these notebooks – Mach's act of writing in them – were epistemological tools. As techniques of science, Hoffmann argues, Mach's writings must be analysed within a knowledge production framework (Hoffman 2003, pp. 183–184). I push Hoffmann's central thesis a bit further. I argue that Mach's unpublished writings – lab notebooks, lecture notes, and correspondence – were the medium through which Mach cultivated his psychophysical imaginarium. Or, at least, we can use these writings to trace, through his use of analogy, some of the inflection points in the development of his psychophysical world view.

Mach's lab notebooks are especially rich resources for illuminating his knowledge-making frameworks, as Hoffman has noted. Mach frequently wrote out research questions, returned to several pages in a row to cross out material, and even wrote down exclamations of frustration; truly a boon for the historian attempting to reconstruct Mach's intellectual craft. In his lab notebooks from the 1870s, the same period in which he was examining accommodation in hearing and lecturing on psychoacoustics, Mach's jottings indicate that he was struggling to understand the movements of the middle ear bones as sound was transmitted through them. These were the accommodation experiments using a Lissajous microscope on a cadaver's ear.[9] Mach sketched out the stroboscopic images and tried to reconcile them with the rotation of the stirrup bone. He noted that the tensor tympani and stapedius muscles reduced this rotation. Then, at the top of the next page, Mach jotted, 'I cannot convince myself of that? Where am I mistaken?' (Mach 1870). In the subsequent lab notebook, after a series of crossed-out derivations, Mach sketched out several images of rotation angles. Then he began investigating the effects of sirens and pipes on the ear muscles. The next pages include sketches of abstract experimental preparations for the series of tests Mach performed with rubber tubes and sounding instruments. We can see that he

---

[9] It is admittedly a little tricky to confirm dates in these materials, but Mach references the accommodation of the stirrup bone near the end of this notebook. The subsequent notebook is dated 1871.

considered using rubber muscles in addition to the experiments performed on human subjects.

Additional drawings of rotation angles, schematics of the middle ear, diagrams of rubber tube preparations, interference waves, and references to Helmholtz, Georg Ohm, André-Marie Ampère, Wundt, Charles Darwin, and Ernst Haeckel populate five more lab notebooks. This chapter is not the place to work out the step-by-step process of Mach's thinking. Rather, I want to show here that the knotty problem of explaining accommodation in hearing dominated Mach's experimental programme. He returned to it again and again over the next decade, each time with a different experimental approach. It became clear to Mach that accommodation in hearing could not be understood via physics or physiology alone. In that sense, it was an emblematic case for the application of the psychophysical parallelism he would propose in a concrete form in 1886.

We can see Mach practicing it himself, however, by as early as 1871 in his university lectures. Mach divided a medium-sized black notebook into pre-paratory notes for two courses of Experimental Physics, one dated to the summer of 1871 and the other to the winter of 1871/1872. The summer course mostly covered optics. The winter course began with mechanics and then moved to oscillators and acoustics. These are neatly written, with derivations and numbered examples of standing waves and so on. At one point in this notebook, seven loose sheets of paper are tucked between the pages. Most of them consists of lists. Here is one (Mach 1871b):

Middle-tones
Tuning-fork research
Piano
Bodies of greater absorption (damping?)
Airspace. Attenuation
Resonators. 1 large pipe.
Tone colour
Cortical Fibres
Sound plates with resonance-tubes
Interference tubes . . . execution via the middle tones
Speed of sound
Application on the forehead
Research on the track

A bit later in the course, Mach discussed Ernst Chladni's vibrating plates and Félix Savart's work on vibrating rods. Following a tidy diagram of longitudinal waves through a vibrating rod, there is another list (Mach 1871b):

Tuning fork research
Tone colour. Partial tones alone.
        Superposition

Spectral analysis analogue. Noise
Resonance tones
König's apparatus   Reed pipes?
Harmonic and unharmonic partial tones of body
Coexistence of the same
Dependence of tone colour of this   drawing*
Independence of tone colour of phase difference
Analogy of spectral analysis . . . Noise.
Resonators *
Organ pipes * open, covered
König's Flame apparatus * vowels

And then on the following page, accompanied by a musical staff with a series of notes labelled with vowel sounds, there is the following (Mach 1871b):

Vowels. Tone colour of the same
König's research (Zenger)*
Artistic representation of vowels with electric tuning forks.
Independence of tone colour of phase difference
Piano research. Very enlightening.
On the other hand it remains a mystery how one is able to sing the same ~~Ton~~
     Vowel of an entire scale.

On another slip of paper that followed a few pages later was the following bracketed list (Mach 1871b):

Partial tones of sound
tone colour
relationship
melody
harmony
ear theory
Euler
Lipps
v. Oettingen

In each of these, we see a seeming jumble of experimental instruments and configurations (tuning forks, resonators, sound plates, a piano, and application on the forehead), physical concepts (attenuation, the speed of sound, resonance, and the theories of Euler, Theodor Lipps, and Arthur von Oettingen), physiological systems (cortical organ and fibres), and musical concepts (tone colour, melody, and harmony). Compared to his laboratory notebooks, which hew closely to physics and physiology, Mach's lecture notes are more wide-ranging. I think we can read this relative freedom of movement between realms – from experimental instruments, to scientific concepts, to musical phenomena, and so on – as a two-step process: first, we can read his formal lecture notes as outlines to constrain his thinking and help him stick to a curriculum of sorts; but second, we should read the inserted lists as additional thoughts and reminders to himself to also talk about his ongoing piano research and so on. These lists were analogies. They functioned to supplement, quite literally, the shortcomings of precise science in explaining the sensory perception of sound. He was already thinking broadly, moving from the acoustics of wave interference to tone colour and then melody and then the structure of the ear, shifting from physics to musical aesthetics to sensory perception.

Acknowledging that the source record is more circumstantial than explicit, I would like to suggest that it was the exercise of articulating his ideas to others – students, friends, and the public – that fuelled the fullest expression of his psychophysical world view. That is, scaling out from his laboratory note-books to his lecture notes to his correspondence to his public lectures, Mach became more bold in the practice and articulation of his monism. Here is one more example, from his correspondence with the music critic Kulke. Through the 1870s, their correspondence returned again and again to the question of an evolutionary theory of musical aesthetics. In 1872, Mach asked Kulke if he believed it possible for listeners in the present to hear what the ancient Greeks had heard. Was it simply a matter of attention (Mach 1872a)? Kulke was at this time working on a Darwinian theory of melody, *Über die Umbildung der Melodie: Ein Beitrag zur Entwickelungslehre*, which was eventually published in 1884 (Kulke 1884; Hui 2014). Mach, we know, was working on an experimental study of accommodation in hearing that was psychophysical in its parallelism, use of analogy, and, ultimately, phenomenology. In an undated letter to Kulke, likely between 1876 and 1878, well before the publication of *Analyse der Empfindungen*, Mach included a sketch that anticipated the famous illustration in that book (see Figure 1.2). It should be noted that, in this iteration, the figure whose perspective the viewer experienced was nursing a coffee and a cigarette; not, as the later image showed, drawing itself and/or the world. Still, it was instructive: 'How to execute the self-perception of "I"'. To underscore my earlier point, for Mach it was the act of articulating his work on accommodation in hearing to others that facilitated increasingly expansive psychophysical positions.

Mach's Piano

I have argued elsewhere that music was a proxy scientific language for Mach, describing it as 'a consequence of his constant and uninhibited engagement with the music world' (Hui 2014, p. 174). Revisiting Mach, and thinking more carefully about the role of parallelism and analogy in his psychophysical experimental programme, I think I can make a more direct claim that

Figure 1.2 Ernst Mach, undated letter to Eduard Kulke (letter no.24, from between 1876 and 1878).
Ernst Mach Papers, Dibner Library of the History of Science and Technology, Smithsonian Institution Special Collections, Washington, DC

Mach's use of music was a by-product of his friendship with musicians. I think we can comfortably understand this use of music to be a deliberate choice by Mach. This was exactly the toggling between realms (from physical to psychological, etc.) that he claimed was necessary for the sciences to move forward (versus atomism, which was stuck in a single realm).

I promised a piano, so let's talk about Mach's piano. Mach had a Bösendorfer grand and was friends with the Bösendorfer family, who were based in Vienna. The 'Bösendorfer sound' was known for its richness and clarity.[10] It was a more brilliant sound than, for example, Steinway pianos, and many preferred this tone colour. This brighter voicing was a consequence of the solid spruce rim, which was jointed together rather than being bent veneer. The entire box of the piano resonated, which threw more upper harmonics into sympathetic vibration when the instrument's keys were struck. The acoustics and aesthetics of Mach's piano should be kept in mind as we think about how he used it in scientific settings both as analogy and experimental instrument.

In 1866, Mach reworked Hermann von Helmholtz's 1863 opus *Die Lehre von den Tonempfindungen* for musicians as *Einleitung in die Helmholtz'sche Musiktheorie: Populär für Musiker dargestellt* (Mach 1866). That is, he eliminated Helmholtz's detailed descriptions of experiments and mathematics and minimised the discussion of physics. Mach disagreed with Helmholtz's understanding of the role of attention in sound sensation as strictly psychological, but otherwise summarised Helmholtz's points in a straightforward way. He embraced Helmholtz's use of the piano as an analogy for the mechanics of the ear. In this text and in popular lectures he gave the year before (1865), Mach described the following scenario to illustrate sympathetic vibration (Mach 1864a, 1864b). Two pianos were placed next to each other, with the dampers lifted on one of them (by pressing on the sostenuto pedal). When a key was struck on the damped piano, the same note rang on the undamped piano. Sounded chords on the damped piano similarly activated the same notes on the undamped instrument. According to Mach, the undamped piano was performing a spectral analysis of sound, separating the sounded tones into individual component parts (Mach 1864a, pp. 23–25).

This was essentially how Helmholtz – and Johannes Müller before him – mobilised this ear-as-piano analogy.[11] Mach expanded on it, however, to also explain accommodation in hearing. In 'Die Erklärung der Harmonie', he described an example of the phenomenon (and possibly performed it during

---

[10] Bösendorfer pianos are also known for their 92- and 97-key models. These were added to the company's series around 1909.

[11] Helmholtz had drawn on Müller's use of the piano as a model for sympathetic vibration. Julia Kursell offers an extensive discussion of the epistemological power of the piano model for Helmholtz (Kursell 2018).

the public lecture) in which a piano sounded two different chords in succession, voiced the same so that all tones were of the same loudness. Then, before
the two chords were played again, the listener was directed to focus their
attention on just the root tones of each chord or just the upper tones. Because
the root tones were the same for both chords, the listener focusing on just the
root tones would experience the sounded chords differently from if they were
focusing on just the upper tones, which changed between the two chords.
This piano demonstration (real or virtual) facilitated the experience of
accommodation in hearing for Mach's audience. Through careful training,
Mach continued, a listener could refine their spectral analyser and differentiate
even a single tone into its constituent fundamental tone and harmonic overtones (Mach 1864b, p. 37). Drawing on Helmholtz, Mach noted that these
overtones contributed to tone colour (timbre) and consonance, and they were
ultimately the root of Western musical harmony. Accommodation informed
the spectral analysis process. So here we slide into Mach's very specific
understanding of the role of analogy as supplement to observed facts.

But the role of attention in altering the individual's experience of sound was
also an observed fact. Mach demonstrated accommodation in hearing again
and again in lectures and his writings so readers could recreate it in their
mind's ear. Mach's piano returns here. The accommodation chord demonstration (he also did a series of demonstrations inverting and reversing
melodies to demonstrate the lack of spatial symmetry in sound), especially
Mach's claim that one could train one's body to experience the phenomenon
in a more pronounced way, would have been additionally amplified by the
Bösendorfer sound (Mach 1864a). Of course, it is unlikely that Mach used his
Bösendorfer piano for his public demonstrations. It is possible, though, that he
used it to work out his thoughts alone, in the laboratory, or even in his lecture
courses (it is unknown where he kept his piano). Again, the bright Bösendorfer
sound was a consequence of the design and voicing that allowed for more
upper harmonics to be sounded. A listener concentrating on discerning
harmonics in a sounded tone would be more likely to hear them on a
Bösendorfer than a Steinway, Pleyel, or Erard. By asking his readers or
listeners to imagine specific psychophysical experiences, Mach established
the broad outlines of a new way of thinking about subjectivity. By walking
his audience through these experiences step by step with a piano, Mach
facilitated for them a psychophysical imaginarium.

Conclusion

In 1872, Mach published *Die Geschichte und die Wurzel des Satzes von der
Erhaltung der Arbeit*, and in it a full articulation of his position that ideas were
bound to specific times and places. It was also in 1872 that Mach asked Kulke
whether they could still hear what the ancient Greeks heard. This was not a

question of whether the sounds of the world changed over time, but whether the perception of these sounds did, for Mach was deep in his studies of how an individual's perception of sound could change moment by moment as they altered their attention. If accommodation in hearing was psychophysical, then it was also historical and historicist. Recall that Mach described his researches in the 1860s as 'alternating study of the physics and physiology of the senses, and through historico-physical studies'. I venture that Mach's early studies of accommodation in hearing were foundational for his later historical epistemology. In describing and explaining the phenomenon of accommodation in hearing, Mach also prompted it in his readers and listeners. Mach moved between realms, from acoustics to physiology to mathematics to musical notation to sound itself, enacting the parallelism he called for. He extended his own feedback loop of psychophysical framework to psychophysical imaginarium to others – to me, and perhaps to you, too.

References

Bohlman, Philip 2000. 'Composing the Cantorate: Westernizing Europe's Other Within', in Georgina Born and David Hesmondhalgh (eds.), *Western Music and Its Others: Difference, Representation, and Appropriation in Music.* University of California Press, pp. 187–212.

Brush, Stephen 1968. 'Mach and Atomism', *Synthese* 18: 192–215.

Daston, Lorraine (ed.) 2000. *Biographies of Scientific Objects.* University of Chicago Press.

Fechner, Gustav T. 1851. *Zend-Avesta oder über die Dinge des Himmels und des Jenseits. Vom Standpunkt der Naturbetrachtung.* Leopold Voß.

  1860a. *Elemente der Psychophysik 1.* E. J. Bonset.

  1860b. *Elemente der Psychophysik 2.* Breitkopf und Härtel.

  1879/1994. *Die Tagesansicht gegenüber die Nachtansicht.* Klotz.

Grey, Thomas 2002. 'Masters and Their Critics: Wagner, Hanslick, Beckmesser, and Die Meistersinger', in Nicholas Vazsonyi (ed.), *Wagner's Meistersinger: Performance, History, Representation.* University of Rochester Press, pp. 165–189.

Grey, Thomas (ed.) 2008. *The Cambridge Companion to Wagner.* Cambridge University Press.

  (ed.) 2009. *Richard Wagner and His World.* Princeton University Press.

Heidelberger, Michael 2004. *Nature from Within: Gustav Theodor Fechner and His Psychophysical Worldview.* Translated by Cynthia Klohr. University of Pittsburgh Press.

Born, Georgina and Hesmondhalgh, David (eds.), *Western Music and Its Others: Difference, Representation, and Appropriation in Music.* University of California Press.

Hoffmann, Christoph 2003. 'The Pocket Schedule', in Frederic Holmes, Jürgen Renn, and Hans-Jörg Rheinberger (eds.), *Reworking the Bench: Research Notebooks in the History of Science.* Springer, pp. 183–202.

Holmes, Frederic and Olesko, Kathryn 1995. 'The Images of Precision: Helmholtz and the Graphical Method in Physiology', in M. Norton Wise (ed.), *The Values of Precision*. Princeton University Press, pp. 198–221.

Hui, Alexandra 2012. *The Psychophysical Ear*. MIT Press.

2013. 'Changeable Ears: Ernst Mach's and Max Planck Studies of Accommodation in Hearing', *Osiris* 28: 119–145.

2014. 'Origin Stories of Listening, Melody, and Survival at the End of the Nineteenth Century', in James Kennaway (ed.), *Music and the Nerves, 1700–1900*. Palgrave, pp. 170–190.

Kessel, Johann 1874. 'Ueber den Einfluss der Binnenmuskeln der Paukenhöhle auf die Bewegung und Schwingungen des Trommelfels am todten Ohre', *Archiv for Ohrenheilkunde* 2: 80–92.

Kulke, Eduard 1884. *Über die Umbildung der Melodie: Ein Beitrag zur Entwickelungslehre*. J. G. Calve'sche K. K. Hof- und Univ.-Buchhandlung.

1906. *Kritik der Philosophie des Schönen*. Deutsche Verlagactiengesellschaft.

Kursell, Julia 2018. *Epistemologie des Hörens: Helmholtz' physiologische Grundlegung der Musiktheorie*. Wilhelm Fink Verlag.

Landecker, Hannah 2007. *Culturing Life*. Harvard University Press.

Mach, Ernst 1861. 'Ueber die Kontroverse Zwischen Doppler und Petzval, bezüglich der Aenderung des Tones und der Farbe durch Bewegung', *Zeitschrift für Mathematik und Physik* 6: 120–126.

1863. *Vorträge über Psychophysik*. Leopold Sommer.

1864a/1896. 'Über die Cortischen Fasern des Ohres', reprinted in Mach, *Populär-Wissenschaftliche Vorlesungen*. J. A. Barth, pp. 17–31.

1864b/1896. 'Die Erklärung der Harmonie', reprinted in Mach, *Populär-Wissenschaftliche Vorlesungen*. J. A. Barth, pp. 32–47.

1865. 'Bemerkungen über die Accommodation des Ohres', *Sitzungsberichte der Kaiserlichen Akademie der Wissenschaften* 51: 343–346.

1866. *Einleitung in die Helmholtz'sche Musiktheorie: Populär für Musiker dargestellt*. Leuschner & Lubensky.

1870. Notizbuch 505. Ernst Mach Nachlass, Deutsches Museum, Munich.

1871a/1896. 'Die Symmetrie', reprinted in Mach, *Populär-Wissenschaftliche Vorlesungen*, J. A. Barth, pp. 101–118.

1871b. Notizbuch 12. Konstanzer Abgabe, Deutsches Museum, Munich.

1872a. Letter to Eduard Kulke. 30 May, letter no. 4. Ernst Mach Papers, Dibner Library of the History of Science and Technology, Smithsonian Institution Special Collections, Washington, DC.

1872b. 'Über die stroboskopische Bestimmung der Tonhöhe', *Sitzungsberichte der Kaiserlichen Akademie der Wissenschaften* 66: 67–74.

1886. *Beiträge zur Analyse der Empfindungen*. Gustav Fischer.

1891. 'Some Questions of Psycho-Physics. Sensations and the Elements of Reality', *The Monist* 1: 393–400.

1896. *Die Prinzipien der Wärmelehre historische-kritisch entwickelt*. J. A. Barth.

1906. 'Eduard Kulke', in Eduard Kulke, *Kritik der Philosophie des Schönen*. Deutsche Verlagactiengesellschaft, pp. x–xiii.

Mach, Ernst and Kessel, Johann 1872. 'Versuche über die Accommodation des Ohres', *Sitzungsberichte der Kaiserlichen Akademie der Wissenschaften* 66: 337–43.

  1874. 'Beiträge zur Topographie und Mechanik des Mittelohres', *Sitzungsberichte der Kaiserlichen Akademie der Wissenschaften* 69: 221–43.

Radano, Ronald, Bohlman, Philip, and Baker, Houston (eds.) 2000. *Music and the Racial Imagination*. University of Chicago Press.

Rheinberger, Hans-Jörg and Fruton, Joseph 1997. *Toward a History of Epistemic Things: Synthesizing Proteins in the Test Tube*. Stanford University Press.

Weber, Ernst Heinrich 1834. *De Pulsu, Resorptione, Auditu, et Tactu*. Prostate Apud C. F. Kohler.

  1846. 'Der Tastsinn und Gemeingefühl', in R. Wagner (ed.), *Handwörterbuch der Physiologie, 3*. Vieweg, pp. 481–588.

2

Mother's Milk and More

On the Role of Ernst Mach's Relational Physics in the Development of Einstein's Theory of Relativity

RICHARD STALEY

Perspectives on the significance of Ernst Mach's work in physics have been dominated by the importance of his denial of absolute time, space, and motion for the subsequent development of the special and general theories of relativity in the hands of Albert Einstein, and by the uncompromising anti-metaphysical, empiricist philosophy his stance seemed to represent. But they have also been troubled by the 1921 publication of an unfinished book, *Optics*, with a preface Mach purportedly wrote in 1913, declining to be associated with the dogmatism of the current theory of relativity. The result is an image of Mach as a physicist for whom the epistemological demands of positivist empiricism trumped theory – even one he had played a large role in instigating.

The most authoritative expressions of this view undoubtedly come from Einstein himself, but it has also underlain many of the principal biographical and philosophical studies of Mach, as well as our major histories of relativity. Characterising it at some length will help develop the possibility of understanding it more fully historically. Einstein's views were expressed in the reflective tones of two extraordinary 'obituaries': first on the occasion of Mach's death in 1916, but also when writing his own 'Autobiographical Notes'. Looking back on the scientists of the nineteenth century in 1949, Einstein described the changing nature of Mach's significance for him personally, singling him out as the one person who had doubted that mechanics could serve as the definitive 'firm and final' foundation for all physics, and even all natural science:

> It was Ernst Mach who, in his *History of Mechanics* [*Die Mechanik in ihrer Entwickelung historisch-kritisch dargestellt*], shook this dogmatic faith; this book exercised a profound influence upon me in this regard while I was a student. I see Mach's greatness in his incorrigible scepticism and independence; in my younger years, however, Mach's epistemological position also influenced me very greatly, a position which today appears to me essentially untenable.
>
> (Einstein 1949, p. 21)

Einstein went on to argue that Mach did not correctly appreciate the 'constructive and speculative nature of thought and more especially of scientific thought', condemning theory 'on precisely those points where its constructive-speculative character unconcealably comes to light', such as atomism (Einstein 1949, p. 21).

Thirty-three years earlier, Einstein's obituary had offered a more extended analysis, suggesting it was not the originality of Mach's epistemological perspective that was most significant, but the way in which he analysed both the connections between facts and concepts and the entrenched points of view sustaining these links, breaking their authority with a demand for new conceptual precision. Referring especially to concepts of time, space, and mechanics that his own work had modified, Einstein wrote, '[N]o-one can take it away from the epistemologists that here they have paved the way for progress; for myself, I know that Hume and Mach have helped me a lot, both directly and indirectly' (Einstein 1916, p. 102). Einstein asked his readers to take *Die Mechanik in ihrer Entwicklung* in hand and consult in particular the sections on Newton's opinions and Mach's critique, quoting liberally from discussions of absolute and relative time and space and the behaviour of water in a rotating bucket to indicate Mach's clear recognition of the weak points of 'classical mechanics', a relatively new term (Staley 2008; Gooday and Mitchell 2013). Einstein even suggested that had physicists been concerned with the constancy of the velocity of light in Mach's youth, he could well have lit upon the special theory himself, while his analysis of the bucket (relating the water's behaviour to the relative rotation of the earth and other heavenly bodies) showed Mach to be very close to demanding a theory of the relativity of acceleration, too. However, Einstein wrote, this was without showing a vivid consciousness of the way in which the equivalence of inertial and gravitational mass elicits this, because experiment cannot decide whether the fall of a body relative to a coordinate system is caused by the presence of a gravitational field or acceleration of the coordinate system (Einstein 1916, p. 103).

Einstein thus identifies specific strengths and limitations when characterising Mach's conceptual contributions to relativity. Just as importantly, his obituary delicately addressed the challenge of a forthright attack Mach had experienced from the prominent Berlin physicist (and Einstein's principal patron), Max Planck. In a lecture published in 1909, Planck had denied both the tenets and the fruitfulness of Mach's stance that the basis for reality subsists in perceptions and that natural science is ultimately an economic adaptation of ideas to perceptions, rather than the striving for a fixed world picture independent of any culture or individual intellect (Planck 1909; Wegener 2010). Privately, Einstein had assured Mach of his support in letters of 9 and 16 August 1909. His obituary noted that Mach's use of the term 'sensation' had led to a misperception of the sober thinker as a solipsist, and Einstein offered a defence of an epistemological approach in the light of specialist disdain. Still more pointedly, he argued that the influence of Mach's historical and critical writings on the evolution of individual sciences was such that even his opponents were unaware of how much of Mach's thinking they had absorbed 'so to speak, with their mother's milk' (Einstein 1916, p. 102).

The portrait that emerges from these comments is of a close, complex, and changing relationship, perhaps summed up most sharply in an exchange between Einstein and his closest friend, Michele Besso, in 1917. As students in Zurich, Besso had introduced Einstein to Mach's history of mechanics twenty years earlier, and he was the one person Einstein acknowledged for assistance in his 1905 paper on special relativity. When Einstein complained in a letter of 29 April that another Machist they had both met in Zurich, Felix Adler, was riding Mach's 'hobby horse to exhaustion', Besso urged consideration for the steed's services in steering through the infernalities of the relativities. Einstein responded that he was not complaining about Mach's little mount (Besso knew what he thought about it); '[H]owever, it can give birth to nothing new, but only stamp out harmful vermin' (see letters of 29 April, 5 May and 13 May in Einstein 1998, pp. 441–442, 444–445, 451–452). When Mach's apparent denial of a link with relativity appeared in 1921, Einstein's exasperation reached a height with his 1922 comment that 'insofar as he was a good student of mechanics, [Mach] was a deplorable philosopher' (Herneck 1959, p. 564).

Thus, to gain a clear perspective on the relations between Mach and Einstein's work, we will need to understand Mach as an epistemologist and as a mechanic; and it should be no surprise that many of our most detailed and influential accounts of their relationship have centred on Mach's philosophical perspective, typically developing more finely grained elaborations of the view Einstein articulated in 1949. This is true of Gerald Holton's influential study of Mach and Einstein, but also John Blackmore's early biography of Mach and John Norton's argument that, in contrast to what he describes as Mach's critically orientated prohibition of the introduction of arbitrary concepts, Hume had more helpfully pointed to the need for invention and freedom in concept formation (Blackmore 1972; Holton 1988a, 1992; but see also Norton 2010; Banks 2012).

Here, I aim to address a limitation to all of these works that follows from their primary orientation around Einstein's achievements, something that risks exemplifying an understanding of the relations between philosophy and science that is in fact an outcome of these developments. Jürgen Renn's otherwise valuable work on the origins of relativity can illustrate both points. Despite identifying a critique of mechanics that was instigated by Mach as one of three principal pathways to relativity, Renn considers Mach's research only through his book on mechanics and describes his stance there as philosophical, without noticing that this is in some contrast to Mach's own view (Renn 2007, p. 21). I will show here that neglecting the twenty years of research that led Mach to his critical reformulations in 1883 has made it difficult to fully understand the character of his epistemology and impoverished our approach to his mechanics. The basis for such a reassessment was laid long ago. In a 1984 paper, Paul Feyerabend turned the tables by suggesting it is not Mach but Einstein whose epistemological writings tend towards positivism. It is revealing that Feyerabend pointed elsewhere in Mach's *Mechanik* (to his discussion of Simon Stevin's account of static equilibrium on an inclined plane) and analysed Mach's 1905 treatment of theory and experiment in *Erkenntnis und Irrtum* (*Knowledge and Error*) in arguing that Mach understood historically grounded principles of physics in the kind of constructive sense Einstein advocated, while avoiding what Feyerabend regards as simplistic references to freely created concepts with no logical link to facts (Feyerabend 1984; but see also Hentschel 1985). Gereon Wolters offered a more comprehensive study of the diverse respects in which Mach's work could help stimulate aspects of Einstein's relativity, including Mach's critique of mechanics as a foundation for the world picture, his distinctive approach to the principle of inertia, critiques of basic concepts of physics, and his methodology (Wolters 1987, 2012). Wolters also gave strong reasons to doubt the episode that seemed to show Mach's abjuration of relativity, arguing that the preface to *Optics*, purportedly written by Ernst Mach, in fact stemmed from his son, Ludwig (but see also Holton 1992). The work of Feyerabend and Wolters can be extended by shifting the direction of comparison from their tight orientation around Einstein and examining Mach's research more developmentally. Building on the detailed and subtle historical understanding of Mach's philosophy that Erik C. Banks and Michael Heidelberger have yielded by beginning with Mach's early writings (Banks 2003; Heidelberger 2010), I will develop a form of historical comparison that considers some similarities and differences in the principal concerns that Mach and Einstein pursued in the periods in which their distinctive approaches were generated. Studying the gradual emergence of Mach's epistemology alongside his research on the physics of time and space will help deepen our historical understanding of the way in which the relations between philosophy, epistemology, and physics were understood in different periods and will draw out several different senses in which Einstein's description of Mach's work as mother's milk is peculiarly apt.

Ernst Mach as an Epistemologist: An Overview

Mach notoriously denied he was a philosopher while also articulating a general approach to scientific knowledge and pursuing clear epistemological demands. Yet, addressing the 1894 *Naturforscherversammlung* in Vienna, he offered a revealing, historically orientated account of tensions between epistemology and physics. Retrospectively, we can see that this lecture helped crystallise a new direction. Mach moved from considering specialised subjects such as mechanics or the nature of sense perception to a more synthetic approach, intellectually and professionally. In the following year, he left his Prague professorship in experimental physics for one in the history and theory of the inductive sciences in Vienna. When he drew together lectures in that subject with several earlier papers, the resulting book bore the title *Erkenntnis und Irrtum* (Mach 1976 [1905]). Thus, late in his career, Mach had become an epistemologist, or *Erkenntnistheoretiker*; his 1894 lecture presaged this change and indicates some of its motivations. Mach began by noting that, twenty years earlier, Gustav Kirchhoff had argued that the object of mechanics is 'description, in complete and simple terms' – and this had produced a 'very peculiar impression' amongst physicists, even astonishment, as Ludwig Boltzmann put it (Mach 1896 [1894], p. 251). Asking how someone so respected for their scientific achievements could be greeted with such scepticism regarding their philosophical views, Mach answered that it might well have to do with scientists' reluctance to enquire into the psychological impulses driving their work (and he gave *Knowledge and Error* the subtitle *Sketches in the Psychology of Research*). Further, Mach suggested it was inevitable that much Kirchhoff had not intended would be read into his rigid words, but also that they would be found to lack ingredients thought essential. Mach was speaking from experience, having anticipated Kirchhoff in describing science as being concerned throughout with the economical description of phenomena; and to his misfortune, Planck's attack later replayed the character and terms of the phenomena Mach diagnosed.

Mach's extended characterisation of description in 1894 has several features worth noting. The first is the importance of its basis in sense phenomena, but also how quickly Mach leaves this. For Mach, it is not sense impression (or fact) which makes science, but communication; and comparison is the most vital element of science because it provides the fundamental condition for communication (something he traced across all concept formation, from the communicative cries of gregarious animals to the shifting terms of abstract theories such as potential theory). Secondly, Mach's portrait of science is focused on process rather than results. By considering change over time and highly diverse stances, Mach builds a complex, subtle image of the benefits and dangers of abstraction and concept formation. He deftly offers examples of respects in which theoretical perspectives such as the emission theory of light by turns hampered and enabled recognition of light's periodicity and polarisation. If the simplicity of Kirchhoff's account expressed the logical and aesthetic sense of the mathematician, Mach's account of description was messier, being based on the practices of the scientist in general and the physicist in particular. He offered an understanding of concepts 'in action' as the trained work of muscles and imagination, whose origins lie in comparing and representing facts with other facts through laws expressed as functional relationships (Mach

1896 [1894], p. 267). This is the basis for a universal comparative phenomenology as the means to advance the diverse specialist sciences, from the physical to the zoological.

Several important implications follow for our understanding of the senses in which Mach, Einstein, and later historians have treated epistemology. The first is methodological: we should look for epistemology in action and as an element of practice, rather than largely as doctrine. Approaching Mach's early research with this aim will deliver quite a different sense of his perspective from the sharply drawn portraits of Planck attacking him or even of Einstein supporting him in 1916.

Secondly, Mach's later phase of work can be described as an attempt to provide a critical historical account of epistemology in scientific practice. Revealingly, when he briefly looked back on his earlier treatment of absolutes and relational knowledge as an epistemologist in *Knowledge and Error*, Mach wrote that he never aimed his arguments against physical working hypotheses, but only against epistemological absurdities. In further contrast to the perspective suggested by Norton and Renn, Mach described himself as always starting from physical details and, from there, rising towards more general considerations, contrasting his concern to address scientists with J. B. Stallo's *Concepts and Theories of Modern Physics*, which had developed similar views but proceeded in exactly the opposite direction, addressing philosophers instead (Mach 1976 [1905], pp. 102–103). Mach's view thus indicates the need to situate 'physics', 'epistemology', and 'philosophy' historically, recognising how fraught these terms may be in periods of contest and change. Although it would be necessary to engage the physics community more broadly to do this fully, developing an understanding of Mach's changing views here will indicate the significance of the task.

Finally, considering the different phases in Mach's work, we can describe his early book on mechanics as providing a critical historical account of the content of a particular field. While this certainly represented his epistemology in practice, the striking terms in which Mach described his epistemological concerns in 1883 have often contributed to others giving them a more partial, selective understanding than Mach's own. In some part, this was a result of one of his most radical aims: Mach sought an account appropriate to all science, not just physics, let alone mechanics. This might represent the most important distinction between Mach's and Einstein's endeavours. Mach's epistemology and his work in mechanics reflected the unusual aim of developing a unified perspective towards science as a whole and the distinct disciplines of physics, physiology, and psychology in particular. In contrast, Einstein's early work in physics aimed to unify what he understood as independent but related fields of mechanics, thermodynamics, and electrodynamics. While each was unusually aware of the limitations of specific fields (especially mechanics), both men proved particularly significant because they understood unification in more comprehensive terms than their contemporaries. However, the form unity could take in these different pursuits was very different, and this has made the character of Mach's work in mechanics difficult to comprehend.

Mechanics in Mach's Early Research

Finishing his studies at the University of Vienna in 1860, Ernst Mach initially aimed to teach in physiologically directed physics, but was instructed by the faculty to habilitate in physics in general instead. He had written his doctoral thesis under Andreas von Ettingshausen on electrical induction, who also suggested he examine the Doppler effect, yet Mach later described his papers on it as suffering from all of the mistakes 'of an autodidact and beginner' (Mach 1873a, p. 3). Developing experiments with sirens and tuning forks attached to wheels or turntables, Mach used rotation to examine the effects of relative motion between an observer and sound source. He was early to propose that Doppler-shifted spectral lines might enable a study of the motion of the stars, but sound was more approachable in the laboratory than the velocities required to realise the Doppler effect in light empirically. Mach also offered a careful account of the different theoretical perspectives and assumptions deployed by Christian Doppler and Josef Petzval (Mach 1873 [1861]). His earliest work thus sought to resolve an ongoing controversy and initiated studies of sound that gave him an exceptionally clear understanding of phenomena propagating in a medium, as Susan Sterrett (1998) has argued.

Yet Mach's strongest intellectual debt in this period was undoubtedly to Gustav Theodor Fechner's 1860 book, *Elemente der Psychophysik* (*Elements of Psychophysics*). This clarified many elements of the relations between psychological and physical effects that Mach had already started examining after reading Johann Friedrich Herbart's account of concepts and mental representations, and he immediately began research on the 'time-sense' of both the ear and the eye (for fuller accounts of Fechner and Mach's early psychophysical research, see Heidelberger 2010; Staley 2018). Mach used a variable pendulum (and later modified a Maelzel metronome) to establish just noticeable differences in the perception of oscillations, publishing the results of experiments he began in the summer of 1860 five years later (Mach 1865c). Similarly, he began research on perceptions of space with a study of the eyes' ability to determine angles and lengths, something he later related to the role of gravity in shaping the visual apparatus of our eyes, leading to significant differences in visual perceptions of vertical and horizontal symmetries (Mach 1861, 1896 [1871]). Perhaps the easiest way of conveying this is to use an illustration Mach favoured: 'It is a fact known to mothers and teachers', he wrote, 'that in their first attempts at writing and reading, children continually confound d and b, as well as q and p, but never d and q or b and p'. Symmetry in the horizontal plane is easily recognised (and similar impressions confused), while symmetry in the vertical plane is less readily perceived (Mach 1896 [1871], pp. 105–106, 1898 [1871], pp. 94–95). This is an interesting and perhaps unexpected respect in which Einstein's term 'mother's milk' provides an apt description of Mach: his research in sense perception often aimed to recover the experience of children, a perspective that guided him writing *Beiträge zur Analyse der Empfindungen* (*Contributions to the Analysis of Sensations*) (Mach 1886).

For Mach, such experiments illustrated significant differences between geometrical, physiological, and physical space. He articulated his views on space and time first in classroom lectures on mechanics in 1864 and on psychophysics in 1864–1865, and he set out their most general results in two papers published in the philosophical journal *Zeitschrift für Philosophie und philosophische Kritik*. Geometrical space was a matter of measurement, while physiological spaces were exemplified in a rich variety of aural, skin, muscle, and visual spaces. Physical space was an epitome of these developments. In the complete physics of the future Mach foresaw, it would be derived from the mutual dependence of phenomena and therefore necessarily involve both time and space. He admitted such reflections were crude and their results might appear trivial, but he affirmed that he had found them very useful in his physical studies (Mach 1866, p. 232; see also Mach 1865a; Banks 2003, pp. 37–38).

Mach's research on perception also led him to explicitly consider relational perspectives, noting in lectures on psychophysics in 1863 that extensive experimental studies of the relations between stimulus and response in perceptual determinations of weight, light, and sound intensity, tone, temperature, distance, and lengths had all shown that it is relative differences, not absolute differences that are critical for our perception. After all, the absolute difference in light intensity between the position of a star and the firmament surrounding it is the same by day and night, but the intensity of reflected sunlight leads us not to notice the difference by day (Mach 1863b, pp. 242–245, on p. 243). In Mach's early work, we can therefore observe him engaging elements that were later to become central to his critical history of mechanics, primarily through studies of perception; and these involved research on organs of perception as well as phenomena, developing a comparative perspective on different forms of perception and diverse spaces. While in 1863 Mach's lectures on psychophysics presented mechanics as the foundation of all physics, by 1867 he had reassessed this perspective and begun to present a markedly critical approach, moving between specific arguments and a much more wide-ranging critique of the mechanical world view.

The Frameworks and Forums for Critique

By the late 1860s, Mach had developed three specific lines of argument against existing mechanical perspectives. The first involved a suspicion of atomistic models, initiated in 1862 when Mach examined the mechanical world view in writing a compendium of lectures on physics for medical students, leading him to refer briefly and inconclusively to the need to reformulate perspectives on the foundation of physics at the book's preface and conclusion (Mach 1863a, pp. vi, 271–272). Mach's work in psychophysics had led him to the view that intuitions of space and time depended empirically upon perception. If this was so, it was not justifiable to assume spatial properties for matter not subject to sense perception. Attempts to represent the spectra of elements mechanically and their failure to match experiment led Mach to think that chemical elements must not be represented in three-dimensional space, yet as he put it later, 'I did not then dare to express this outright before orthodox physicists' (Mach 1872, p. 55, 1911 [1872], pp. 86–87).

Mach's second line of argument sought to resolve difficulties in the customary definition of mass, but it already linked acceleration and the law of inertia, the subject of his third line of argument. Later, Mach related that he first spoke of each of these critical points in the lecture courses he gave each year on different subject matters (such as psychophysics or mechanics) or on the principal questions of physics, but the paper he wrote on mass was rejected by the *Annalen der Physik* in 1867. Although it was published the following year in *Carl's Repertorium*, Mach hesitated to bring his other arguments before the wider community of physicists (Mach 1872, pp. 47–48, 50, 1911 [1872], pp. 75–76, 80). His classroom lectures thus became an extremely important forum for the gestation and articulation of Mach's views, a semi-private space at a critical distance from his discipline. However, in 1871, Mach used the opportunity of a lecture before the Royal Bohemian Society of Sciences to develop a pointed critique of the mechanical world view, which he published as a booklet in 1872 – so no journal editor would need to take responsibility for its content.

First, consider the paper that initiated his critiques of the foundations of the current world view (Mach 1868). Mach's article on mass began by outlining his view of a 'properly scientific' approach to the basis of mechanics. This would regard those theorems that experience has suggested as hypotheses (rather than as a priori principles) and then show how the rejection of these hypotheses would lead to contradictions with our best-constructed facts. The paper's aim was to clearly establish what should be regarded as a priori – which, in Mach's view, was only the principle of causation – and what should be treated as hypothetical or a matter of experience (Mach 1868, 1872, p. 50, 1911 [1872], pp. 80–81). His treatment of the concept of mass began with the empirical theorem that bodies experienced a mutual acceleration in the line of their conjunction, to then provide a definition of mass through empirically determined ratios – with the implication that the mass of a single body is undefined and never considered alone (in contrast to Newton's definition in terms of quantity of matter). Similarly, when he outlined a short set of theorems of experience and definitions at the end of the paper to offer the most 'scientific' arrangement of the theorems of mechanics, Mach began with the following: 'Theorem of experience. – Bodies placed opposite to one another communicate to each other accelerations in opposite senses in the direction of their line of junction. The law of inertia is included in this' (Mach 1868, p. 359, 1872, p. 53, 1911 [1872], p. 84). Together with its economy of expression, the latter inclusion also illustrates a different kind of relational argument tracing the interdependence of phenomena: the thought that the law of inertia should be considered within the dynamical framework of the force laws and gravitation, considered in terms of (directly measurable) accelerations and their absence.

Both his insistence on formulating the laws of mechanics on an empirical basis and a radical form of conceptual unification were to be leitmotifs of Mach's work; one can see each strategy in terms of the economic description of nature that Mach began articulating publicly as a goal of science in his lecture. Mach included this paper in the notes to its 1872 publication, and he also elaborated on his account of mass in the second chapter of *Die Mechanik in ihrer Entwickelung* in 1883. He developed the argument at length in section 5, immediately preceding the first of the two sections to which Einstein pointed in 1916, while Mach's empirical theorems and definitions were incorporated in the core of his formulation of the essential laws of mechanics (now described as simpler, methodologically better arranged, and more satisfactory than Newton's), which concluded section 7 (Mach 1883, pp. 202–207, 227). Einstein's notes show that in his 1909/1910 lectures on mechanics, he modelled his introduction of the concept of mass on Mach's, and he introduced the concept of electronic charge in an analogous way (Einstein 1993, pp. 5, 9).

Mach later wrote that 'all the essential features' of his *Mechanik* had been stated in this paper, while his epistemological position had been outlined for science in general and for physics with special exactness in his lecture on the conservation of work (which he reprinted when challenged by Planck; Mach 1901, p. 276; 1909 [1872]; 1919 [1902], p. 555). The observation highlights the importance of recognising how many years it took for Mach's work to win wider recognition. Although he held an academic professorship, Mach was an intellectually marginal figure for much longer and in a more significant disciplinary sense than Einstein, although both men went through similar stages of iconoclasm, acceptance, and late disciplinary distance, and each had a trusted confidant (Josef Popper-Lynkeus was Mach's Besso; see Mach 1872, p. 2). We should also consider the extent to which Mach's presentation of the 'essential features' depended on the forum in which his work was delivered, and how it changed over this substantial period. What is decisive in this respect is that, in contrast to the discussion of Newton and absolutes that Einstein picked out of Mach's 1883 book, his 1871 lecture focused on the principle of excluded perpetual motion and the conservation of energy; Mach critiqued specific laws of mechanics only in extensive supplementary notes to its publication. I will first draw out the contrast between the body and notes to Mach's lecture on work and then show that between his two rather different presentations of similar conceptual points in 1871 and 1883, Mach had tested the basic theorems of mechanics bodily.

Mach began his lecture to the Royal Bohemian Society with an account of what his mother had never been able to teach him. As a child, he had wondered why we should suffer the rulership of a king and why there was such a difference between rich and poor, questions his mother could never answer adequately. Mach found you could grow so accustomed to such puzzles that they no longer troubled you, or you could learn to understand them at the hands of history and thereby be able to approach them without hatred – for in the hands of history one could find the tools to doubt, test, and change the apparently self-evident, as he aimed to do in physics (Mach 1872, pp. 1–4, 1911 [1872], pp. 15–18). The principal argument Mach developed in 1871 was a three-stage discussion of the grounds for the mechanical view of nature, carefully distinguishing a limited mechanics from a full physics. Mach first demonstrated that the principle of excluded perpetual motion was not a fruit of recent arguments about the conservation of energy from Hermann von Helmholtz and others, but historically preceded them. Stevin, Galileo, and Christiaan Huygens had all relied on a prescription against unceasing motion in developing different specific principles. Mach also examined the current role of the mechanical view of nature to argue that it should be regarded as limited in scope: attempts such as those of Wilhelm Wundt to trace all phenomena to motion should be regarded as mechanical only and not phys-ical. Here, Mach raised his earlier queries about atomism, and he argued more clearly than previously that science was concerned with the connections between appearances, and what was represented behind appearances must be regarded as hypothetical only – but also that there was no need to impose on the creations of thought the limitations of the tangible and visible (Mach 1872, p. 27, 1911 [1872], p. 50). Mach's final step was to show that the principle of excluded perpetual motion was in fact a special form of the law of causation (adapting a formulation due to Fechner), which was a presuppos-ition for all scientific investigation – and since it rested on such wide experi-ence, it was therefore independent of the mechanical world view (Mach 1872, 1911 [1872], pp. 69–71).

Why did Mach's account of the logical and empirical roots of the mechan-ical world view begin with his mother? The end of his introduction and the book's concluding note offer clues. Introducing his study, Mach wrote that history had made all and could alter all: referring to his mother and social inequality expresses an important respect in which Mach sought a critical approach capable of understanding all. This surely helped win him followers amidst the circles of socialists in Vienna and Zurich, men such as Adler and Besso (Feuer 1974, pp. 11–22). After its final endnote, the book finished with a brief 'General remark' noting that we soon learn to distinguish our conceptions from sensations (perceptions). The task of science can then be split into three parts involving: establishing the laws of the connection of conceptions, which is psychology; finding the laws of the connection of sensations (perceptions), or physics; and clarifying the laws of the connections between sensations and conceptions, or psychophysics (Mach 1872, pp. 57–58, 1911 [1872], p. 91). These remarks express both the experiential and disciplinary sides of Mach's epistemology. He sought an approach capable of linking inner and outer and distinct disciplines – a form of unity still more radically extensive than Einstein's unified physics. Mach's endeavour to encompass a mother's knowledge or its lack offers a second respect in which Einstein's term 'mother's milk' offers a peculiarly apt description of Mach's work – he would offer an epistemology capable of shifting between home economics and the national economy, the soul and matter and interrelated scientific disciplines – all knowledge.

If the text of his lecture insisted on distinguishing between physics and the mechanical world view and he concluded with a general remark, Mach's notes more commonly took up critical particulars. Notably, the first offered his critique of customary understandings of the law of inertia, which is undefined if the law does not specify in relation to which frame of reference a body remains at rest or moves in uniform motion. Mach pushed this point to a paradox: there is no difference between the relative motion when the earth is considered at rest with the stars revolving around it, or conversely the earth rotates and the stars remain fixed. Geometrically the same, current physical understandings admit inertial effects such as the bulging of the earth at the equator in the second case alone – a result Mach thought had to be avoided. While Carl Neumann had noted this problem and introduced a hypothetical body at rest in absolute space to which other motions could be referred, Mach thought the law of inertia should be reformulated, and he indicated the significance of the actual bodies through which systems of coordinates are determined. Usually, the corner of a room or a tower would do, but in an earthquake or if the stars were to shake and move, how would one apply and express the law of inertia? Mach's point was that the bodies to which one must appeal explicitly or implicitly in describing a phenomenon should be regarded as belonging to the essential conditions, the causal nexus; and he sought a form of expression of the law of inertia that would recognise the share every mass has in the determination of the direction and velocity in the law of inertia. Mach admitted that no definite answer could currently be given to this by experience, and he promised to return to it in the future (Mach 1872, pp. 47–50, 1911 [1872], pp. 75–80).

That return came for the first time *in extenso* in Mach's 1883 treatise on mechanics, where he gave an alternative expression to Newton's in terms of the average acceleration in relation to distant masses. Limited by the fact that one could not sum the infinite masses of the universe it engaged, this at least showed the possibility of formulating the law of inertia without referring implicitly to absolute space (Mach 1883, pp. 218–222). This is where Einstein – after 1907 – took the most concrete inspiration for what in 1918 he christened 'Mach's principle', a generalisation of the demand to trace inertia back to the interaction of bodies that he first pursued without clearly distinguishing it from the generalised principle of relativity (Einstein 1918). Einstein initially saw this as a central goal and believed that general relativity exemplified it, yet he later came to think he was mistaken. Others have argued that he misunderstood Mach's view, but in fact general relativity realises it appropriately (Barbour 1990, 1989, 2007; Barbour and Pfister 1995). It is usually thought to be the most specific sense in which Einstein drew on Mach. There are two respects whereby the forum in which Mach discussed the law of inertia in 1883 subtly but importantly changed the sense in which his arguments would be understood. The first concerns the nature of *Mechanik* as a text, while the second concerns the way in which its chapter 2, section 6 presented a concentrated form of Mach's aims, and has often been taken to define them.

Mach's *Die Mechanik in ihrer Entwickelung*: Mother's Milk and More

As a treatise, Mach promised that his *Mechanik* was no textbook to inculcate mechanical principles, and also that its mathematics was secondary. Instead, its aims were anti-metaphysical: 'to expose the scientific content of mechanics, how we have come to this, from what sources we have created it, and how far it can be regarded as a certain possession' (Mach 1883, p. v). The participatory tone was central. It is appropriate that Mach won his reputation in good part through a book that emerged from classroom lectures and combined pedagogical aims with thoroughgoing critique; this is the most obvious respect in which Einstein's term 'mother's milk' is particularly apt. But it is notable that Mach felt that before readers were brought to criticise its foundations they needed to appreciate the character and tools of mechanics, as well as the imagination required for its development in the hands of Newton and others. Thus, it was only after 190 pages of conceptual analysis, mathematical development, diagrams of forces, and experimental apparatus, and after treating Newton's achievements at length, that Mach turned to critique (Mach 1883, p. 191).

Following Einstein's views, we have often thought that Mach principally trained readers in criticism, but I want now to raise the possibility that his text may also have helped train their creative imaginations. This was surely an aim of Mach's ceaseless investigations of subtly different conceptual perspectives (often while identifying common assumptions), but I will develop my argument by noting one particularly pertinent example. Before pivoting towards the critique of action and reaction and mass in which he developed further the content of his 1868 publication, in 1883 Mach offered some illustrative physical examples of the principle of reaction, including a revealing thought experiment in which he considered a load on a table. Mach noted that the table is pressed by the load only as much as the load presses on it, and if the table were to fall vertically in free descent, all pressure on it ceases, for as Mach wrote, the pressure on the table is determined by the relative acceleration of the load with respect to the table. The relative acceleration is determinative. The point is simple, but it was accompanied by an illustration depicting a table and load, and Mach went on to embed it in a rich context, pointing to Galileo's thought experiments on fall, discussing the behaviour of a pendulum in accelerated motion or free fall, and the peculiar sensations we feel jumping or falling from a height, or would feel if transported to other planets (Mach 1883, pp. 191–192).

Mach's thought experiment can be traced back to an occasion on which he had jumped into water and noted that the pressure of a load in his hand ceased (Mach 1865b, p. 329). Still more importantly, in 1873, Mach noticed a peculiar visual phenomenon while seated in a train banking through a corner, and he set out to investigate the physiological phenomena that allow us to determine the vertical, pairing laboratory studies of the bodily and visual perceptions of rotation with investigations of fall. He built a four-metre diameter rotating frame in which one could sit and be spun on an axis or in a circle; sat on a trolley rolling down an inclined plane constructed through three rooms of his laboratory; studied the perception of people in mineshaft elevators or heaving ships; and explored the experience of rise and fall possible under the controlled conditions of a see-saw. Mach's bodily experience of the laws of motion – of Galileo's experiments on an inclined plane and especially of Newton's spinning bucket experiment – enabled him to discover the mechanical system in the inner ear that leads to our perception of rotation and the vertical (Mach 1873b, 1875; Staley 2013, 2017).

In 1920, Einstein described the first decisive step he had taken towards general relativity in 1907. His publication had offered an abstract treatment, formally considering two frames of reference – one accelerated and the other located in a homogeneous gravitational field – and noting that, as far as was known, the laws of physics were the same in each. This allowed him to 'assume the complete physical equivalence of a gravitational field and a corresponding acceleration of the reference system', before exploring the consequences for clocks and spectral lines in a gravitational field (Einstein 1907). Einstein wrote that this happiest thought of his life had occurred to him while working in the Patent Office in Bern, when he realised the following:

The gravitational field has only a relative existence to an observer, similar to the electric field produced by magneto-electric induction. *This is because for an observer falling freely from the roof of a house there exists during his fall* – at least in his immediate environment – *no gravitational field*. If the observer lets a body go, it will remain at rest relative to him or in uniform motion, independent of its chemical and physical nature. The observer is justified in thinking of his state as being 'at rest'.

(Einstein 2002[1920], pp. 20–21, emphasis in original)

Though it has since been celebrated and elaborated pedagogically, historians have rarely given Einstein's comments on fall a history. Michel Janssen, for example, has written that here 'Einstein had only his imagination to go on' (Janssen 2014, p. 174). Yet, in fact, Einstein's thought experiment was a variation on Mach's thought experiment, which was in turn a variation on his extensive experimental work on the physiology of rotation and fall.

We can gain an understanding of why Einstein was not conscious of this imaginative debt by considering the sections in which Mach developed his central arguments about the law of inertia in 1883. After discussing action and reaction and mass, Mach turned to Newton's views on time, space, and motion, quoting at length from the passages in the *Principia* in which Newton distinguishes absolute, relative, and mathematical time and space and states his endorsement of absolutes. Declaring instead that his own book has shown that all of the theorems of mechanics concern experience of the relative positions and motions of bodies, after considering the need for multiple bodies in order to determine the direction of inertial forces, Mach turns to those points with which Newton appears to distinguish between relative and absolute with most justice: his discussion of rotation. Now, Mach considers the example of Newton's treatment of the rotating bucket (as well as the difference between a stationary earth and the Copernican perspective that he had described in 1872), arguing for a single interpretation since the relative motion is the same and stating that Newton had neglected to consider the effect of the earth and the distant stars in understanding the inertial effects of rotation (Mach 1883, pp. 207–222, on 216–217).

In 1883, Mach's arguments about inertia were thus incorporated in a critique of absolutes in general and Newton in particular, as Einstein's 1916 obituary recalled; its strong argument against Newton is surely one reason that this particular stance against metaphysics attracted so much attention. These passages provide the original source for the questions about absolute motion that August Föppl and Henri Poincaré extended into their treatments of electrodynamics, both of which were significant for Einstein; and here Einstein might also have found an exemplar for his more distinctive insistence that a single interpretation be given for the same relative motion between a magnet and a conductor, the opening argument of his 1905 paper

(Einstein 1905; Holton 1988b). Historians such as Renn have sometimes seen Mach's approach as philosophical (Renn 2007, p. 21), but while it undoubtedly reflected his epistemology, Mach certainly did not understand it as philosophically motivated. Nevertheless, while offering a study of physical space and time and charting new possibilities for the law of inertia, it is also true that in this section Mach refrained from considering the perceptual phenomena, psychophysical investigations, and physiological research that had in fact stimulated much of his early work on space and time and been integral to his epistemology. While a footnote drew attention to the exclusion of physiological space and time (Mach 1883, p. 210), few who read this account would have realised the extent to which bodily experience underwrote Mach's abjuration of metaphysical absolutes and recognise that Mach had *been* Newton's bucket before he critiqued it.

Mach was surprised by the success of *Die Mechanik*, and ultimately he watched it go through seven editions in his lifetime, with each edition noting specific responses to his work and responding both to support and criticism. Several of his additions indicate that Mach's attack on absolutes found wider favour, and by 1908, he was even writing of a small but growing group of decided 'relativists'; his work provides the basis for what Renn has described as a 'third way' to relativity through the critique of mechanics (Mach 1908, p. 257). Although focused on mathematicians and physicists who had discussed relative motion in particular, Mach's list of eight relativists began with J. B. Stallo and W. K. Clifford, who had both incorporated relational approaches in the context of much more wide-ranging abjurations of absolutes; this may be a reason for Mach's readiness to use the personal term. Ironically, he was then unaware that another small group of physicists and mathematicians had begun to think of themselves as adherents of the principle of relativity in electrodynamic theory, or of the Lorentz–Einstein theory of relativity, including Einstein, Planck, Max Laue, and Jakob Laub. When Hermann Minkowski's famous lecture on space and time made this new development apparent, Mach sought someone to tutor him in the theory; and when he republished his 1872 lecture in response to Planck's attack on him in 1909, Mach incorporated a note to affirm that he upheld the principle of relativity (Mach 1911 [1872], p. 95). Soon he wrote that space and time were, for him, problems that were being approached by Einstein, Minkowski, and Hendrik Lorentz.

Mother's Milk and More

In the course of this study of the relations between Mach's epistemology and his mechanics, I have identified several different senses in which Einstein's term 'mother's milk' offers a peculiarly apt description. First, we have seen that Mach's studies of sense perception sought to recover basic perceptual experience sensitive to the relations between different sensations of space and time – the vision of a child, for example. This helped initiate his work on physical space and time. Second, we have noted that Mach sought an epistemology suitable for examining conceptions and perceptual experience, psychology, physics, and psychophysics, but also the social world, incorporating the unfinished lessons of his mother and the sciences in the same framework of the economic description of phenomena. Often this has been seen in simply positivist terms stressing empirical measurement, but Mach equally sought knowledge of the soul and matter – of all experience. In a third, more familiar sense, Mach's writings provided a critical pedagogy, unfolding the relations between experience and the theorems of mechanics so as to open them to new approaches; mother's milk for the physicist. Further, I have shown here senses in which his published studies of mass, action and reaction, and inertia were all linked conceptually, but that they were also derived and emerged from perceptual studies and bodily experience. It is perhaps especially appropriate that we can identify an unrecognised imaginative resource for Einstein in Mach's pairing of a focus on relative acceleration, falling bodies, and bodily experience, bringing elements of his earlier phase of work into his 1883 discussion of mechanics; mother's milk and more. Finally, I would suggest that when Mach later turned to give an account of epistemology in practice, his discussions of thought experiments and the psychology of research offered tools we might use to understand not only Mach himself, but also Einstein's work still more fully than we do at present.

References

Banks, Erik C. 2003. *Ernst Mach's World Elements: A Study in Natural Philosophy.* Kluwer.

   2012. 'Sympathy for the Devil: Reconsidering Ernst Mach's Empiricism', *Metascience* 21: 321–330.

Barbour, Julian B. 1989. *Absolute or Relative Motion? A Study from Machian Point of View of the Discovery and the Structure of Dynamical Theories.* Cambridge University Press.

   1990. 'The Role Played by Mach's Principle in the Genesis of Relativistic Cosmology', in B. Bertotti, R. Balbinot, S. Bergia, and A. Messina (eds.), *Modern Cosmology in Retrospect.* Cambridge University Press, pp. 47–66.

   2007. 'Einstein and Mach's Principle', in J. Renn and M. Schemmel (eds.), *The Genesis of General Relativity, Volume 3: Gravitation in the Twilight of Classical Physics.* Springer, pp. 569–604.

Barbour, Julian B. and Pfister, Herbert (eds.) 1995. *Mach's Principle: From Newton's Bucket to Quantum Gravity.* Birkhäuser.

Blackmore, John T. 1972. *Ernst Mach: His Work, Life, and Influence.* University of California Press.

Einstein, Albert 1905. 'Zur Elektrodynamik bewegter Körper', *Annalen der Physik* 17: 891–921.

1907. 'Über das Relativitätsprinzip und die aus demselben gezogenen Folgerungen', *Jahrbuch der Radioaktivität und Elektronik* 4: 411–462.

1916. 'Ernst Mach', *Physikalische Zeitschrift* 17: 101–104.

1918. 'Prinzipielles zur allgemeinen Relativitätstheorie', *Annalen der Physik* 55: 241–244.

1949. 'Autobiographical Notes', in P. A. Schilpp (ed.), *Albert Einstein: Philosopher–Scientist*. The Library of Living Philosophers, pp. 2–94.

1993. *The Collected Papers of Albert Einstein, Vol. 5: The Swiss Years: Correspondence, 1902–1914*. Martin J. Klein, A. J. Kox, and Robert Schulmann, eds. Princeton University Press.

1998. *The Collected Papers of Albert Einstein, Vol. 8: The Berlin Years: Correspondence, 1914–1918. Part A: 1914–1917*. A. J. Kox, Martin J. Klein, Robert Schulmann, and Józef Illy edn. Princeton University Press.

2002 [1920]. 'Fundamental Ideas and Methods of the Theory of Relativity', in *The Collected Papers of Albert Einstein, Vol. 7: The Berlin Years: Writings, 1918–1921*, English translation of selected texts ed. Princeton University Press, pp. 113–150.

Feuer, Lewis S. 1974. *Einstein and the Generations of Science*. Basic Books.

Feyerabend, Paul K. 1984. 'Mach's Theory of Research and Its Relation to Einstein', *Studies in the History and Philosophy of Science* 15: 1–22.

Gooday, Graeme and Mitchell, Daniel 2013. 'Rethinking "Classical Physics"', in R. Fox and J. Z. Buchwald (eds.), *Oxford Handbook of the History of Physics*. Oxford University Press, pp. 721–764.

Heidelberger, Michael 2010. 'Functional Relations and Causality in Fechner and Mach', *Philosophical Psychology* 23: 163–172.

Hentschel, Klaus 1985. 'On Feyerabend's Version of "Mach's Theory of Research and Its Relation to Einstein"', *Studies in History and Philosophy of Science Part A* 16: 387–394.

Herneck, Friedrich 1959. 'Zu einem Brief Albert Einsteins an Ernst Mach', *Physikalische Blätter* 15: 563–564.

Holton, Gerald 1988a. 'Mach, Einstein, and the Search for Reality' (orig. 1968), reprinted in his *Thematic Origins of Scientific Thought: Kepler to Einstein*. Harvard University Press, pp. 237–277.

1988b. 'On the Origins of the Special Theory of Relativity', as reprinted in his *Thematic Origins of Scientific Thought: Kepler to Einstein*, Harvard University Press, pp. 191–236.

1992. 'More on Mach and Einstein', in J. T. Blackmore (ed.), *Ernst Mach – A Deeper Look: Documents and New Perspectives*. Kluwer, pp. 263–276.

Janssen, Michel 2014. '"No Success Like Failure . . .": Einstein's Quest for General Relativity, 1907–1920', in M. Janssen and C. Lehner (eds.), *The Cambridge Companion to Einstein*. Cambridge University Press, pp. 167–227.

Mach, Ernst 1861. 'Über das Sehen von Lagen und Winkeln durch die Bewegung des Auges', *Sitzungsberichte der Mathematisch-naturwissenschaftlichen Classe der Kaiserlichen Akademie der Wissenschaften Wien* 43: 213–224.

1863a. *Compendium der Physik für Mediciner.* W. Braumüller.

1863b. 'Vorträge über Psychophysik', *Österreichische Zeitschrift für praktische Heilkunde* 9: 146–148, 167–170, 202–204, 225–228, 242–245, 260–261, 277–279, 294–298, 316–318, 335–338, 352–354, 362–366.

1865a. 'Bemerkungen zur Lehre vom räumlichen Sehen', *Zeitschrift für Philosophie und philosophische Kritik* 46: 1–5.

1865b. 'Über Flussigkeiten, welche suspendirte Körperchen enthalten', *Annalen der Physik* 126: 324–330.

1865c. 'Untersuchungen über den Zeitsinn des Ohres', *Sitzungsberichte der kaiserlichen Akademie der Wissenschaften in Wien. Mathematisch-naturwissenschaftliche Classe* 51: 133–150.

1866. 'Bemerkungen über die Entwicklung der Raumvorstellungen', *Zeitschrift für Philosophie und philosophische Kritik* 49: 227–232.

1868. 'Über die Definition der Masse', *Repertorium für physikalische Technik, für mathematische und astronomische Instrumentenkunde (Carl's Repertorium der Physik)* 4: 355–359.

1872. *Die Geschichte und die Wurzel des Satzes von der Erhaltung der Arbeit.* J.G. Calve.

1873a. 'Einleitung', in *Beiträge zur Doppler'schen Theorie der Ton- und Farbenänderung durch Bewegung. Gesammelte Abhandlungen.* J. G. Calve, pp. 3–4.

1873b. 'Physikalische Versuche über den Gleichgewichtssinn des Menschen', *Sitzungsberichte der Kaiserlichen Akademie der Wissenschaften Wien* 68: 124–140.

1873 [1861]. 'Über die Controverse zwischen Doppler und Petzval, Bezüglich der Aenderung des Tones und der Farbe durch Bewegung', in *Beiträge zur Doppler'schen Theorie der Ton- und Farbenänderung durch Bewegung. Gesammelte Abhandlungen.* J. G. Calve, pp. 21–28.

1875. *Grundlinien der Lehre von den Bewegungsempfindungen.* Engelmann.

1883. *Die Mechanik in ihrer Entwickelung historisch-kritisch dargestellt.* F. A. Brockhaus.

1886. *Beiträge zur Analyse der Empfindungen.* Fischer.

1896 [1871]. 'Die Symmetrie', in *Populär-wissenschaftliche Vorlesungen.* J. A. Barth, pp. 100–116.

1896 [1894]. 'Über das Prinzip der Vergleichung in der Physik', in *Populär-wissenschaftliche Vorlesungen.* J. A. Barth, pp. 251–274.

1898 [1871]. 'On Symmetry', in *Popular Scientific Lectures,* 2nd edn. Open Court/Kegan Paul, Trench, Trübner & Co., pp. 89–106.

1901. *Die Mechanik in ihrer Entwickelung historisch-kritisch dargestellt,* 4. verbesserte u. vermehrte edn. F. A. Brockhaus.

1908. *Die Mechanik in ihrer Entwickelung historisch-kritisch dargestellt.* 6. verbesserte u. vermehrte edn. F. A. Brockhaus.

1909 [1872]. *Die Geschichte und die Wurzel des Satzes von der Erhaltung der Arbeit*. 2. unveränderter Abdruck nach der in Prag 1872 erschienenen 1. Aufl. edn. J. A. Barth.

1911 [1872]. *History and Root of the Principle of the Conservation of Energy*. Translated by P. E. B. Jourdain. Open Court/Kegan Paul, Trench, Trübner & Co.

1919 [1902]. *The Science of Mechanics: A Critical and Historical Account of its Development*. Translated by T. J. McCormack, 4th edn, with additions through the 4th German and 2nd English edn. Open Court.

1976 [1905]. *Knowledge and Error: Sketches on the Psychology of Enquiry*. D. Reidel.

Norton, John D. 2010. 'How Hume and Mach Helped Einstein Find Special Relativity', in M. Dickson and M. Domski (eds.), *Discourse on a New Method: Reinvigorating the Marriage of History and Philosophy of Science*. Open Court, pp. 359–386.

Planck, Max 1909. 'Die Einheit des physikalischen Weltbildes', *Physikalische Zeitschrift* 10: 62–75.

Renn, Jürgen 2007. 'The Third Way to General Relativity: Einstein and Mach in Context', in J. Renn and M. Schemmel (eds.), *The Genesis of General Relativity, Volume 3: Gravitation in the Twilight of Classical Physics*. Springer, pp. 21–75.

Staley, Richard 2008. *Einstein's Generation: The Origins of the Relativity Revolution*. University of Chicago Press.

2013. 'Ernst Mach on Bodies and Buckets', *Physics Today* 66: 42–47.

2017. 'Beyond the Conventional Boundaries of Physics: On Relating Ernst Mach's Philosophy to His Teaching and Research in the 1870s and 80s', in F. Stadler (ed.), *Integrated History and Philosophy of Science: Problems, Perspectives, and Case Studies (Vienna Circle Institute Yearbook)*. Springer, pp. 69–80.

2018. 'Sensory Studies, or When Physics was Psychophysics: Ernst Mach and Physics between Physiology and Psychology, 1860–71', *History of Science* doi:10.1177/0073275318784104.

Sterrett, Susan G. 1998. 'Sounds Like Light: Einstein's Special Theory of Relativity and Mach's Work in Acoustics and Aerodynamics', *Studies in History and Philosophy of Modern Physics* 29: 1–35.

Wegener, Daan 2010. 'De-Anthropomorphizing Energy and Energy Conservation: The Case of Max Planck and Ernst Mach', *Studies in History and Philosophy of Modern Physics* 41: 146–159.

Wolters, Gereon 1987. *Mach I, Mach II, Einstein und die Relativitätstheorie: Eine Fälschung und ihre Folgen*. Walter de Gruyter.

2012. 'Mach and Einstein, or, Clearing Troubled Waters in the History of Science', in C. Lehner, J. Renn, and M. Schemmel (eds.), *Einstein and the Changing Worldviews of Physics*. Springer, pp. 39–57.

3

Meaningful Work

Ernst Mach on Energy Conservation

DAAN WEGENER

Introduction

By the final decades of the nineteenth century, scientists generally agreed that the law of energy conservation was science's highest-level generalisation. What the law meant, however, was a highly contested issue. Even its name – and therefore, in a sense, its very identity – was subject to debate. Around this time, in the early 1870s, Ernst Mach gave his lecture *History and Root of the Principle of the Conservation of Energy* (or rather 'work' – *Kraft*) (Mach 1872/1910). This is widely regarded as one of Mach's key publications.[1] In the preface to the second edition, Mach said that it laid the foundation for his oeuvre.

This paper discusses the role of energy conservation in Mach's thought. For a proper understanding of it, we need to compare his statements about this law both with those of his contemporaries *and* with his own philosophical outlook – 'system' seems too strong a word. By comparing Mach with his contemporaries, we can gauge his originality and situate him within nineteenth-century debates about energy conservation. Mach's particular position within these debates can be understood more fully once we recognise how his statements about energy exemplify recurring themes in his work. At the same time, a close study of Mach's understanding of energy will contribute to our knowledge of the history of energy conservation in the nineteenth century, as well as deepen our understanding of Mach's overall philosophy.

It is, of course, impossible to cover the entire history of the law of energy conservation in the nineteenth century within the space of a single chapter. Fortunately, we can draw on existing studies of this subject. Four aspects of its history will be highlighted: its status within science as a fundamental law; the priority disputes that ensued in part as a consequence of this; terminological confusion related to disputes about the law's proper meaning; and finally the widespread use of this law, ranging from physics and physiology to psychology and sociology. Mach was simultaneously a participant in this debate

---

[1] Blüh (1970, p. 14), Blackmore (1972, p. 86), Haller and Stadler (1988, p. 20), Banks (2003, pp. 24, 180), and Staley (2018, p. 3).

(interpreting the law of energy conservation) and a commentator on it (interpreting debates on the law of energy conservation).

The law of energy conservation was especially praised for its unifying power. Hermann Helmholtz's *On the Conservation of Force* (1847) described it as a guiding thread that could be pursued in all branches of physical science. In an analogous fashion, one could use the law of energy conservation as a tool to explore and trace the connections between the many compartments of Mach's thought. Although some of these will be addressed in passing in this chapter, only one will be worked out: Mach's understanding of scientific meaning. Meaning, according to Mach, did not exist in disembodied minds, nor could it be captured by timeless and formal definitions. More down to earth, Mach understood meaning in terms of communication, practice, and history. Energy conservation was a prime example.

Mach in the Context of Nineteenth-Century Debates on Energy Conservation

In order to situate Mach with respect to nineteenth-century debates about the law of energy conservation, it is useful to enquire when and where he first heard about this law. Mach's own testimony is at once vague and revealing. In a footnote to his *Principles of the Theory of Heat*, he recalled:

> When I was at school, about 1853, I read somewhere the expression 'mechanical equivalent of heat'. By repeated occupation with mechanical constructions, the impossibility of constructing *mechanically* a *perpetuum mobile* had long become clear to me. The above expression at once made it subjectively certain to me that such a construction is impossible in any other way as well. When later I became acquainted with the principle of energy, this principle appeared to me trustworthy and almost self-evident.
>
> (Mach 1896/1986, p. 434, emphasis in original)

Interestingly, the passage distinguishes three stages: his practical experience with mechanical constructions; encountering the phrase 'mechanical equivalent of heat' and immediately recognising that it had implications beyond mechanics; and getting to know the energy principle – which felt more like recognising a known or familiar truth than discovering a new one. Thus, on the basis of this passage, we cannot pin down the precise moment when Mach first learnt the law of energy conservation. That may well be intentional. As we shall see, he believed that scientific concepts are fundamentally historical. They come into being gradually – both in science and in the individual mind.

The vagueness of the passage ('about 1853', 'somewhere', 'later') is striking. To appreciate it, one must consider the larger point that it served to illustrate: that in science it is difficult – indeed impossible – to think of ideas as personal property. Why? Because all scientists share in the convictions common to their time, which facilitates the quick exchange of ideas between them (Mach 1896/ 1986, p. 227). Once we think of scientific knowledge in terms of (oral) communication, we can appreciate that even individual accomplishment depends on countless intractable influences: 'We ought also to consider more the ready stimulus to independent inquiry that comes from conversation, even from a single word, or from hearsay' (Mach 1896/1986, p. 227). The lesson here is that to assess Mach's own understanding of the law of energy conservation, we must first listen to what his contemporaries had to say about it.

From the start, the law of energy conservation had been considered *a fundamental principle* – at least by those who introduced it. Thus, in different ways, both Robert Mayer and Helmholtz linked the law to a general principle of causality. When William Thomson first employed the term 'energy' in 1849, he did so in a grand and yet surprisingly casual manner (apparently assuming that the word itself required no prior clarification): 'Nothing can be lost in the operations of nature – no energy can be destroyed.'[2] In popular scientific lectures, such claims were sometimes repeated and inflated. In his infamous 1874 Belfast Address, for example, John Tyndall declared that the energy doctrine '"binds nature fast in fate," to an extent not hitherto recognised, exacting from every antecedent its equivalent consequent, from every consequent its equivalent antecedent' (Tyndall 1910, p. 478). Emil du Bois-Reymond stressed in 1882 that the law was more than a fact of experience; it agreed with the fundamental conditions of our intellect (du Bois-Reymond 1912, vol. II, pp. 145–146). In effect, the aims and methods of physical science were reformulated in terms of energy and the law of energy conservation. As James Clerk Maxwell opined in the mid-1870s:

> The principle of the conservation of energy is our unfailing guide. It gives us a scheme by which we may arrange the facts of any physical science as instances of the transformation of energy from one form to another. It also indicates that in the study of any new phenomenon our first inquiry must be, How can this phenomenon be explained as a transformation of energy? What is the original form of the energy? What is its final form? and What are the conditions of transformation?
>
> (Niven 1890, vol. II, pp. 594–595)[3]

Moreover, Maxwell conceptualised scientific experiments in terms of energy with its tripartite division of energy source, transport of energy, and measurement.[4] Max Planck took the law as a symbol of external and unchanging reality itself and effectively made it the shibboleth of science: 'Is there today a

---

[2] Thomson (1849/2015, vol. I, p. 118), Smith (1998, p .94), and Harman (1982, p. 51).
[3] Cf. Smith (1998, p. 126).
[4] Niven (1980, pp. 518–519). See Galison (1987, pp. 23–27).

single physicist worthy of serious consideration who doubts the reality of the energy principle? Rather, the recognition of this reality is nowadays a prerequisite for winning any scientific respect.'[5]

Mach also considered the law of energy conservation a fundamental principle, connecting it to the principle of causality and to scientific thought in general: 'We shall now attempt to show that the broad view expressed in the principle of the conservation of energy, is not peculiar to mechanics, but is a condition of logical and sound scientific thought generally.'[6] Yet it would be rash to equate Mach's understanding of the law with Tyndall's. While it is interesting to note the general tendency of linking the law of energy conservation to the principle of causality, it makes all the difference *how* this principle of causality is understood and used.

Mach famously defined causality as a functional relationship: '*the mutual dependence of phenomena*' (Mach 1872/1910, p. 61, emphasis in original). It should be noted that this notion of causality differs from more common definitions of causality that refer to succession in time (the effect has to occur at a later time than the cause) and some form of physical interaction (often mechanical). The law of causality or the interdependency of all phenomena (α, β, γ, ...) could be given various formal expressions, depending on convention:

$f(\alpha, \beta, \gamma, \ldots) = 0$, $a = \psi(\alpha, \beta, \gamma, \ldots)$, $F(\alpha, \beta, \gamma, \ldots) = \text{const.}$ (Mach 1872/1910, p. 73)

According to Mach, the law of energy conservation was an instance of the third expression. Indeed, he claimed to have deduced the law from his definition of causality: 'I believe that I have shown that the theorem of excluded perpetual motion is merely a special form of the law of causality, which law results immediately from the supposition of the dependence of phenomena on one another' (Mach 1872/1910, p. 73). It should be noted that Mach does not explicitly mention the conservation of work (or energy) in this quote, but only the impossibility of perpetual motion. The two are not identical. Yet in his 1871 lecture Mach did not explicitly distinguish the two. In fact, the impossibility of perpetual motion is initially presented as one of two common expressions of the law of energy conservation (Mach 1872/1910, p. 19). Finally, energy conservation follows directly from Mach's principle of causality if, following Erik C. Banks, we interpret the phenomena α, β, γ as work elements (Banks 2003, p. 182). This interpretation is supported, at least partially, by Mach's *Principles of the Theory of Heat*, where the energy principle is regarded 'as a contribution to the direct description of an extensive domain of facts' (Mach 1896/1986, p. 367).[7]

---

[5] Planck (1908/1970, p. 25) and Wegener (2010, pp. 146–159).
[6] Mach (1883/1974, p. 604, cf. p. 606).
[7] I would like to thank Kenneth Caneva for drawing my attention to this passage.

It would seem, therefore, that for Mach the law of energy conservation was indeed fundamental. For a correct understanding of Mach, however, some qualifications are necessary. First, he stresses in the 1871 lecture that he cannot become enthusiastic 'for the mysticism which some people love to push forwards by means of this theorem' (Mach 1872/1910, p. 73). Indeed, Mach can be read as saying that the law of energy conservation is merely a special form of the causal principle, nothing more or less. Other laws of nature may be expressed in a similar manner and are similarly linked to the causal principle. Second, for Mach, the causal principle describes what science is about but does not have any specific empirical content. He stressed this point repeatedly, and we will return to it later. Third, Mach was an anti-dogmatic thinker. For Mach, the law of energy conservation never became an article of scientific faith. Ever the opponent of dogmatism, Mach warned against statements such as, '"Man *must* not be descended from monkeys," "The earth *ought* not to rotate," … "Energy *must* be constant," and so on' (Mach 1886/1914, p. 40, emphasis in original). Fourth, Mach rejected cosmological claims about 'the energy of the *universe*' as nonsensical, because it cannot be measured (Mach 1986/1986, p. 439, emphasis in original).[8]

At times, Mach even seems to say that the law is not always valid: 'The principle of energy consists in a special form of viewing facts, but its domain of application is not unlimited' (Mach 1896/1986, p. 319). This striking conclusion follows upon a discussion of irreversible phenomena, associated with the second law of thermodynamics. The problem (identified by Thomson and repeated by Mach; Mach 1896/1986, p. 252) is how to reconcile energy conservation with the fact that when heat is conducted from a hotter to a cooler body some of the capacity to do work is lost forever. It is senseless, Mach said, to attribute a 'work value' to energy if it cannot actually be used (Mach 1896/1986, pp. 318–319). From the context of the discussion, however, it appears that Mach is specifically targeting the substance interpretation (or view) of energy here. Mach, of course, thought in terms of functional relations, not of substances. Taking a more general perspective, Mach claimed that the distinction between the first and second laws of thermodynamics was merely a historical coincidence. Indeed, he suggested that 'a full insight into the conservation of energy' would include both laws of thermodynamics (Mach 1896/1986, pp. 310–311).[9] To conclude, Mach's statements about the relationship

---

[8] These points are also addressed in Banks (2003, pp. 131–132, 218).

[9] Mach himself noted that there are analogous examples in the history of physics, where a single fact had initially been expressed (incompletely) by different principles. Compare Mach's *Mechanics*: 'The reason the perception of these facts was embodied in so great a number of principles is wholly an historical one; the perception was not reached at once, but slowly and by degrees. In reality only *one* great fact was established' (Mach 1883/1974, p. 306, emphasis in original). See also Preston (2008, p. 93).

between the law of energy conservation and the second law of thermodynamics seem contradictory. I would suggest the solution is that, according to Mach, the narrow substance conception of energy conservation was irreconcilable with the second law of thermodynamics, whereas his functional understanding of energy conservation included it.

The rising status of the law of energy conservation made its discovery a valuable property and so *priority disputes* were predictable. There were multiple potential candidates for the title 'discoverer of energy conservation'. As Thomas Kuhn argued in a classic paper, the law of energy conservation (or rather the elements required for its full articulation) was discovered by at least a dozen scientists around 1830–1850. The list can be extended. Indeed, nineteenth-century physicists were imaginative when it came to identifying discoverers of the law of energy conservation. Since the 'discoverers' came from different nations, the priority disputes often had a strong nationalistic undertone. Kuhn, of course, did not take sides. His whole point is that scientific discoveries are not made by individuals – they are not events that can be localised in space and time. For our purposes, it is interesting that Kuhn explicitly stated that 'no two of our men even said the same thing' (Kuhn 1977, p. 70). This created another ground for controversy. It was not simply a matter of who came first, but who found the 'correct' expression first.

Mach contributed to the priority disputes by criticising them. First of all, his historical-critical analyses of science – including but not limited to the history of energy conservation – reach back far in time. He sketched a history of the development of the law of energy conservation that stretched over 'many centuries'.[10] The principle of causality, to which the law of energy conservation is closely connected, is even older. There are no ultimate beginnings.[11] Secondly, Mach's comparative analysis of Mayer, Helmholtz, and James Prescott Joule (the main protagonists of the priority dispute) argued that each had made his own contribution to the energy principle (Mach 1896/1986, pp. 224–250). It should be noted, however, that Mach's sympathies went out to Mayer, whose words inspired him and in whose work he read some of his own ideas. Thirdly, as indicated earlier, for Mach, science was a collective enterprise in which individual accomplishment rested on countless untraceable influences. From this perspective, priority disputes are simply misguided.

Terminological Confusion and Widespread Use

Terminology was intimately bound up with the 'correct' expression of the law of energy conservation. So far, we have spoken of 'the law of energy

---

[10] Mach (1872/1910, p. 40. Cf. Mach (1896/1986, p. 226).
[11] Cf. Cohen (1970, p. 155): 'it idealises all reality backwards into the *Urzeit*'.

conservation', and for convenience will continue to do so, but it should be observed that contemporary usage was far more varied. There were 'alternatives' for all of its three components (law, energy, and conservation). Instead of 'law', some spoke of 'doctrine' or 'principle'; instead of 'energy', some spoke of 'force' or 'work'; and instead of 'conservation', some spoke of 'correlation', 'interaction', or 'persistence'. Minor other variations existed, but these were the most common ones. The phrase 'law of energy conservation' was thus only one of many ways of combining these components.

Kuhn emphasised that shifts in terminology can reveal underlying conceptual shifts.[12] As Kenneth Caneva and others have noted, Helmholtz's 'conservation of force' cannot simply be rendered as conservation of energy.[13] Interestingly, although his presentation of this law changed significantly after his visit to England in 1853 (implicitly equating it with the law of energy conservation, stressing the importance of the concept of work or mechanical effect, and firmly rooting it in experience), he continued to speak of the conservation of force. From the 1850s onwards, one finds explicit reflections on the varying terminology whereby alternatives are sometimes introduced and legitimised.

In a 1858 lecture, for example, the Dutch physicist Johannes Bosscha Jr introduced the phrase 'conservation of work capacity', explaining that 'conservation of force' is misleading because the 'force' in 'conservation of force' is conceptually different from the term 'force' in Newton's laws of motion (Bosscha Jr 1858).[14] He gave the example of Michael Faraday who, evidently confused by the terminology, had argued that gravity violated the conservation of force, because as an object falls its force of motion or *vis viva* increases, and at the same the force of attraction also increases (as the distance to the centre of attraction decreases). The terms 'actual' and 'potential' energy, introduced and propagated by William Macquorn Rankine and William Thomson, had the benefit of being derived from the Greek and were therefore easily translatable into different languages, but Bosscha Jr thought 'energy' would not remind readers enough of its physical meaning. In his view, 'work capacity' was more evocative.

For some physicists, energy or work was more than an additional physical concept to be distinguished from force. In a sense, energy increasingly threatened to displace force from its central position in physics. Physicists such as Gustav Kirchhoff and Heinrich Hertz had serious reservations about the concept of force. Kirchhoff, for this reason, preferred to speak of the

---

[12] For example, see Kuhn (1978, p. 356).

[13] See Elkana (1970, pp. 263–298), Smith (1998, pp. 126–138), Wegener (2009b, pp. 265–287), and Cahan (2012, pp. 55–68). By far the most comprehensive discussion of terminology related to energy conservation can be found in Caneva (2021).

[14] Cf. Wegener and van Lunteren (2012, pp. 384–399).

'constancy of work' (Kirchhoff 1865, p. 20). Some, such as the energeticists Wilhelm Wien and Wilhelm Ostwald, went further and saw force as a derivative concept to which no reality whatsoever ought to be attributed. To them, everything was energy.

Mach's terminology changed over time. Initially, he had used Helmholtzian terminology and treated the 'conservation of force' as a law of mechanics (Mach 1863, p. 49). In his 1871 lecture, however, Mach preferred to use the term 'work'. He explained that this was 'understood by all' and less likely to lead to misunderstandings than 'force' (Mach 1872/1910, p. 19). Furthermore, Mach's terminological preference reflected his criticism of the mechanical world view. One of the main points of the lecture is that the mechanical world view – based on point-particles governed by forces acting at a distance – was not a scientific necessity, but the product of 'convenience, history, and custom' (Mach 1872/1910, p. 57). More specifically, Mach stressed that although Helmholtz had shown how the law can be derived from the mechanical world view, it is actually much deeper, with historical roots predating even the science of mechanics (Mach 1872/1910, pp. 38–39). Notwithstanding his well-known criticism of the mechanical world view, Mach did not say that the concept of force had no scientific legitimacy. As Banks has shown, Mach did not believe that the concept of energy could fully replace that of force. Moreover, as John Preston has pointed out, Mach's *The Science of Mechanics* 'contains nothing one could call a critique of the concept of force' (Preston 2008, p. 100). The main reason for using the term 'work' rather than 'force', then, was that different concepts require different words to avoid confusion. Interest in terminology is also evident in Mach's other writings. The historical part of his *Principles of the Theory of Heat* devoted attention to the various phrases used by Mayer, Helmholtz, and Thomson (Mach 1896/1986, p. 238). In its critical part, addressing the limits of the energy principle, he noted that 'a better terminology here appears desirable' (Mach 1896/1986, pp. 38–39). It should be emphasised that, for Mach, terminological reform was never a goal in itself. As far as possible, he used ordinary language (Mach 1886/1914, p. 47). It might be that this is why, in 1871, he preferred the term 'work' to 'energy'. That he generally referred to the energy principle in his later writings may simply reflect that the language of energy had become more conventional at the end of the century.

Mach recognised that terminological consensus does not mean uniformity of use. The *widespread use* of the law of energy conservation partly accounts for divergence in how it was interpreted. Moving from one disciplinary context to another, concepts are transformed. As Mach observed:

> [The] fermentation of ideas gives rise to very remarkable phenomena. While many physicists are concerned to purify physical conceptions by psychological, logical and mathematical methods, other physicists,

mistrustful of this tendency, and more philosophical than the philosophers themselves, are coming forwards as advocates of the old
metaphysical conceptions which the philosophers have already largely
abandoned. Philosophers, psychologists, biologists, and chemists, all make
the most widely extended applications of the principle of energy and of
other physical concepts, with a freedom which the physicist would hardly
venture to use in his own field. We may almost say that the customary
roles of the special departments have been interchanged.

(Mach 1886/1914, p. 84)

This seems so outlandish that one must go a little into its history and
philosophy. As Timothy Lenoir has shown, Du Bois-Reymond's and
Helmholtz's deployment of the law of energy conservation was part of their
campaign for reductionist physiology and against vitalism (Lenoir 1997,
pp. 75–95). The argument came down to this: if there is a 'life force', distinct
from all other forces of nature, then this comes into being and ceases to exist
with each individual living thing. But the law of energy conservation prohibits
such a spontaneous (dis)appearance of energy. Although the logic of this
reasoning can be contested, it seems that the argument was effective (at least
in combination with the other 'boundary work' done by the Berlin
physiologists).

In similar fashion, an argument could be constructed against the intervention of mind (human or divine) in nature.[15] In a nutshell, the argument was
that Cartesian dualism cannot be reconciled with the following three
assumptions:

(1) Energy can only be carried by matter.
(2) Causal interaction involves the exchange of energy.
(3) Energy is conserved.

If the mind were to act on the brain, such as by setting a brain molecule in
motion, then according to (2), the mind would have to add a bit of energy to
the brain. This would either violate (1), because the immaterial mind would
then have to be a carrier of energy, or (3), because energy would be created out
of nothing. The power of the argument was its simplicity. But the first two
assumptions were often challenged.

One can challenge the first assumption by saying that potential energy is an
immaterial form of energy and then suggesting that the mind could also be seen
as an immaterial form of energy. More generally, energy itself was sometimes
portrayed as an immaterial substance. In this case, (1) would prove to be false,
while (2) and (3) could be maintained. In practice, this defensive strategy did not

---

[15] The following discussion draws on Wegener (2009a, pp. 121–151).

really favour dualism. Instead, it was the energetic monists such as Ostwald who tended to emphasise that energy is an immaterial substance.

Dualists denied the second assumption, typically using two counterexamples. The first was the 'normal force', a force that is perpendicular to the motion of the object it is acting on. A normal force changes the direction of an object but leaves the magnitude of its velocity unchanged. Because the kinetic energy of an object is indifferent to direction (is not a vector quantity), a normal force can act on an object without changing its energy. 'Unstable equilibrium' was their second counterexample. In unstable systems, minor actions may have major effects, as when a single spark causes a huge explosion, liberating large amounts of energy. In theory, there were systems so unstable that an arbitrarily small amount of energy could set them in motion. Dualists suggested that the mind was like a normal force, or that the brain was an unstable system. In both cases, (2) would prove to be false, while (1) and (3) could be maintained.

Although the implications of energy conservation for the mind–body problem were debated by philosophers, psychologists, and physicists of varying stature, not all were convinced of the importance of this question. Mach, for example, was quite amazed by the debate: 'I cannot refrain from here expressing my surprise that the principle of the conservation of energy has so often been dragged in in connection with the question whether there is a special psychical agent' (Mach 1886/1914, p. 55, note 1). Why the surprise? In the first place, the philosopher 'misses the point' of the law of energy conservation (Mach 1886/1914). The energy principle does not uniquely determine what is going to happen. There is room for other factors, such as normal forces, that do not expend or create any energy. Moreover, the debate presupposed a substantial conception of energy, whereas Mach had a relational conception of energy. Secondly and similarly, the mind–body problem as conceived in this debate – which focused on the possibility of physical interaction between mind and body – was a non-issue when viewed from Mach's radically different, 'sensational' perspective. (It is remarkable, therefore, that fragments from his notebooks show that Mach occasionally did take the debate seriously.[16]) Finally, Mach noted that 'the stock reply of the physicist has no intelligible meaning in relation to a case so far removed from the scope of his ideas' (Mach 1886/1914, p. 55).

Fresh Perspectives on (the) Meaning (of Energy Conservation)

More than just a participant in nineteenth-century debates about energy conservation, Mach was an interpreter of them. Although scientists agreed

---

[16] Banks (2003, p. 119). Cf. Mach (1905/1976, p. 195).

that the law of energy conservation was of fundamental importance, there was no agreement over its meaning. Mach could make sense of the situation by taking a more general perspective. In fact, he introduced an original conception of scientific meaning itself, grounding it in communication, practice, and history.

The history of statics clearly illustrated 'the process of science generally'. Its beginnings pointed to the importance of the 'experiences of the manual crafts'. The next step – crucially, the one to which science owed its origin – came with 'the necessity of putting these experiences into *communicable* form' (Mach 1883/1974, p. 89, emphasis in original). This form, however, is necessarily partial and speculative; it can be no more than a sketch or caricature, a 'rough *outline*' of the 'entire fact, in all its infinite wealth, in all its inexhaustible manifoldness' (Mach 1883/1974, p. 90, emphasis in original).[17] As he put it in *Knowledge and Error*: 'If we mix up ideas or concepts with facts, we identify what is poorer and subservient to special purposes with what is richer and indeed inexhaustible' (Mach 1905/1976, p. 102). Mach's economy of thought was thus not one of cataloguing 'simple facts', the prosaic work of a librarian, but of talking about inherently complex experience.[18] It is a by-product of the desire to communicate.

The fundamental role of communication in Mach's thinking has been overlooked by most commentators.[19] In his *Principles of the Theory of Heat*, Mach expanded on this. A scientific theory, he explained there, is the result of two processes: processing sense experience and actively reproducing the facts in thought. 'This reproduction, if it is to have a scientific character, must be communicable' (Mach 1896/1986, p. 116). Given the modern division of scientific labour, the limited horizon of experience of the individual researcher needs to be expanded by '*linguistic communication [sprachlichen Mitteilung]*' (Mach 1896/1986, p. 336).[20] And again: 'Science can grow vigorously only by the fusion of the experiences of many men, by language [*sprachlichen*

---

[17] Compare '*Wir denken ja auch nicht, wenn ich so sagen darf, in den natürlichen Verhältnissen, sondern übertreiben in Gedanken um zu verstehen*' ('We also do not think, if I may say so, in natural conditions, but exaggerate in thought to understand') (Mach, cited in A. Hohenester, 'Ernst Mach als Didaktiker, Lehrbuch- und Lehrplanverfasser', in Haller and Stadler (1988, pp. 150–151)). And compare: 'One could say that the retina schematises and caricatures.... It is an analog of abstraction and of the formation of concepts' (Mach, cited in Ratliff (1970, p. 39)).

[18] Neither Mach's notion of concepts nor his notion of facts has received much attention. For a comparison between Hertz's theory of scientific 'images' and Mach's notion of scientific 'concepts', see Preston (2008, pp. 93–94). For a comparison between Mach's and Einstein's views about 'facts', see de Waal and ten Hagen (2020).

[19] The importance of 'sharing' is noted, however, in Cohen (1970, p. 127).

[20] See Mach (1896/2016, p. 411, emphasis in the original).

Mitteilung]' (Mach 1896/1986, p. 396).[21] The original German seems to suggest that Mach specifically had oral communication in mind. Mach's argument that conversation, a single word, and hearsay can move us into action exemplifies his overall emphasis on the importance of (oral) communication in science.

Scientists use a wide variety of linguistic tools to communicate, including words and formulas. Whence do they derive their meaning? How is it that they are understood? Interestingly, Mach firmly grounded meaning in practical contexts of use. In the first place, historically speaking, scientific concepts emerged from the experiences of manual labour. The history of the science of mechanics furnishes an example, Mach said, adding that we should devote more attention to the role of material culture in science (Mach 1883/1974, p. 104). It is even true on a personal level. Recall that when Mach heard the phrase 'mechanical equivalent of heat', he found this neatly captured his own experience with constructing mechanical engines. Secondly, as we have seen in the previous paragraphs, scientific descriptions are neither complete nor neutral; instead, they serve what 'is of importance for the given technical (or scientific) aim in view' (Mach 1883/1974, p. 90, cf. p. 578). Finally, and perhaps most importantly, understanding a scientific concept and knowing how to use it amount to the same thing. Although the principles of causality, of sufficient reason, and of the impossibility of perpetual motion may be excellent instruments 'in the hands of an experienced investigator', they become 'empty formula[s] in the hands of even the most talented people in whom special knowledge is lacking' (Mach 1872/1910, p. 69, see also p. 72). In his 1871 lecture on the conservation of work, Mach's main point is that these principles require experience to acquire empirical content. This has not escaped the attention of Mach's commentators. But the repeated reference to 'hands' seems to anticipate a more original point, made explicit in later writings, that hands-on experience is required for a full understanding of the principles. In *The Science of Mechanics*, he said:

> If we know principles ... only in their abstract mathematical form, without having dealt with the palpable simple facts, which are at once their application and their source, we only half comprehend them ... We are in a position of a person who is suddenly placed on a high tower but has not previously travelled in the district round about, and who therefore does not know how to interpret the objects he sees.
>
> (Mach 1883/1974, pp. 394–395)

---

[21] See Mach (1896/2016, p. 444, emphasis in the original). Cf. Mach (1896/1986, p. 237).

And, in *Knowledge and Error*, we are told:

> [P]ractice in enquiry, if it can be acquired at all, will be furthered much
> more through specific living examples, rather than through pallid abstract
> formulae that in any case need concrete examples to become intelligible.
>
> (Mach 1905/1976, p. xxxi)

Mach repeatedly compared science to craft knowledge, playing the piano, and
cooking a dish.[22] 'A concept cannot be passively assimilated; it can be acquired
only by doing, only by concrete experience in the domain to which it belongs
[*nur durch* Mittun, Mitleben *in dem Gebiet, welchem der Begriff angehört*].
One does not become a piano player, a mathematician, or a chemist, by
looking on; one becomes such only after constant practice of the operations
involved.'[23] (The original German is suggestive of the communal nature of
science.) And elsewhere: 'Long and thoroughly practised *actions* ... are thus
the very kernel of concepts.'[24] Concepts are not images – they are activities, or
an integral part thereof.[25] The concept of work and the law of energy
conservation may serve as illustrations:

> Of course the development of the concept of force [*Arbeit*] has a long
> history ... However, if you are conscious that you can always carry out
> this test [calculating the work for a given displacement] knowing that a
> case of equilibrium will yield zero or a negative sum while a case of
> motion yields a positive, then you have a concept of work and can by
> means of it distinguish static from dynamic cases.
>
> (Mach 1905/1976, p. 97)

> That general concepts are not ... mere words, clearly emerges from the
> fact that very abstract propositions are understood and correctly applied
> in concrete instances, witness the countless applications of the propos-
> ition 'energy remain constant'.
>
> (Mach 1905/1976, p. 92)

Mach's criticism of the viewpoint that concepts are mere words or images is
in line with his rejection of mind–body dualism. Specifically, as Richard Staley

---

[22] Cf. Cohen (1970, p. 138): '[F]or [Mach] laws are *recipes for producing* individual descriptive statements ... Our analogy to the kitchen is fair because recipes are to be practiced'
(emphasis added). Cohen does not seem aware that Mach basically made the same
analogy. Also compare the Carl Runge quote in Warwick (2003, p. 227).

[23] Mach (1896/1986, p. 382). Mach (1896/2016, p. 468), with original emphasis on *Mittun*
and *Mitleben*. Also see Mach (1905/1976, p. 95) and Mach (1896, p. 368).

[24] Mach (1896, p. 252). See also Mach (1896/1986, p. 369). For connections with Mach's
influence and relevance for education, see Zudini and Zuccheri (2016, pp. 651–669).

[25] As a 1910 notebook briefly puts it: '*Begriffe* = *Tätigkeitsimpulse und
Denktätigkeitsimpulse*' (Haller and Stadler 1988, p. 209).

has recently argued, for Mach, the senses already do part of the mental labour (Staley 2018). 'The sense-organs themselves are a fragment of soul; they themselves do part of the psychical work, and hand over the completed result to consciousness' (Mach 1886/1914, p. 71). We are convinced of the principle of inertia not just because we understand it on reflection, but because we *feel* it. This is not detrimental to our theoretical knowledge, but complements it because the senses and indeed our entire organism can also be seen as a piece of machinery, acting like mechanical measuring instruments.[26] Shifting his perspective, but with the same effect of blurring the distinction between mind and body in the production of knowledge, Mach described prejudice as 'a sort of reflex motion in the province of intelligence'.[27] Mach's view that feeling, action, and thought cannot be entirely separated is consistent with his understanding of concepts as well-practiced activities. The same goes for his argument that there is continuity between instinct and scientific knowledge.[28] The fact that Mach praised Mayer's 'powerful formal instinct' should thus be understood against the background of his acknowledgement of the epistemological function of instinct (Mach 1896/1986, p. 231).

Mach's understanding of meaning also incorporated a historical dimension. The belief that scientific concepts, laws, and theories are fundamentally historical entities underpinned Mach's historical-critical analyses. One can gain a fuller understanding of the content of science – to the point of being able to criticise it – by studying its history. As he repeatedly pointed out, the religious connotations (God's works are indestructible) and the economical connotations (balancing accounts) of the law of energy conservation make sense from a historical perspective. In general, Mach believed that everything that we seek behind the relations between phenomena is liable to reflect and change with cultural circumstances (Mach 1872/1910, p. 49). This might suggest that we can liberate ourselves from history by studying it in order to identify and then eliminate what is metaphysical or arbitrary, but this is not quite true. In the first place, Mach famously said that convenience, history, and custom dictate which facts we regard as fundamental (Mach 1872/1910, p. 57). Consequently, the foundations of science have an ineluctable historical component. Secondly, the concepts of science are so soaked in history that their full meaning can be grasped neither by a timeless formal definition (which requires interpretation anyway) nor by a snapshot in time, but only by a historical narrative. To repeat and extend a previous quote on energy conservation:

> That general concepts are not ... mere words, clearly emerges from the
> fact that very abstract propositions are understood and correctly applied

---

[26] Mach (1883/1974, p. 394). For a similar comparison, see Ratliff (1970, p. 40).
[27] Mach (1896, p. 232); Cohen (1970, p. 144).
[28] Mach (1905/1976, pp. 17, 200); Mach (1883/1974, p. 394).

in concrete instances, witness the countless applications of the proposition 'energy remain constant'. It would however be idle to attempt to find
a clear, momentary conscious idea which would exactly cover the sense of
the sentence as it is spoken or heard. The difficulty disappears if we
recognise that concepts are not momentary entities ... every concept
has its sometimes long and eventful formative history, and its content
cannot be explicitly be expounded by a transient thought.

(Mach 1905/1976, p. 92)

It is fitting that he chose energy conservation as an example. He had
discussed its very long and complex 'formative history' in three of his historical-critical studies (of the conservation of energy, of mechanics, and of
thermodynamics). Moreover, nineteenth-century discussions about the proper
meaning of the law of energy conservation can be understood and overcome,
from a Machian perspective, once one realises that meaning is determined by
communication, practice, and history. Even the development of Mach's own
understanding of the law of energy conservation, at least as he recalled it,
conveniently confirmed his own down-to-earth meta-theory.

Conclusion

The law of energy conservation played a central role in Mach's thought. We
have contextualised Mach's statements on it in two different ways: firstly, by
placing them against the background of the nineteenth-century history of the
law of energy conservation; and secondly, by relating them to Mach's theory of
(scientific) meaning. The two are linked, I would suggest, because the meaning
of the law of energy conservation was highly disputed. Mach's theory of
meaning addressed this issue from a more general perspective.

Four aspects of the nineteenth-century history of energy conservation have
been addressed in this chapter. The first aspect is the fundamental status of the
law as a principle of causality, as a guide in science, and as something of a
scientific creed. Mach's position with respect to this issue was ambivalent: on
the one hand, he linked the law of energy conservation to scientific thought
'per se', while on the other hand, he criticised the dogmatism surrounding it.
The second aspect regards priority disputes about the true discoverer of the
law. From Mach's perspective, by emphasising the historical and collective
nature of science, these debates were misguided. It is worthwhile, however, to
see how different individuals made their own personal contributions to the law
of energy conservation. The third aspect relates to terminological confusion
and attempts to avoid it. In his 1871 lecture, Mach preferred 'conservation of
work' to 'conservation of force'. Later, he also employed the phrase 'energy
principle'. He called for further terminological refinement to deal with conceptual problems related to the first and second laws of thermodynamics.

The fourth and final aspect involves applications of the law in different disciplinary contexts. Neither the property of a single individual nor of any particular discipline, the meaning of the law of energy conservation could shift depending on its context of use. Mach – equally at home in the fields of physics, physiology, psychology, and philosophy – had a particularly clear view of the situation and appears to have been amused by the abiding confusion.

When compared to his contemporaries, Mach held somewhat aloof. He was an astute observer of his time, not just a 'representative' thereof. Examples of scientific theory and practice were always connected to the larger issue of how science works. To fully appreciate the significance that the law of energy conservation had for Mach, it should therefore be connected to more general concerns in his philosophy of science. His account of meaning is particularly suitable for this, because it neatly captures his meta-perspective on nineteenth-century debates about the meaning of the law of energy conservation. Unsurprisingly, the law of energy conservation was one of his main examples. I have argued that, in Mach's analysis, meaning is grounded in communication, practice, and history. These three dimensions of meaning are closely related. Separating them is artificial, as many passages in Mach's work can be used to illustrate several of them at once. Science is a collective enterprise driven by the merger of horizons of experience through verbal communication. Concepts and their meanings could not be the property of a single individual. Indeed, as Mach saw it, historically, concepts were born out of the need to communicate and tailored to particular contexts of use. To grasp their proper meaning, years of practice and participation in the (communal) life of the profession were necessary. Without routine use, laws such as energy conservation had no meaning. It follows that their meaning cannot be grasped instantaneously or covered by a static picture. Both in the individual mind (and hands!) and in science, laws develop in time – that is, historically. Hence the necessity of approaching the content of science through historical narratives. They alone are flexible enough to capture the protean nature of scientific meaning.

In light of this, there is something paradoxical in speaking of Mach's understanding of the law of energy conservation. The more singular, abstract, and unchanging his conception of it appears, the more it contradicts his own theory of scientific meaning. Of course, nothing requires us to apply Mach (consistently) to Mach. A Machian outlook would nevertheless suggest that we should portray him in constant dialogue with his contemporaries, pay attention to how he put the law to use in different contexts, and be open to the idea that his ideas continued to develop. In this respect, Mach's recollections of how he first learnt the law of energy conservation can serve as a reminder.

References

Banks, Erik C. 2003. *Ernst Mach's World Elements: A Study in Natural Philosophy.* Kluwer.

Blackmore, John T. 1972. *Ernst Mach: His Work, Life, and Influence.* University of California Press.

Blüh, Otto 1970. 'Ernst Mach – His Life as a Teacher and Thinker', in Robert S. Cohen and Raymond J. Seeger (eds.), *Ernst Mach: Physicist and Philosopher.* D. Reidel, pp. 1–22.

Bosscha Jr, Johannes 1858. *Het Behoud van Arbeidsvermogen in den Galvanischen Stroom. Eene Voorlezing.Voorgedragen in het Natuurkundig Gezelschap te Utrecht.* Sythoff.

Cahan, David 2012. 'Helmholtz and the British Scientific Elite: From Force Conservation to Energy Conservation', *Notes and Records of the Royal Society*, 66: 55–68.

Caneva, Kenneth 2021. *Helmholtz and the Conservation of Energy: Contexts of Creation and Reception.* MIT Press.

Cohen, Robert S. 1970. 'Ernst Mach: Physics, Perception and the Philosophy of Science', in Robert S. Cohen and Raymond J. Seeger (eds.), *Ernst Mach: Physicist and Philosopher.* D. Reidel, pp. 126–164.

Cohen, Robert S. and Seeger, Raymond J. (eds.) 1970. *Ernst Mach: Physicist and Philosopher.* D. Reidel.

de Waal, E. and ten Hagen, S. (2020). 'The Concept of Fact in German Physics Around 1900: A Comparison between Mach and Einstein', *Physics in Perspective*, 22: 55–80.

du Bois-Reymond, E. (ed.) 1912. *Reden von Emil du Bois-Reymond*, 2 vols. Von Veit.

Elkana, Yehuda 1970. 'Helmholtz' "*Kraft*": An Illustration of Concepts in Flux', *Historical Studies in the Physical Sciences*, 2: 263–298.

Galison, Peter 1987. *How Experiments End.* Chicago University Press.

Haller, Rudolf and Stadler, Friedrich (eds.) 1988. *Ernst Mach. Werk und Wirkung.* Hölder-Pichler-Tempsky.

Harman, Peter M. 1982. *Energy, Force and Matter: The Conceptual Development of Nineteenth-Century Physics.* Cambridge University Press.

Kirchhoff, Gustav R. 1865. *Ueber das Ziel der Naturwissenschaften.* Mohr.

Kuhn, Thomas S. 1977. 'Energy Conservation as an Example of Simultaneous Discovery', in his *The Essential Tension: Selected Studies in Scientific Tradition and Change.* Chicago University Press, pp. 66–104.

   1978. *Black-Body Theory and the Quantum Discontinuity 1894–1912. With a New Afterword.* Chicago University Press.

Lenoir, Timothy 1997. *Instituting Science: The Cultural Production of Scientific Disciplines.* Stanford University Press.

Mach, Ernst 1863. *Compendium der Physik für Mediciner.* Braumüller.

1872/1910. *History and Root of the Principle of the Conservation of Energy.* Open Court.

1883/1974. *The Science of Mechanics: A Critical and Historical Account of Its Development.* Open Court.

1886/1914. *The Analysis of Sensations and the Relation of the Physical to the Psychical.* Open Court.

1896. *Popular-Scientific Lectures.* Open Court.

1896/1986. *Principles of the Theory of Heat, Historically and Critically Elucidated.* Kluwer.

1896/2016. *Die Prinzipien der Wärmelehre. Historisch-kritisch entwickelt,* eds. M. Heidelberger and W. Reiter. Xenomoi.

1905/1976. *Knowledge and Error.* D. Reidel.

Niven, William D. (ed.) 1890. *The Scientific Papers of James Clerk Maxwell,* 2 vols. Cambridge University Press.

Planck, Max 1908/70. 'The Unity of the Physical World-Picture', as translated in S. E. Toulmin (ed.), *Physical Reality: Philosophical Essays on Twentieth-Century Physics.* Harper, pp. 3–27.

Preston, John M. 2008. 'Mach and Hertz's Mechanics', *Studies in History and Philosophy of Science* 39: 91–101.

Ratliff, Floyd 1970. 'On Mach's Contributions to the Analysis of Sensations', in Robert S. Cohen and Raymond J. Seeger (eds.), *Ernst Mach: Physicist and Philosopher.* D. Reidel, pp. 23–41.

Smith, Crosbie 1998. *The Science of Energy: A Cultural History of Energy Physics in Victorian Britain.* Athlone.

Staley, Richard 2018. 'Sensory Studies, or When Physics Was Psychophysics: Ernst Mach and Physics between Physiology and Psychology, 1860–71', *History of Science,* doi:10.1177/0073275318784104.

Thomson, William 1849/2015. 'An Account of Carnot's Theory of the Motive Power of Heat; with Numerical Results Deduced from Regnault's Experiments', in J. Larmor (ed.), *Sir William Thomson, Baron Kelvin: Mathematical and Physical Papers,* 6 vols. Cambridge University Press, pp. 113–155.

Tyndall, John 1910. *Fragments of Science. Being a Series of Detached Essays, Addresses and Reviews,* 6th edn. Burt.

Warwick, Andrew 2003. *Masters of Theory: Cambridge Mathematics and the Rise of Mathematical Physics.* Cambridge University Press.

Wegener, Daan 2009a. *A True Proteus: A History of Energy Conservation in German Science and Culture.* PhD thesis, Utrecht University.

2009b. 'Science and Internationalism in Germany: Helmholtz, Du Bois-Reymond and Their Critics', *Centaurus* 51: 265–287.

2010. 'De-anthropomorphizing Energy and Energy Conservation: The Case of Max Planck and Ernst Mach', *Studies in History and Philosophy of Modern Physics* 41: 146–159.

Wegener, Daan and van Lunteren, Frans 2012. 'Verspreiding en ontwikkeling van de wet van behoud van energie in Nederland, 1844–1900: een terreinverkenning', *Tijdschrift voor Geschiedenis* 125: 384–399.

Zudini, Verena and Zuccheri, Luciana 2016. 'The Contribution of Ernst Mach to Embodied Cognition and Mathematics Education', *Science and Education* 25: 651–669.

4

Mach on Analogy in Science

S. G. STERRETT

Introduction

Many nineteenth-century natural philosophers and scientists employed analogy, and some (e.g. James Clerk Maxwell, John Herschel) discussed it as a subject in its own right, too. Analogy plays a role in Ernst Mach's philosophy as well as in his scientific work. What I want to do in this chapter, however, is examine Mach's views on how analogy is used in natural science. I think the uses Mach saw for analogy in natural science is nothing short of majestic and that, when properly understood, his views on analogy help us to see the roles appropriately played by logic, psychology, and scientific principle when analogy is used in natural science. Unfortunately, some reprints and translations of Mach's essay on the topic contain omissions whose presence and placement are crucial to understanding the surrounding text. These textual inadequacies have undoubtedly contributed to a lack of understanding – and thus of appreciating – Mach's views. (I describe and discuss the associated textual corrections in Footnotes 1 and 2 to this chapter for the benefit of the reader who wants to get a clear idea of the nature and significance of these textual inadequacies.)

Until it is fully understood, what Mach says about analogy can come across as unexceptional and, at times, even contradictory. Neither is true. In fact, some points that are today regarded as discoveries about, or recent advances in, the study of analogical reasoning are seen to be not only already articulated by Mach, but taken to a higher level of sophistication. Inasmuch as is possible, though, I aim to avoid using anachronistic terminology to express Mach's points about analogy, and I will focus simply on sorting out and presenting Mach's view.

The Distinctiveness of Analogy

Mach says that analogy is a *special case* of *similarity* ('*Die* Analogie *ist jedoch ein* besonderer *Fall der* Ähnlichkeit'[1]), and that there are good reasons for

---

[1] 'Die Aehnlichkeit und die Analogie als Leitmotiv der Forschung' in the journal *Annalen der Naturphilosophie* (Mach 1902). The third sentence of the paper is '*Die* Analogie *ist*

regarding analogy that way. Analogy is a special case of similarity in that what matters in the kind of similarity we call an analogy are 'conceptual relations': both the conceptual relations that concepts have to each other and the relations that concepts have to the objects between which the analogy is drawn. One way in which the kind of similarity that is an analogy differs from other cases of similarity, he notes, is as follows: 'Not a single immediately perceptible feature of one object need be found to be a feature of the other object' (*'Nicht ein einziges unmittelbar wahrnehmbares Merkmal des einen Objektes braucht mit einem Merkmal des anderen Objektes übereinstimmender, identischer Weise wiedergefunden werden'* (Mach 1902, p. 5). This is, I think, meant to be arresting, for it was fairly common to portray an analogy as requiring – or even as consisting in – the fact that two objects have some of the same immediately perceptible (or at least observable) features. This idea is still commonly found today in introductory textbooks on logic and scientific reasoning as well (e.g. Salmon 2012).

The way Mach presents analogy in the outset of the essay is thus as a contrast case to other cases of similarity. That is, first he characterises similarity in terms of how it compares to identity as far as having features in common: 'Similarity is partial identity: the characteristics of similar objects are in part identical and in part different' (Mach 1976, p. 162). Mach then emphasises that two objects related by analogy might not have *any* immediately perceptible features in common. So analogy is an extreme case – we might say it is a limiting or degenerate case – of similarity; Mach says just that it is a special, or very particular (*besonder*) case of similarity. Analogy, he is highlighting here, is about sameness (identity) of *relations between*, not sameness (identity) of *features of* objects. Further, recognition of an analogy consists in determining that the same relations exist between the features of one object as exist between the homologous features of another object. We shall see that when it comes to proffering his own definition of analogy,

---

jedoch ein besonderer *Fall der* Aehnlichkeit'. This sentence appears in both the version in the journal (1902) and in the German-language edition of the anthology *Erkenntnis und Irrtum* (1905). It does not appear in the English translation by Thomas J. McCormack in *Knowledge and Error* (Mach 1976), which was translated from the 5th edition of *Erkenntnis und Irrtum*, 1926. Since Mach also indicates later in the paper that his view is that analogy is a special case of similarity, there is no question that he endorses the statement. Excising the statement from the essay, however, as in the English translation, or even moving it from its position as the third sentence in the paper, changes the emphasis of the sentence that follows it. In the original version published in the journal in 1902 and the anthology in 1905, the statement that begins 'Not a single observable mark ...' is clearly meant to be bringing out a point about analogy; in the English translation (Mach 1976), where the sentence *'Die Analogie ist jedoch ein* besonderer *Fall der* Aehnlichkeit' has been excised, it appears to be a comment about similarity. My thanks to Karin Krauthausen for urging me to consult the version that originally appeared in *Annalen der Naturphilosophie* and providing me with a copy of it.

though, Mach does not speak in terms of objects or, even, features of objects. Rather, he speaks of an analogy holding between systems of concepts. Why is he doing so here, then, at the very outset of the essay?

As I see it, the reason that Mach is highlighting the fact that these two different objects between which an analogy is drawn might not have any (immediately perceptible) features in common is not so much to endorse a definition of analogy as holding between objects, but rather to draw attention away from the *sharing of features* and towards what is essential to an analogy on his account of analogies in natural science: *relations between concepts*. At this point of the discussion in the essay, he does so from *within* the presuppositions and terminology of the existing discourse about similarity, which is similarity of objects. Discussions later in Mach's 'Similarity and Analogy' may appear inconsistent with this early portion of the essay, unless one takes account of that.

It is the common usage of the term 'analogy', on which an analogy is drawn in terms of objects and their features, that is under discussion when, later in the same essay, Mach points out that the expectations generated by analogies are not logically justified. ('*Diese Erwartung ist logisch nicht berechtigt ...*'; Mach 1902, p. 9) On this kind of characterisation, 'Inferences from similarity and analogy are not strictly matters of logic, at least not of formal logic, but only of psychology' (Mach 1976, p. 166). Analogy is distinguished from similarity here, too, and on the same general basis that we will see Mach emphasise on his own account, on which analogy holds between systems of concepts: relations rather than relata. Considering the case of an analogy between two objects M and N where 'an object M has marks a, b, c, d, e, and another object N agrees with it as regards a, b, and c,' he writes that 'If a, b, c, d, and e above are directly observable, we speak of similarity; if they are conceptual relations between marks, analogy is closer to normal usage' (Mach 1976, p. 166). If d and e are 'indifferent', the analogy merely makes us *associate* d and e with the object N. If d and e are especially 'useful or noxious' properties, he says, we may go on to investigate further. But Mach does not say that those investigations are carried out using analogy: it seems that they would be carried out using whatever methodologies for finding or figuring things out one would normally use: 'by simple sense observation or by means of complex technical or scientific conceptual reactions' (Mach 1976, p. 166). This discussion seems to be reviewing and describing contemporaneous accounts of analogy, on which analogies are heuristics for suggesting possibilities to investigate. However such investigations turn out, our knowledge is extended, he says. Therefore, this use of analogy results in extending our knowledge, even though the inference from the analogy itself is not logically justified.

In what follows, I will lay out Mach's account of analogy as well as his views about the significance of analogy in many advances in natural science. Some apparent inconsistencies may arise from what he says about analogy in that context and what he says about analogy in the context just quoted above in discussing the term in common usage. Hopefully, what I point out about the two uses will show that the inconsistency is merely apparent. The uses of the term 'analogy' in the different contexts are actually referring to different kinds of analogy: in the common use of the term, an analogy is drawn between two *objects* and is feature-based. In the powerful use made of it in the historical case studies he has in mind when talking about analogy in natural science, an analogy is drawn between two *systems of concepts*. The latter kind of analogy opens up a role for scientific laws and principles to play in analogies.

Mach's Account of Analogy

Since Mach characterises similarity as 'partial identity' and includes analogy as a special case of similarity, not all cases of similarity are cases of analogy, and so some things can be said of analogies that are not true in general for cases of similarity. Both identity and analogy are special cases of similarity. They are extreme cases: at one extreme (identity), all of the features *must* be the same; at the other extreme (analogy), none of them *need* be the same. To put it another way: what we can say about every case of identity that we cannot say about every case of similarity is that all of the features between the two things being compared are the same. What we can say about every case of analogy that we *cannot* say about every case of similarity is that a specified set of connections or relations – connections or relations that may not be immediately perceivable – are the same.

When highlighting how analogy differs from other kinds of similarity, Mach spoke of the marks (*Merkmale*), or features, of objects, but, as noted above, I think the terminology (marks, or features) used there when discussing analogy was for consistency with the terminology already in use when talking about similarity, of which analogy is a special case. It makes sense to distinguish between those contexts in which he is using the terminology in common use when discussing analogy and similarity in order to contrast his view with it from contexts in which he is presenting his own definition of analogy.

When it comes to proffering *his own* definition of analogy, Mach defines analogy as holding between 'systems of concepts' rather than between objects or their features. It is a significant difference that, in the first place, the things that an analogy holds between are *systems*, because a system consists of, or contains, interrelated items. Secondly, he says they are systems of *concepts* – thus, the relations are between interrelated concepts (rather than features, or marks, of objects). They are *logical* relations.

More specifically, Mach defines analogy as 'a relation between *systems of concepts*, in which both the *difference between* two homologous concepts, and the *identity of the logical relations* of each pair of homologous concepts come to clear consciousness' (emphasis added).[2] That is, in an analogy, we can see clearly that the concept in one system and the homologous concept in the analogous system of concepts are different (when they do differ), while, at the same time, we are also aware that, for each pair of homologous (corresponding) concepts in the two analogous systems, the logical relations associated with one are the same as with the other. What are identical in an analogy are these logical relations between concepts.

I think it is worth noting that Mach does not say that more is required regarding the logical relations than just being conscious *that they are identical*. Often a determination of identity can be made on partial information, as when we can see that the levels in two glasses are the same without having to measure what those levels are, or showing that two shapes are the same without having to quantify or identify the shape of either by overlaying each over the other. Likewise, it seems that what is required is just to show that the logical relations in one of the systems of concepts work in the same way as the logical relations in the analogous system of concepts work with the homologous concepts.

I pointed out above that, on Mach's definition of analogy in natural science, there are two aspects that distinguish an analogy from other cases of similarity: (1) an analogy is drawn between *systems of concepts*, whereas this is not true for all cases of similarity and (2) the logical relations in each of the two systems between which the analogy is drawn are identical. We might wonder what each of these two aspects contributes to his definition of analogy independently.

Although Mach defines analogy as holding between systems of *concepts*, the provision in his definition about identity of relations could be illustrated with systems of spatially interrelated things rather than systems of logically interrelated concepts. I find it helpful to consider how this provision would go for a concrete example before considering how it would go for a system of concepts. This might be illustrated, it seems to me, by the following simple example: consider a sketch that has been constructed to record and display observations of the location of a system of things. Here, a sketch made of astronomical observations provides a good example. Each of the points in the sketch bears spatial relations to other points on the sketch. Each mark on the sketch is unlike the body in space that it is homologous to, though, in terms of what is immediately perceptible about it. Yet, we can point out that certain spatial relations that each mark on the sketch has to other marks on the sketch are *the same as* the relations that the observations in the sky that a certain mark is

---

[2] The original in German reads: '*eine Beziehung von Begriffsystemen, in welcher sowohl die Verschiedenheit je zweier homologer Begriffe als auch die Uebereinstimmung in den logischen Verhältnissen je zweier homologer Begriffspaare zum klaren Bewusstsein kommt*'.

homologous to has to the other observations in the sky that the other marks on the sketch are homologous to.

Now, as I have described it, the above example of a sketch and some observations in the sky is an example of a similarity that illustrates aspect (2) of Mach's definition of analogy in natural science, but it would *not* count as an analogy on Mach's view, since it is not a case of a similarity holding between systems of *concepts*. This is important to keep in mind so as to fully appreciate what he is saying in discussing examples of the use of analogy in mathematics, especially ones that involve geometrical figures.

What does Mach cite to show the use of analogy in mathematics, then? First, his general remarks on the physical application of mathematics involve analogies of *operations* rather than *objects*: it is not just the correspondence of *mathematical entities* or marks with *physical entities* that makes the application of mathematics an analogy. Mach writes, 'Every physical application of mathematics rests on taking note of analogies between facts and mathematical operations' (Mach 1902, p. 7). When Mach says that Hermann Grassman's mechanics or vector theory makes use of an analogy between 'lines and forces, areas and torques, and so on', he makes it clear that he would spell this out as an analogy between operations defined on lines and areas, as well as facts about forces and torques. I take him to be saying that there is an analogy between two systems: one a system containing lines and areas and an operation that relates them; and the other a system containing forces and torques and some facts that relate them. In addition, the logic of the operation in the system relating the lines and areas is identical to the logic of the facts in the system relating the forces and torques. Algebra can be used to formulate the relation in such a way that the identity is clear (i.e. 'the logical relations come clearly to consciousness').

Note that Mach's analogy is not the only kind of mathematical analogy possible; other mathematicians opted for other ones. Others writing contemporaneously on Grassman's mechanics sometimes spoke of analogies in terms of objects and their properties. One explains the fact that 'the summation of sects . . . corresponds completely to the discussion of the resultant of a system of forces in a plane' as following from the observation that 'a sect possesses the exact geometrical *properties* of a force, namely, magnitude, direction, and position' (Hyde 1905, p. 31, emphasis added). Mach's choice to pick out *operations* rather than the *things* being operated on or *features* of the things being operated on in the analogies he draws in the physical application of mathematics is due to his view of analogy as a relation between *systems of concepts*. The operations in Grassman's abstract treatments of mechanics were often spatial transformations (rotations, translations, etc.).

Secondly, the analogy that Mach chooses in order to illustrate 'the great value of geometry in cognition' is more involved and contains subtleties, but in it we see the same point. The case deals with the 'optical properties' of

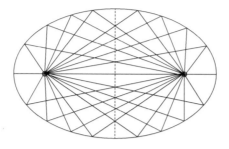

Figure 4.1 Foci of an ellipse. The two points of intersecting rays shown are called foci. Every light ray sent out from one of the foci will be reflected off the surface of the ellipse on a path that intersects the other focus. The same holds for sound waves, too, and this is the phenomenon behind a 'whispering gallery' – a room with an elliptical surface in which a whisper spoken at one of the ellipse's foci can be heard by someone located at the other focus of the ellipse

curves – specifically, 'logical properties' of conic sections. 'Optical properties' of conic sections arise from the study of how light rays are reflected off and travel within spaces bounded by surfaces shaped like those conic sections: probably the most familiar 'optical property' of a conic is the pair of foci of an ellipse. In a space bounded by a surface shaped like an ellipse, light rays emanating from one focus, no matter how they are orientated, will pass through the other focus (see Figure 4.1).

Considered *with respect to the optical properties* of a conic section, the *focus (of an ellipse)* is a concept bearing a logical relation to the concept of *ellipse*. It was in studying the *conceptual relations* of foci (of various particular conic sections) to particular conic sections that Johannes Kepler employed the reasoning Mach cites here. Thus, on my reading of Mach, this *would* count as an analogy, and Mach does refer to it as an analogy, too. But he is *not* referring, as above, to the physical application of the concepts of conic section, but rather to an analogy that Kepler finds among conic sections. As we will see in what follows, it is a case of analogy used within mathematics itself to generate a result in abstract mathematics.

Kepler explains how the concepts of *foci* and *conic section* figure in the characterisations of a circle, ellipse, parabola, hyperbola, and point (the degenerate case). Mach quotes him as saying (in Latin in the original): 'The one focus of a circle is A, namely at the centre; in the ellipse there are two foci, A and B, equidistant from the centre of the figure in the more pointed part.' Kepler then examines how the location of a focus or foci with respect to the curve formed by the outer surface of the cone and the cutting plane varies: 'In the circle, the focus is thus at the centre, as far from the circumference as possible, in the ellipse already less and in the parabola much less, and finally in

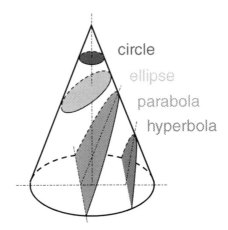

circle

ellipse

parabola

hyperbola

Figure 4.2 Conic sections. The conic sections are the various shapes formed by the intersection of a solid cone and a plane cutting through it at various angles

the straight line it is at minimal distance, that is, it falls on the line.' The straight line is not obvious in a progression of conic sections (indeed, it is not generally included), but Kepler remarks on the apparent misfit in a parenthetical remark: '[W]e speak of straight lines not so much in the ordinary way but rather to complete analogy.' Using the graphical means of a plane intersecting a conic, though, as is seen in Figure 4.2, a straight line can be seen as being at the other end of the progression that begins with a horizontal plane intersecting a cone to form a circle. A line would in fact result from the intersection of the plane and the cone when the plane's orientation coincides exactly with the angle of the cone. How to treat the *focus* of the conic when the conic is a straight line? This is a matter of using the analogy: 'It follows by analogy [with the hyperbola] that in a straight line either focus … falls on the line: there is but one [focus], as in the circle' (Mach 1976, p. 164). The circle and the line are the two extremes. Mach attributes Kepler's ability to grasp these 'deep-seated' analogies to his use of 'the principle of continuity', which I take to refer to how Kepler has organised the progression of conic sections from circle to line. The circle can be continuously transformed into an ellipse, the ellipse into a parabola, the parabola into a hyperbola, and then the hyperbola into a line, and the foci continuously change their locations accordingly. In fact, thinking of these figures – circle, ellipse, parabola, and hyperbola – as conic sections is associated with imagining a cutting plane being continuously rotated from a beginning position in which it cuts out a circle through cutting out an ellipse, parabola, hyperbola, and line. Significantly, Kepler's explanation of his reasoning shows that the foci associated with each of these conic sections figures in the comparisons and transformations. In regarding them as 'optical properties' of the conic sections, the foci here are characterised in terms of *facts about how rays of light behave*, so this example illustrates how Mach's approach to analogy provides a role for scientific principles to play.

Mach on Analogy in Natural Science

We have seen that Mach's definition of analogy ('a relation between systems of concepts, in which both the difference between two homologous concepts, and the identity of the logical relations of each pair of homologous concepts come to clear consciousness') provides a role for scientific principles and laws to play in analogy. When it comes to explaining the significance of analogy in natural science, Mach specifies no set way in which analogy is used. In the historical cases he discusses, there is always some area of science about which a lot is already known, but analogies are generated and contribute to extending knowledge in a variety of ways. What comes through in reading his discussions of cases in the history of science is the need for agility and flexibility in recognising the possible analogies one can draw upon and being able to adapt them to the needs of the situation at hand.

As might be expected considering the era, Mach describes cases of analogies between different kinds of waves, different kinds of currents, and different kinds of fields. But he also mentions Galileo's discovery of the moons of Jupiter as an important case of analogy in natural science, writing that Galileo's discovery was 'more powerful than any other arguments [by analogy][3] in supporting the Copernican system'. We might ask what kind of analogy that is, or how analogy is involved in that case, as it seems to concern two particular *physical* systems (i.e. the system of the planet Jupiter and its moons and the system of the sun and the planets in our solar system). Galileo's discovery of the moons of Jupiter, reported in *The Starry Messenger*, arose from his handwritten sketches of their positions at successive points in time: he came to see that the bodies he was observing must have been moving in orbits around Jupiter. Mach's succinct comment about the kind of analogy between the system of Jupiter and its moons and the solar system he is referring to as so powerful an endorsement of the Copernican view is telling: '[W]e have here a small scale model of the solar system' (Mach 1976, p. 167). The reason that the phrase 'small scale model' is telling is because Mach certainly knew something about methodologies of modelling (what we would

---

[3] The journal version of Mach's essay seems to say this is the strongest of all the arguments by analogy, rather than the strongest of all arguments. '*Die Entdeckung der Jupiter Trabanten durch Galilei hat das Copernikanische System mächtiger als alle anderen Argumente durch die Analogie gestutzt. Das Jupitersystem stellte ein verkleinertes Modell des Planetensystems dar*' (Mach 1902, p. 11). Hence, I have indicated the qualification in brackets in my English rendition here.

call) kinematic similarity used to model one physical system with another. In fact, he wrote about them, though he did not use the terminology of kinematic similarity or physical similarity. In *The Science of Mechanics*, Mach discusses Newton's notion of similar systems, which Newton uses to indicate when the motions of the bodies in one physical system of bodies will be homologous to those in another. As I wrote on an earlier occasion in discussing the history of the concept of physical similarity:

> After generalizing one of his own conclusions, Mach remarks: 'The considerations last presented [on similarity and similar systems] may be put in a very much abbreviated and very obvious form by a method of conception first employed by Newton.' He does not quite accept Newton's use of the term similar system there, though:
>
> 'Newton calls those material systems *similar* that have geometrically similar configurations and whose homologous masses bear to one another the same ratio. He says further that systems of this kind execute similar movements when the homologous points describe similar paths in proportional times' (Mach 1960, p. 203).
>
> Mach admires Newton's methodology here, but he points out an issue with Newton's use of the term *similar* . . .'
>
> (Sterrett 2017, p. 379)

Mach's care in distinguishing the use of geometrical similarity from uses of other closely allied kinds of comparisons is evident in this discussion in *The Science of Mechanics* (Mach 1960) – as it is in his essay on similarity and analogy (Mach 1902, 1976). I went on to describe Mach's attempt to rescue Newton's analysis here:

> However – and what is significant and interesting – Mach does not say that Newton is wrong here; rather, what he says is that what Newton was doing is better understood in Mach's day in terms of affine transformations:
>
> 'The structures might more appropriately be termed *affined* to one another. We shall retain, however, the name phoronomically [kinematically] *similar* structures, and in the consideration that is to follow leave the masses entirely out of account (Mach 1960, p. 204).'
>
> (Sterrett 2017, p. 380)

Now, I suggest, the way in which Mach proposes Newton's work should be understood fits Mach's definition of an analogy in natural science, for it turns out that 'phoronomically similar' structures relate homologous concepts (rather than bodies) and that the relations between them are identical. As I explained previously:

> . . . Mach shows how to understand phoronomically [kinematically] similar structures for the topic of oscillation he has been discussing:

'In two such similar motions, then, let
the homologous paths be s and αs,
the homologous times be t and βt;
whence the homologous velocities are $v = s/t$ and $\alpha v = \alpha/\beta$ s/t,
the homologous accelerations $\varphi = 2s/t^2$ and $\varepsilon\varphi = \alpha/\beta^2$ $2s/t^2$
Now all oscillations which a body performs under the conditions above
set forth with any two different amplitudes 1 and α, will be readily
recognised as *similar* motions (Mach 1960 p. 204).'

(Sterrett 2017, p. 380)

So, if path, time, velocity, and acceleration can be considered concepts,
what Mach describes above are two systems of interrelated concepts. He
states the logical relations between them in the excerpt shown above: we
can immediately see that the relations between the interrelated concepts
in the two systems of concepts are identical. Hence, it is a relation
between two systems of concepts that meets Mach's definition of analogy
in natural science. His assessment of the value of such methods of
investigation is thus an assessment of the value of analogy in natural
science, too, and about this he is positively effusive: 'After showing how
elegantly theorems about centripetal motion can be obtained by such
means', Mach remarks:

It is a pity that investigations of this kind respecting mechanical and
phoronomical *affinity* are not more extensively cultivated, since they
promise the most beautiful and most elucidative extensions of insight
imaginable (*The Science of Mechanics*, p. 205).

Thus Mach sees the great power of the notion of similar systems.

(Sterrett 2017, p. 380)

As Mach made a point about the crucial importance of Kepler's use of
continuity along with his use of analogy in developing his account of conic
sections, it is noteworthy that Mach's discussion about this 'scale model'
kind of analogy in *The Science of Mechanics* also employs continuity. The
transformations he describes between the two systems of bodies (which is
what would be relevant to the case of seeing the system of Jupiter's moons
as a scale model of the solar system) are *continuous* transformations, and
many of the concepts related by the analogy likewise include continuity
(e.g. time is continuous, paths and velocities are at least part-wise continuous). What is significant about that with respect to scale models and similar
systems is that, when these ideas were later formalised in the landmark
paper of 1914 by Edgar Buckingham, the idea of a similar system (and so of
a scale model) was presented in terms of one system undergoing continuous changes to create the other, all the while obeying applicable physical
laws and relations at every point in the transformation (Buckingham 1914).
Thus, continuity is important in the case of the kind of analogy that

underwrites use of a scale model – even in the later formulation of it that appeared in a physics journal.

Mach also cites analogies between different kinds of waves in illustrating how important analogies have been in the development of natural science: surface (water) waves, sound waves, and light waves. A favourite theme of his in both his popular lectures and his scientific publications is the analogy between sound and light; in his essay 'Similarity and Analogy' he writes, 'As to light, the appropriate ideas were developed from the case of sound' (Mach 1976, p. 167). This is the kind of statement that might make Mach look like an uncritical user of wave analogies, or at least analogies that are out of date and irrelevant to the knowledge of today. Yet that judgement is mistaken, for as I pointed out in 'Sounds Like Light: Einstein's Special Theory of Relativity and Mach's Work on Acoustics and Aerodynamics':

> Commentators on Einstein and special relativity tend not to look in the direction of work in acoustics for conceptual precursors to the special theory of relativity; expositions on special relativity that compare light and sound tend to associate the insight of special relativity with the contrast between light and sound, and the similarities between them with the (discredited) classical wave theory of light.
>
> ... Mach was exceptional here in that, in the very context of drawing an analogy between sound and light, he explicitly freed the notion of a wave from the necessity of having a mechanical basis.

> (Sterrett 1998, p. 2)

The occasion for Mach's own substantive scientific work on wave analogies was determining the correct explanation of the Doppler effect. He used acoustic experiments to prove that the observed effects (difference in pitch of sound) arose from the relative motion of the sound source and the observer of the emanating sound wave, and then concluded that, based on the analogy between light waves and sound waves, the explanation of the Doppler effect for light (difference in colour of light) was likewise due to the relative motion of light source and observer. To develop the analogy, he looked at many different kinds of waves and identified what he thought they had in common that also captured what was essential to waves. This included being propagated in time, having spatial periodicity, having temporal periodicity, and being able to be algebraically summed. Mach says something very bold and striking about what his experiment showed: 'It is of absolutely no significance for the question of whether Doppler's principle applies to light', he says, 'whether or not light is a mechanical wave motion [like sound]. One could just as well think of light as chemical oscillation, for many of the appearances, such as anomalous dispersion and fluorescence, can be better understood, in many respects, under such a notion of light' (Sterrett 1998, p. 22). In that earlier work on Mach, I commented that Mach confidently states that the Doppler principle can be applied to light in the same way as for sound, on the basis that light and sound are propagated in time, have spatial and temporal periodicity, and can be algebraically summed. This is in keeping with his earlier remarks on the Doppler effect for sound, in which he kept clear of appealing to any causes arising from the mechanical nature of sound waves, and stuck to kinematical considerations.

(Sterrett 1998, p. 22)

The above quote indicates Mach's use of an analogy between light and sound; it is drawn in terms of an analogy between different systems of (interrelated) concepts, in that homologous concepts such as *pitch* (frequency of sound) and *colour* (frequency of light) are different, yet their relations to other concepts such as (sound/light) wave velocity and (sound/light) wavelength are the same between homologous concepts.[4] It certainly fits Mach's definition of an analogy in natural science.

There is another noticeable pattern, a trend we see as we consider more and more of Mach's discussions of the use of analogy in historical examples: a trend towards eliminating the more material-laden aspects of a situation. Often, material aspects of an area are responsible for a point of disanalogy between two areas of science, but they make no difference to many other points of analogy. Leaving the material medium out of the account of waves seems to be part of this pattern, just as, in Mach's discussion of Newton's work on 'similar systems' above, Mach proposed to 'leave masses out of the account'. In applying his 'method of physical analogy', James Clerk Maxwell likewise would leave masses and materiality of a fluid out of his account. Mach shows his admiration for Maxwell's successful investigations in natural science, and especially for his 'method of physical analogy' throughout his essay on similarity and analogy. Maxwell, Mach says, 'describes analogy as that partial similarity between the laws in one field and those in another, so that each illustrates the other' – a view, he says, from which his own view of analogy is 'not different' (Mach 1976, p. 162). Later, he uses Maxwell as an illustration of the fact that the use of analogy tends to lead to being able to use abstraction.

From his ability to see an analogy between different kinds of physical phenomena in electrostatics and electrodynamics with fluids, Maxwell is able to use both abstraction (i.e. finding features common to various phenomena in physics and imagining a fluid that has just those and not other features such as mass) and physical intuition: '[W]e do not take it as real and we know precisely how it coincides conceptually with the facts to be represented' (Mach 1976, pp. 168–169). Mach, too, employed analogy in somewhat the same way in investigating different kinds of waves, in that one goal of his

---

[4] This is explained in greater detail in Sterrett (1998).

comparisons was to separate off certain facts about waves in order to use them in reasoning what could be deduced more generally about any given wave.[5]

In saying that his view of analogy is not different from Maxwell's, and throughout the entire essay, Mach seems to be supportive of other scientists and natural philosophers writing on analogy. His general approach in this essay seems to be to find common ground, rather than to find fault with the views of other natural philosophers and scientists on the topic. Yet we may wonder what points of disagreement there might be; philosophers often find it illuminating to identify them in order to better understand someone's views. Here, I would suggest there is a point of disagreement with Herschel, in spite of Herschel clearly articulating some of the most significant points about analogy that Mach wished to emphasise. Mach certainly recognised Herschel's strength in identifying and reasoning with analogies; he mentions Herschel's striking success in predicting experimental results on the basis of analogy before there was direct experimental evidence for them (e.g. the case of polarisation of light (Mach 1902, p. 225), among others), but he does not endorse Herschel's philosophical views on the matter. Yet Herschel's views in *Preliminary Discourse* seem at times akin to Mach's, such as when Herschel writes of the 'general resemblance between the two sciences of electricity and magnetism' and notes that 'many of the chief phenomena in each were ascertained to have their parallels, mutatis mutandis, in the other' (Herschel 1845, section 85, p. 94), which he explains as follows: 'If we encounter the same elementary phenomena in the analysis of several composite ones, it becomes still more interesting, and assumes additional importance: while at the same time we acquire information respecting the phenomenon itself, by observing those with which it is habitually associated ...' (Herschel 1845, section 85, p. 93). There is a basic difference, though, in that Herschel consistently conceives of and applies analogies in terms of a 'cause' that they have in common, whereas Mach's definition and use of analogies is in terms of systems of concepts and relations. The difference between Mach's and Herschel's uses and conceptions of reasoning by analogy shows up strikingly in how they conceive of and make use of the analogy between sound and light, as Herschel writes that 'an analogy between sound and light has been gradually traced into a closeness of agreement, which can hardly leave any reasonable doubt of their ultimate coincidence in one common phenomenon, the vibratory motion of an elastic medium' (Herschel 1845, section 85, p. 94). In contrast, the value of Mach's application of the analogy between sound and light in explaining the Doppler effect is that it *did* allow for agnosticism regarding the existence of a medium for the transmission of light – which,

---

[5] This is laid out in Sterrett (1998), which explains the momentous significance for twentieth-century physics of Mach's theoretical and experimental work in his explanation of why the Doppler effect and the phenomenon of shock waves hold for waves of any kind.

I have argued (Sterrett 1998), was later crucial for the development of the special theory of relativity. Thus, the difference between Mach and Herschel on the use of analogy is significant, not just philosophically, but in terms of the scientific conclusions that their methods underwrite. It is not surprising, therefore, that the enthusiasm and admiration Mach shows Maxwell for his method of physical analogy does not extend to Herschel.

Conclusion

I have not done justice here to all of the aspects of Mach's rich and subtle discussion of analogy. There is much more he says to open his readers' eyes to uses of analogies in natural science. We now take stock of a few basic points we have been able to touch upon in this chapter.

In his short essay on similarity and analogy, Mach explores the use of analogy in natural science, while at the same time recognising a more common usage of the term. He distinguishes analogy from similarity for the more common usage of the term on the basis of an emphasis on identity of relations (between features of objects) rather than relata (features and objects). This basis for identifying what is distinctive about analogy is formulated in such a way that it later carries rather smoothly over into his own account of analogy in natural science. Mach discusses limitations of the common usage of analogy (i.e. inferences are not logically justified), as well as virtues (it can stimulate investigation, extend our knowledge, and give a biological and physiological account of why we value analogies) (Mach 1976, p. 166). These points about the value of analogy are being rediscovered. An especially striking example of a discipline in which analogy is indispensable, in spite of being recognised as not providing logical justification of inferences, is ethnographic archaeology.

Alison Wylie (1985) and Mads Ravn (2011, 2018) each give a historical narrative of how views of the use of analogy in ethnographic archaeology have changed. At first, the power of analogy to extend knowledge was embraced: knowledge about certain present societies that were considered representative of past ones in other places, it was argued, could be used to extend the sparse evidence from those past societies. When the bases for making such comparisons between present and past societies were later recognised as faulty, and thus the conclusions obtained by the use of analogy untrustworthy, the use of analogy was disavowed, even disparaged. However, Wylie showed that even the methods proposed as alternatives to analogy by those who openly disavowed analogy were actually analogical methods of reasoning, too. Instead of disavowing the use of analogy, she argues, legitimate critiques of the use of analogy can help guide more appropriate use of analogy in ethnographic archaeology. These critiques direct our investigations into both the source-side domain and the subject-side (target) domain of knowledge. These points are in line with Mach's views on what I have called the common view of analogy above.

However, Mach went further than the question of how analogies can be licensed and talked about the value of analogy even when the analogy is shown to be undermined by a negative analogy (i.e. when we find features that the two objects being compared do not have in common). Mach said the cases where there is such a negative analogy are equally important (Mach 1902, p. 10). A recent in-depth study by Ravn argues this, too. In 'Roads to Complexity: Hawaiians and Vikings Compared' (Ravn 2018), Ravn explicitly highlights the role that negative analogies play in making the point that complexity in a society does not necessarily depend on certain specific features. Ravn's study is a very detailed use of analogy in which comparisons and analogies are drawn using the 'long view' of societies, rather than snapshots of them at only certain points in time, and in which processes, rather than only features, of societies are considered. What makes Ravn's study notable here is that archaeology is often cited for its use of feature-based analogies. Here, it is fitting to recall Mach's admiration of Kepler's use of continuity in drawing analogies between the conic sections. Thus, we see in these case studies in archaeology that delving seriously into valuing and making use of analogy in the more common sense of the term eventually leads towards the kind of analogy that Mach meant when talking about its use in the natural sciences: the analogy is between systems of concepts, and what is being equated are logical relations between concepts.

Mach's real interest in this essay lies in the notion of analogy as used in natural science. On Mach's account, the notion of analogy in natural science is a relation between *systems of (interrelated) concepts*, and what makes the relation an analogy is the identity of logical relations between homologous concepts. Mach saw analogies used in a variety of ways in natural science, so that even though he said his account was 'not different' from Maxwell's view of analogy, he had a lot to say about how and where analogies had been, and could be, used in natural science. Maxwell's description of analogy as a 'partial similarity between the laws in one field and those in another', Mach said, brought to light what was most valuable about analogy for scientific enquiry: each of the two laws illustrates the other. He discusses how powerful this method can be in extending knowledge, and he gives some unusual and surprising examples from the history of science.

We have also seen that Mach's appreciation of the power of analogy is evident in other works, albeit not always described explicitly as an analogy. For Mach, natural science is more variegated than just scientific enquiry into the unknown. His discussion indicates that sometimes the use of an analogy is about understanding one known thing in terms of another known thing (Mach 1976, pp. 167–168), but he also said that sometimes new analogies and questions arise that were not the target of enquiry (Mach 1976, p. 165). He noted that there may be many different areas one could draw upon in using analogy: 'Several equally known areas M, N, O, P may enter into analogy, in groups of two or more', so that there may be 'different analogies, each justified in its setting' (Mach 1976, p. 167).

Mach felt that there was much to be gained in using analogy, and that opportunities to do so were being wasted. He warned about how much could be lost in not seizing the opportunity to do so, as evidenced by his closing anecdote in the essay. The anecdote is meant to have a moral of epic significance: Newton's failure to consider analogies other than the one he was using in planning his experimental investigations – analogies that were easily available to him – meant that he ignored easily available but crucial observations, and this failure left him clinging to the wrong theory of light (ibid., p. 169).

Mach clearly had a mission in writing about analogy: I hope that the work I have done here, although just touching on a few of Mach's key ideas about analogy, will help us better understand what that mission was.

References

Buckingham, Edgar 1914. 'On Physically Similar Systems: Illustrations of the Use of Dimensional Equations', *Physical Review* 4: 345–376.

Herschel, John 1845. *Preliminary Discourse on the Study of Natural Philosophy*. Longman, Greens, et al.

Hyde, Edward W. 1905. *Grassman's Space Analysis*, 4th edn. John Wiley & Sons.

Mach, Ernst 1902. 'Die Aehnlichkeit und die Analogie als Leitmotiv der Forschung', *Annalen der Naturphilosophie* 1: 5–14.

   1905. *Erkenntnis und Irrtum: Skizzen zur Psychologie der Forschung*. J. A. Barth.

   1960. *The Science of Mechanics: A Critical and Historical Account of its Development*. Trans. Thomas J. McCormack, 6th edn, with Revisions through the 9th German edn. Open Court.

   1976. *Knowledge and Error*. Trans. Thomas J. McCormack. D. Reidel.

Ravn, Mads 2011. 'Ethnographic Analogy from the Pacific: Just as Analogical as any other Analogy', *World Archaeology* 43: 716–725.

   2018. 'Roads to Complexity: Hawaiians and Vikings Compared', *Danish Journal of Archaeology* 7: 119–132.

Salmon, Merrilee 2012. 'Arguments from Analogy', in *Introduction to Logic and Critical Thinking*. Cengage Learning, pp. 132–142.

Sterrett, S. G. 1998. 'Sounds Like Light: Einstein's Special Theory of Relativity and Mach's Work in Acoustics and Aerodynamics', *Studies in History and Philosophy of Modern Physics* 29: 1–35.

   2017. 'Physically Similar Systems: A History of the Concept', in L. Magnani and T. Bertolotti (eds.), *Springer Handbook of Model-Based Science*. Springer, pp. 377–411.

Wylie, Alison 1985. 'The Reaction against Analogy', *Advances in Archaeological Method and Theory* 8: 63–111.

Ernst Mach's Enlightenment Pragmatism

History and Economy in Scientific Cognition

THOMAS UEBEL

Ernst Mach's philosophy of scientific knowledge is profitably understood as an original form of pragmatism. Far from endorsing a simple copy-theory of empiricism, he insisted that all knowledge claims go beyond what is immediately 'given' and can only be understood and legitimated by taking their context into account. His broadly naturalistic approach was informed as much by his awareness of cultural development as it was by evolutionary theory. Mach recognised science itself as a deeply historical phenomenon and scientific knowledge as path dependent, thoroughly fallible, and far from ever closed. Conceptual perplexities, he held, can only be resolved by historical-comparative investigations. What merits thinking of Mach as a pragmatist, I will argue, is his insistence, as a philosopher, on the ultimately practical orientation of all thought as a matter both of fact and norm, and, as a historian of science, on the need to investigate the specific problem situations out of and in response to which concepts and theories developed. Last but not least, the practical orientation of his philosophy found expression in his allegiance to the ideal of enlightenment. The aim of the present chapter is to make these claims perspicuous.

To appreciate the pragmatist nature of Mach's conception of scientific knowledge, we must focus not on his *Analysis of Sensations* (1886), but on the more philosophical chapters of his large historical works, *The Science of Mechanics* (1883), with related essays, and *Principles of the Theory of Heat* (1896), as well as his early *History and Root of the Principle of the Conservation of Energy* (1872), and, of course, his *Knowledge and Error* (1905).[1] There, the supremely practical office of all thought is argued to be reflected still in the theoretical abstractions of natural science, however far removed from the exigencies of everyday life they may appear to be. I begin with remarks on Mach's understanding of enlightenment. These are meant to elucidate the distinctive historical perspective that allowed him to avoid the

---

[1] See Mach (1872/1911, 1882/1986, 1883/1960, ch. 4, sect. 4, 1884/1986, 1896/1986, chs 25–34, 1905/1976, passim).

simplistic biologism of which he is sometimes accused and to recognise instead the cultural dimension of science against the background of which his pragmatism unfolds.[2]

Mach's Ideal of Enlightenment

Mach invoked the idea of enlightenment repeatedly in his historical-physical studies. The notion figures prominently in the introductions not only of his history of mechanics (*The Science of Mechanics*) and *Principles of the Theory of Heat*, but also of the earlier *History and Root of the Principle of the Conservation of Energy* – as well as his *Contributions to the Analysis of Sensations*. What did he mean by it?

It is well known, thanks to his own later recollections, that having been deeply impressed in his youth by Immanuel Kant's critical philosophy, Mach turned away from it most decisively.[3] Less often noted is that while Mach rejected transcendental idealism in order to embrace what has been called 'critical empiricism', he continued to affirm the Enlightenment ideals of which, in German-speaking lands, Kant was the best-known advocate.[4] Mach tended not to mention other philosophers, so it is notable that the two philosophers mentioned as influences in the introduction to his early *History and Root* are Kant and Johann Friedrich Herbart.[5] Given that the influence of the former's 'Copernican Revolution' was largely suspended, only his partisanship for enlightenment merited his mention here.

The sense in which Mach invoked the notion is very much that of Kant's definition of the process of enlightenment, not the historical period (though the process may serve to characterise an ideal that was given currency then). Kant's definition may be translated as 'Enlightenment is mankind's exit from self-incurred dependence.'[6] Even in the original, both of the last two words are

---

[2] This perspective is sharply at variance with Paul Pojman's judgement on Mach's adoption and employment of the principle of economy: 'As is typical for Mach, widely accepted biological principles became uncritically held cornerstones of his epistemology' (2009/ 2018).

[3] For example, see Mach (1910a/1992, p. 6). This passage is not contained in the excerpts from this paper translated by John Blackmore (1992), but a translation of it can be found in footnote 3 of von Mises (1938/1970).

[4] The term 'school of critical empiricism' is used by Hermann Lübbe (1972/1978, p. 90) for the philosophy of Mach and Richard Avenarius; his further attempt to relate their efforts to the tradition of phenomenology is bracketed here.

[5] Prominent exceptions are his retrospective remarks on the reception of his work as in Mach (1910a) and the prefaces and additions to later editions of his main works, such as Mach (1872, 1883, 1886).

[6] See Kant (1784/2009, p. 1). I prefer 'dependence' for '*Unmündigkeit*' to the more commonly employed 'tutelage', whose legal meaning appears to have fallen out of use (the original term literally means being unable to speak for oneself in legal matters). For

problematical, and Kant sought to clarify their meaning further. Given our concern, we may abstract from the undesirable state being one's own fault and focus on the fact that the dependence is intellectual: it is 'the inability to use one's reason without guidance by another'.

Mach retained the central idea of autonomy, but his conception was not deontological. In general, he did not publish on ethical matters.[7] On occasion, however, he let his desire for an ethical world order shine through clearly and, late in his career, he closed *Knowledge and Error* with a passionate statement.[8] Mach's own ethical views were described by Anton Lampa, a close associate, as broadly utilitarian, albeit with a very characteristic difference:

> His utilitarianism is determined by the highest ethical ideal and therefore freed from all remains of elementary instincts. For the ideal that guided him is that of a moral world order. The sober criticism which dethroned venerated opinions is not his end but only a means to the highest that we humans may aspire to. He does not draw a utopia but only place an aim before us of the achievability of which we may be convinced, for we already hold in our hands the means to realise it: science.
>
> (1912, p. 58)

Contemporary readers may differ as to whether Mach's optimism – shared by his later admirers in the Vienna Circle under the heading 'scientific humanism'[9] – was realistic, but it was an indispensable ingredient of his philosophical outlook.

---

remarks on the difficulty of translation, see https://persistentenlightenment.com/2013/05/28/translatingkant1/, dated 28 May 2013, by James Schmidt (accessed 26 January 2019).

[7] One exception is an intervention on racism. See Mach (1907) and, for discussion, Uebel (2003).

[8] Most hints appear in his *Popular Scientific Lectures*, but there, too, less in the English version – he deplored the 'club law' governing international relations and 'the hate of races and nationalities' at the end of his lecture on projectiles (1898/1986, pp. 335–337) – than in the later German version, where one can in addition find critical remarks on imperialism (albeit not under this name), endorsements of Josef Popper's plan for a mixed economy ('*Allgemeine Nährpflicht*'), and of the unrestricted education of women (1910b/2014, pp. 287–288, 361, 366; on Popper's plan, see also 1905/1976, p. 63, note 21). For the clear expression of his humanist credo, see the closing sentences of *Knowledge and Error*: 'Let us remember what miseries our forebears had to endure under the brutality of their social institutions, their laws and courts, their superstition and fanaticism; let us consider how much of these things remains as our own heritage and imagine how much of it we shall experience still in our descendants: this should be sufficient motive for us to start collaborating eagerly in realizing the ideal of a moral world order, with the help of our psychological and sociological insights. Once such an order is established, nobody will be able to say that it is not in the world, nor will anybody need to seek it in the heights or depths' (1905/1976, p. 361).

[9] For example, see Carnap (1963, pp. 82–84) and Feigl (1949).

Mach's historical-physical works transposed the idea of autonomy to the intellectual-scientific sphere. Mach did not hold any one individual or ruling elite responsible for failing the rationality of science, but rather credited lapses to the vicissitudes of cultural and intellectual development in general, and particularly to an all-too-common acquiescence in common conceptions and inattention to the use of language across different contexts and periods. Consider the 'Introduction' to the early *History and Root* which, notably, begins with recollection of his puzzlement, as a child, of why people should want to be ruled over by a king and why there existed such inequality of wealth in society. Mach continued:

> There are two ways of reconciling oneself with actuality: either one grows accustomed to the puzzles and they trouble one no more, or one learns to understand them by the help of history and to consider them calmly from that point of view.
>
> Quite analogous difficulties lie in wait for us when we go to school and take up more advanced studies, when propositions which have often cost several thousand years' labour of thought are represented to us as self-evident. Here too there is only one way to enlightenment: historical studies.
>
> (1872/1911, p. 16)

Note that, for Mach, history not only held the key to understanding how we got to where we are, but also to discerning the way out of current predicaments. We are not condemned to remain stuck where we are. 'Let us not let go of the guiding hand of history', Mach advised. 'History has made all; history can alter all' (1872/1911, p. 18).

It would be difficult to exaggerate the importance of history to Mach's philosophy.[10] We will consider the mechanics of his historical enlightenment below. First, let us note that we will find history also centrally involved when we ask why, eleven years later, Mach specified that the 'enlightenment tendency' of his book on mechanics meant that it was 'anti-metaphysical' (1883/1960, p. xii), and why, another five years on, the first chapter of his *Analysis of Sensations* was titled 'Introductory Remarks. Anti-Metaphysical' (1886/1897, p. 1). The reason is this: 'We are accustomed to call concepts metaphysical, if

---

[10] I am far from the first to notice this, though it bears reiterating. Compare Erwin N. Hiebert: 'The success of Mach's philosophical enterprise, if we call it that, rests upon the feasibility of using the history of science to illuminate specific problems to which Occam's razor can be applied to remove metaphysical ballast and inherited anthropomorphisms and ambiguities' (1970, p. 189). Unfortunately, Hiebert gave Mach's philosophy a traditionalist reading, unlike Paul Feyerabend in the same volume, who noted that 'Mach develops the outlines of a knowledge without foundations' (1970, p. 178). Rudolf Haller (1988) combined the appreciation of Mach's historicism with that of his naturalistic anti-foundationalism.

we have forgotten how we reached them. One can never lose one's footing or come into collision with facts, if one always keeps in view the path by which one has come' (1872/1911, p. 17). Again, Mach's optimism is striking, but so is his analysis of what he called 'metaphysics'. Metaphysics for him was what we place beyond the reach of experience – indeed, far above it, what we cannot remember ever having learnt. That is why historical-physical works provide enlightenment and why such enlightenment is needed: the study of their history brings these runaway conceptualisations, whether mere concepts or whole theories, back down to earth and makes them available for critical discussion. Mach also called 'metaphysical' ideas that have become absolutised: '[I]f from history we learned nothing else but the variability of views, it would be invaluable.' To excise metaphysics, then, is an eminently practical affair of enlightenment import: 'Whoever knows only one view or one form of a view does not believe that another has ever stood in its place, or that another will ever succeed it: he neither doubts nor tests' (1872/1911, p. 17).

So historical studies for Mach were far from merely antiquarian. Rather, they intended a 'critical epistemological enlightenment of the foundations' of scientific disciplines or sub-disciplines (1896/1986, p. 1, trans. amended) and so were essential to sustaining further progress. But precisely how were they to achieve this? It is important to note that Mach did not offer an anticipation of George Santayana's thought that those who cannot remember the past are condemned to repeat it, true as that thought may be; nor did he merely advertise the beneficial consequences of broader viewpoints. Rather, Mach had a far more specific point in mind.

A Dialectical Conception of Enlightenment

Philipp Frank dug deeply in his 1917 appreciation of Mach when he characterised him as a thinker whose distinction lies in having equipped us to continue the work of the Enlightenment.[11] Mach achieved this by drawing attention to a certain dynamic in scientific theorising that demanded constant vigilance if any advance achieved was not to abet stagnation or even regression.[12]

---

[11] Compare Richard von Mises's statement that Mach is 'recognised today as *the most powerful and, for our own time, the most typical philosopher of enlightenment in the last generation*' (1938/1970, p. 249, emphasis in original). Although von Mises held Mach's historical perspective responsible for this judgement and soon added that, for Mach himself, unlike for Kant, 'the Copernican doctrine is likewise only a certain phase in our knowledge of the universe which may be modified again by assimilation of new facts, perhaps even by findings in an entirely different field' (1938/1970, p. 251), it remained Frank's merit to have pinpointed the dialectical character of Mach's conception.

[12] Feyerabend spoke of Mach's 'dialectical rationalism' (1985, title of §5) but did not specify it other than to point to his understanding of the close relation between observation and

As Frank put it, an 'essential characteristic' of the Enlightenment was its 'protest against the misuse of merely auxiliary concepts in general philosophical proofs' (1917/1949, p. 73). In doing so, it reveals a 'tragic feature': 'It destroys the old system of concepts, but while it is constructing a new system, it is already laying the foundation for new misuse. For there is no theory without auxiliary concepts, and every such concept is necessarily misused in the course of time' (1917/1949, p. 78). With Mach, Frank here pointed to nineteenth-century materialism as an example and concluded that 'in every period a new enlightenment is required to abolish this misuse' (1917/1949, p. 73). It follows that 'progress in science proceeds in an eternal struggle' and that 'it is this restless spirit of enlightenment that protects science from petrifying into a new scholasticism' (1917/1949, p. 78, trans. amended). As we shall see, this was no mere fanciful interpretation by Frank. Aware of the fact that new conceptualisations can advance knowledge as much as they thereby limit further advance, Mach himself advised their periodic reassessment.

It may be asked, how might such reassessments be possible? After all, it is impossible to check any correspondence of our conceptual frameworks with what they seek to comprehend. But that is not what Mach required: it was a distinction of his enlightenment perspective that no Archimedean points were available. What was to lead us out of '*Unmündigkeit*' was precisely the abandonment of unrealistic ideas of knowledge, even as mere limit concepts. Rather than search for timeless standards, Mach advised us to consult the contingent history of our very own projects and find their measure of success and failure in what we ourselves can discern. To make '*Mündigkeit*' possible, these measures must relate to the point of the projects in question. To make a difference, reason had to adapt to the human scale: the '*Selbstbesinnung*' that enlightenment demanded required its naturalisation and pragmatisation.

In following this route, Mach's enlightenment quite effectively and intentionally sought to escape the hubris that postmodern critics are fond of accusing the enlightenment of.[13] In fact, it is tempting to speak of a 'dialectic of enlightenments' here that Mach clearly perceived – and to claim that the implicit pun here is by no means achieved only for the price of an anachronism. There are two components to this view that need to be noted explicitly.

The first is that the plural just used is no slip-up, but reflects Mach's own usage and indicates, to start with, that for him enlightenment was not just one historical period, but an ongoing project involving many distinct episodes,

---

theory. My explication of Mach's 'dialectical' conception of enlightenment builds on Frank (1917), who does not, however, use that term. On the significance of this conception for some of Mach's readers in the Vienna Circle, see Uebel (1996).

[13] Or, indeed, in this respect their precursors in critical theory; see Horkheimer and Adorno (1947).

edging uncertainly forwards. Here is Mach, revisiting more than twenty years later the topic of his first major 'historical-critical' publication: 'The fate of all momentous enlightenments is very similar. On their first appearance they are regarded by the majority of people as errors' (1894/1986, p. 138, trans. amended).[14] While enlightenments may be relatively local events, they still upset established ways of looking at things and are vilified for precisely this. The second component is the distinctive one. Mach continued:

> The majority of people … take success for proof. *So it can happen that a view which has led to the greatest discoveries*, like Black's theory of caloric, *in a subsequent period in a province where it actually does not apply may actually become an obstacle to progress by its blinding our eyes to facts which do not fit in with that view.* If a theory is to be protected from this dubious role, the grounds and motives of its development and persistence must be examined from time to time with the utmost care.
>
>   (1894/1986, p. 138, trans. amended, emphasis added)

Here, the dialectic becomes plain. While the new enlightenment upsets the old order, it tends to establish a new one that, if knowledge is to progress, needs to be upset in due course just as much as the old one needed to be.

It is worth breaking down Mach's remark even further. Note, to begin with, what is implied: success is *not* proof of truth. (It is not a vulgar pragmatism equating truth with its cash value that Mach is envisaging here.) Then note wherein the dialectic consists. Enlightenments turn into their opposite and then, rather than help cognition, hinder it. The insight once gained, against resistance to be sure, is now being applied too widely – it overreaches itself and so turns from what enlightens into what obscures. Once this dialectic is pointed out, it is of course readily evident – on reflection, examples of it multiply.[15] It is difficult not to draw another conclusion which Mach also

---

[14] Unfortunately, Thomas J. McCormack translated this as: 'The fate of all momentous discoveries is similar.' His choice is understandable, given the peculiarity of a plural for 'enlightenment', but it is nevertheless unfortunate, for it helps blind readers to Mach's distinctive conception.

[15] Mach gives the following illustration: 'Habitual judgment, applied to a new case without antecedent tests, we call prejudgment or prejudice. Who does not know its terrible power! But we think less often of the importance and utility of prejudice … no one could exist intellectually if he had to form judgments on every passing experience, instead of allowing himself to be controlled by the judgments he has already formed … On prejudices, that is, on habitual judgments not tested in every case to which they are applied, reposes a goodly portion of the thought and work of the natural scientist … Not until the discrepancy between habitual judgments and facts becomes great is the investigator implicated in appreciable illusion. Then tragic complications and catastrophes occur in the practical life of individuals and nations … The very power which in intellectual life advances, fosters, and sustains us, may in other circumstances delude and destroy us' (1884/1986, pp. 232–233).

drew: that scientific reason must remain self-consciously incomplete and oppose any claim anywhere to have spoken the last word.[16]

The Mechanics of Mach's Historical Enlightenment

How, then, did Mach propose to overcome this tendency of liberating insights to turn into shackles of progress – in particular, why should history be of help here? Consider again his advice (emphases now added): 'If a theory is to be protected from this dubious role, the *grounds* and *motives* of its *development* and *persistence* must be examined from time to time with the utmost care.' To be explained, then, are both the *origin* and *persistence* of a doctrine; both, moreover, must be considered with regard to what *prompted* and what *legitimated* it. What Mach proposed as remedy against fossilisation, therefore, was that periodically any conceptual entity – whether concept or theory: I will use 'doctrine' for short – should be investigated twice over and each time with two different concerns in mind. A fourfold enquiry was to lay bare the limits of the conceptualisation at issue and thereby open a way forwards.

The comparison of doctrine then and now was the key. Here is how Mach also put the matter:

> Historical studies are a very essential part of a scientific education. They acquaint us with other problems, other hypotheses, and other modes of viewing things, as well as with the facts under conditions of their origin, growth and eventual decay. Under the pressure of other facts which formerly stood in the foreground other notions than those obtaining today were formed, other problems arose and found their solutions, only to make way in their turn for the new ones that were to come after them. Once we have accustomed ourselves to regard our conceptions as merely a means for the attainment of definite ends, we shall not find it difficult to perform, in the given case, the necessary transformation in our own thought.
>
> (1896/1986, p. 5)[17]

---

[16] 'Physical science does not pretend to be a complete view of the world; it simply claims that it is working toward such a complete view in future. The highest philosophy of the scientific investigator is precisely this toleration of an incomplete conception of the world and the preference for it rather than an apparently perfect, but inadequate conception' (1883/1960, p. 559).

[17] Mach here picked up the thread from his early remark that, given that 'the essence of classical education is historical education' – teaching 'the point of view of another eminent nation, so that we can, on occasion, put ourselves in a different position from that in which we have been brought up' – 'there is, for the investigator of nature, a special classical education which consists in the knowledge of the historical development of his science' (1872/1911, pp. 17–18).

Note the key move Mach effected: 'regard our conceptions as merely a means for the attainment of definite ends'. The tendency to misleading absolutisations was to be counteracted by keeping clearly in mind the very specific problematics our doctrines address.

Broadly speaking, of the two angles of the investigation Mach envisaged, one is causal ('motives') and one is normative ('grounds'). These angles are to be applied to both the original problem situation and the current one. The insight sought was to emerge from the comparison of the fit of doctrine to the phenomena then and the possible misfit to phenomena now. Clearly, such a comparison has no need for an Archimedean point outside – it has to be achieved from the inside, as it were, given the point and purpose of the doctrine at issue. Unlike the correspondence of our conceptual schemes with their objects and unlike any teleological histories, Mach's insight possessed empiricist credentials. This is precisely what the pragmatic turn of his historical enquiries was meant to ensure: that both the motive and the ground of a doctrine were in principle accessible to intersubjective investigation.

What for Mach made possible the overcoming of the apparently inevitable but hopefully non-terminal hindrances to further knowledge expansion by a doctrine's fossilisation was placing it in a practical context. We have to ask what the problem was that the doctrine was meant to solve in the first place, what conditioned its acceptance, and whether these grounds were adequate. Then, we have to ask what the problems are that the doctrine is confronted with today and where it shows difficulties. Finally, we have to investigate wherein the difference between the original and the new problem situation consists. The hope is that doing so will make plain, sooner or later, just why the doctrine no longer works – to be precise, what the step of generalisation was or is that brought on the failure now faced. The key, in short, lies in looking at the matter historically and doing so from a pragmatic perspective.

Mach's Non-Reductive Naturalism

It may not require the foregoing considerations, of course, to prepare the grounds for what we will turn to next: the demonstration of Mach's pragmatism as deeply anchored in his philosophy of scientific knowledge – indeed, his conception of human cognition in general. But it is helpful to see his pragmatism enabling the agenda of enlightenment. His pragmatism, too, had a point!

Mach's was an epistemological or methodological naturalism.[18] It should not be surprising, then, that this entailed a pragmatic orientation. One

---

[18] This was argued convincingly by Haller (1980, 1988). Without stressing the naturalism, Erik C. Banks' important work (2014) also opposed the phenomenalist reading of Mach.

important point to be made here is that Mach's naturalistic perspective on cognition was by no means owed solely to the influence of Darwin's evolutionary theory, though of course it fed off it and cohered with it. To be sure, the biological perspective on mental life was of central importance for his epistemology. What is notable, however, is that the formulation of the economy principle (about which more below) preceded his embrace of evolutionism (if not by much).[19] Even more notable is that Mach's biologism was clearly not reductive in intent.

Importantly, his category of 'biological' events includes 'consciously aware theoretical activity' as much as 'physical motoric, practical activity' (1910b/ 2014, p. 372); in other words, it includes all human intentional activity. Beyond this broad use of the term 'biological', we may also note that, for Mach, the division between 'instinctual' and consciously rational thought was by no means sharp or even categorical. Two phenomena here are of relevance: first, that scientific thought itself has roots in pre-scientific mental habits;[20] and second, that scientific thought prompts new 'intuitive' modes of mental operation.[21] 'Instinctual' and rational thought mark but gradations in a continuum of mental activity. But even here – or perhaps especially here – the

---

[19] About the development of his epistemological views, Mach stated that 'while I was studying the achievements of the scientists I lectured on at the beginning of my teaching activity as a Privatdozent in physics, I spotted that what most distinguished their approach was using a selection of the simplest, most economical and most purpose-suited means. Through contact with the economist Emmanuel Hermann in 1864 … I accustomed myself to designate the mental activity of scientists as economical … The more one goes into the methods of science … the more one recognises scientific activity as economical.' Mach went on to state that he had been prepared for Darwin's evolutionary theory by his study of Jean-Baptiste Lamarck in high school, and that he had begun to assimilate them already when he lectured in Graz from 1864 to 1867 and conceived 'the competition of scientific thoughts as a life and death struggle resulting in the survival of the fittest. This view does not contradict my theory of economy, but helps enlarge it by merging it into a biological-economical representation of economy. Expressed in the briefest way, the task of scientific knowledge becomes *the fitting thoughts to facts and thoughts to each other* … All beneficial knowing consists of special cases or parts of biologically helpful processes. This is because the physical and biological behavior of the more highly organised species of being are co-determined and added to by an inner process of knowing, of thinking' (1910a/1992, pp. 133–134, emphasis in original, trans. amended; cf. 1896/1986, p. 350).

[20] 'The impulse to complete mentally a phenomenon that has only been partially observed, has not its origin in the phenomenon itself; of this fact we are fully sensible. And we well know that it does not lie within the sphere of our volition. It seems to confront us as a power and a law imposed from without and controlling both thought and facts' (1884/ 1986, pp. 220–221).

[21] Thus, Mach called Newton's generalisation of Galileo's insight that gravity determines acceleration '*an instinct acquired in the course of research* [which] teaches a feeling for what is important and its separation from the incidental and irrelevant' (1910b/2014, p. 376, emphasis added).

force of scientific enlightenment can be felt: 'The greatest advances of science have always consisted of some successful formulation, in clear, abstract, and communicable terms, of what was instinctively known long before, and of thus making it the permanent property of humanity' (1882/1986, p. 191).

Another respect in which the non-reductive nature of Mach's biologism is of great importance concerns his recognition of the social dimension of human life both as precondition for the very emergence of scientific thought and for its continued institutional development:

> The first real beginnings of science appear in society, particularly in the manual arts, where the necessity for the communication of experience arises. Here, where some new discovery is to be described and related, the compulsion is first felt of clearly defining in consciousness the important and essential features of that discovery, as many writers can testify. The aim of instruction is simply the saving of experience; the labour of one man is to take the place of that of another.
>
> (1882/1986, p. 191)

Mach's recognition of the decisive role of the manual arts in forcing the articulation and more or less formal documentation of practical knowledge, thereby setting science on its path, is striking in light of some twentieth century approaches to a sociological history of science (e.g. Edgar Zilsel, Boris Hessen). His own concern was to stress the roots of this development with reference often also to the anthropology of his day (the terminological shortcomings of which he shared):

> Pressed by the goal of self-preservation to adopt practical-economical behaviours, man originally reacts wholly instinctively to favorable and unfavorable circumstances. Only as soon, however, as the social development, the division of labour and the emergence of craftsmen, forces the individual to turn his attention to the intermediate means, the intermediate goals for the satisfaction of his needs, does the intellect self-consciously play a prominent role. Practical discomfort is replaced by a pressing intellectual discomfort. The voluntarily chosen intermediate goal is now pursued with the same energy and means as previously the stilling of hunger. The instinctive movements of a savage, the semi-consciously learnt touches of craftsmen are the precursors of the concepts of researchers. The conceptions and the despised philistine arts of craftsmen develop imperceptibly into the conceptions and techniques of the physicists and the economy of action expands to become the intellectual economy of researchers which may also express itself in the pursuit of the most ideal goals.
>
> (1910a/1992, p. 6)

Consciously directed, instrumentally rational thought emerged only with the division of labour, and the emergence of science itself appeared to require still further social differentiation: 'Cultural progress is conceivable only when there is a certain audacity and thus can generally be set going only by those who are partly relieved from toil' (1905/1976, p. 58). Yet in Mach's view, it should be added, science was by no means condemned to serve the ruling classes only. 'The business of science has this advantage over every other enterprise, that from its amassment of wealth no one suffers the least loss. This, too, is its blessing, its freeing and saving power' (1982/1986, p. 198).[22] Just how the emancipatory potential of science was to be set free was not discussed, however.

The Economy of Thought

Already in this early work, then, Mach concluded about scientific theories in general that they operate according to a principle of mental economy:

> If all individual facts – all the individual phenomena knowledge of which we desire – were immediately accessible to us, a science would never have arisen. Because the mental power, the memory, of the individual is limited, the material must be arranged.... [A] 'law' has not in the least more real value than the aggregate of the individual facts. Its value for us lies merely in the convenience of its use: it has an economical value.
>
> (1872/1911, pp. 54–55)

Science operated within a practical context where it was constrained by what is thinkable and doable with the resources at hand. To render comprehensible the otherwise incomprehensible multitude of phenomena was precisely its task. And still before the task of economic systematisation came that of economical description.

Yet description was not a matter of unadulterated reporting: 'In the reproduction of facts in thought, we never reproduce the facts in full, but only that side of them which is important to us, moved to this directly or indirectly by a *practical interest*. Our reproductions are invariably abstractions' (1883/1960, pp. 578–579, emphasis added). Moreover, interest-relativity was not the only challenge. 'Observation and theory too are not sharply separable, since almost any observation is already influenced by theory and, if important enough, in turn reacts on theory' (1905/1976, p. 120) Given this 'positivist' anticipation of the theory-ladenness of observation, it is clear that achieving economy in thought extended deeply into theory.

Evolutionary in origin, the principle of economy of thought pertained to the material means of representation as much as to its content. Mach explored

---

[22] On this point, Mach was criticised by his American associate Paul Carus, who defended all business enterprises. But Mach stood firm; see Mach (1896/1986, pp. 361–362).

both. To begin with, language was the 'instrument' of the communication of scientific results across individuals and entire generations and was 'itself an economical contrivance'.[23] In terms of the content of our representations, the principle of economy can be traced throughout the development of science, both in specific doctrines and in general features of theory formation. Concerning the principles of the determination of laws of nature, Mach gave an evolutionary-economical rendition of David Hume's critique of the idea of causation as necessary connection (1883/1960, p. 581), and later provided a dissolution of the nominalism–realism dispute (1896/1986, p. 383). As regards the notion of law itself, Mach wrote of his favourite example: 'In nature there is no law of refraction, only different cases of refraction. The law of refraction is a concise compendious rule, devised by us for the mental reconstruction of a fact' (1883/1960, p. 582; cf. 1882/1986, p. 193; 1884/1986, p. 231; 1896/1986, p. 57). In sum: 'Science itself . . . may be regarded as a minimal problem, consisting of the completest possible representation of facts with the *least possible expenditure of thought*' (1883/1960, p. 586, emphasis in original).[24]

So whatever the practical purposes of enquiry may be that set the parameters of convenience now, all cognition and science bears the traces, deep in its conceptual make-up, of this origin. 'In the economical schematism of science lie both its strength and its weakness. Facts are always represented at a sacrifice of completeness and never with greater precision than fits the need of the moment' (1882/1986, p. 206) In sum:

> The character and course of development of science becomes more intelligible if we keep in mind the fact that science has sprung from the needs of practical life, from provision for the future, from techniques.... The investigator strives for the removal of intellectual discomfort; he seeks a *releasing thought*. The technician wishes to overcome a practical discomfort; he seeks a *releasing construction*. Any other distinction between discovery and invention can scarcely be made.
>
> (1896/1986, p. 407, emphasis in original)

The extent to which Mach's 'positivism' was indeed a 'pragmatism' as far as science was concerned can hardly be rendered plainer.

---

[23] Thus Mach claimed that 'the first and oldest words are names of "things"', even though 'the thing is an abstraction, the name a symbol, for a compound of elements from whose changes we abstract. The reason we assign a single word to a whole compound is that we need to suggest all the constituent sensations at once.' It follows that 'the whole operation is a mere affair of economy. In the reproduction of facts, we begin with the more durable and familiar compounds and supplement these later with the unusual by way of corrections' (1883/1960, pp. 578–579).

[24] For an analysis of Mach's economy of science in action, see Chapter 8 in this volume,

The Principle of Scientific Significance

To round off the picture, consider that Mach also formulated a maxim for scientific reasoning that has a good claim of being placed next to what William James called 'Peirce's principle'. In *The Science of Mechanics*, Mach wrote:

> The function of science, as we take it, is to replace experience. Thus, on the one hand, science must remain in the province of experience, but, on the other, must hasten beyond it, constantly expecting confirmation, constantly expecting the reverse. *Where neither confirmation nor refutation is possible, science is not concerned.*
>
> (1883/1960, pp. 586–587, emphasis added)

Compare now Peirce's principle in his 'How to Make Our Ideas Clear':

> Consider what effects, which might conceivably have practical bearings, we conceive the object of our conception to have. Then, our conception of those effects is the whole of our conception of the object.
>
> (1878/1923, p. 132)

The process by which science was distinguished by Mach – that it is possible to test its claims – is, broadly speaking, what taking account of 'practical bearings' involves.[25] This suggests a deep affinity between their conceptions.

Chronologically, Peirce's principle appears to predate Mach, but there are no indications that Mach ever read Peirce (unlike, in later years, the other way around) or that James tipped him off to this on his visit to Prague in 1882.[26] Moreover, already in *History and Root* Mach had written:

> If the hypotheses are so chosen that their subject matter can never appeal to the senses and therefore also can never be tested, as is the case with the mechanical molecular theory, the investigator has done more than science, whose aim is facts, requires of him – and this work of supererogation is an evil.
>
> (1872/1911, p. 57, trans. amended)

To be sure, here Mach appeared to focus on the verifiability of what is directly observable, while the 'confirmation [or] refutation' which he wrote of in *Mechanics* may not be so strictly limited (it is arguable that 'the province of experience' exceeds that of the 'senses'). In any case, however, the later passage also exhibits the anticipatory, future-orientated nature of his conception of

---

[25] Already Feyerabend once noted that Mach used 'a vague form of the verifiability principle' (1970, p. 180).

[26] In *Pragmatism*, James wrote that Peirce's principle 'lay entirely unnoticed by any one for twenty years' (1907/1991, p. 24) until he 'brought it forward again' in a paper of his own in 1898.

scientific knowledge that distinguishes it as properly pragmatist from mere positivism.[27] It would appear, therefore, that the pragmatic principle was discovered, independently, twice over.

In *Pragmatism*, James repeated his paraphrase of Peirce's principle from his 1898 paper in which he had revived it:

> To attain perfect clearness in our thoughts of an object, then, we need only consider what conceivable effects of a practical kind the object may involve – what sensation we are to expect from it, and what reactions we must prepare. Our conception of these effects, whether immediate or remote, is then for us the whole of our conception of the object, so far as that conception has positive significance at all.
>
> (1907/1991, pp. 23–24; cf. 1898/1963, p. 13)

For present purposes, we need not go into the differences between the pragmatic maxims of Peirce and James or theirs and Mach's principle of scientific significance. What matters now is wherein they all agree. Mach's maxim for scientific theorising specifies in broadly logical terms Peirce's 'practical bearings, we conceive the object of our conception to have' and James's 'conceivable effects of a practical kind [an] object may involve – what sensation we are to expect from it, and what reactions we must prepare'. All three agree that only differences that make a discernible difference matter, for only they can make a difference to how we deal with the world and the challenges it presents to us.

It is this tying of significance to the possibility of human action and intervention in the world that renders all three criteria pragmatist. This is no coincidence. It is as characteristic of the pragmatism of Peirce as of James that belief is not regarded primarily as a vehicle of representation of how the world is independently of what our concerns may be. Peirce called belief 'a habit which will determine our actions' (1877/1923, p. 15) and, more carefully, attributed to it 'just three properties. First, it is something that we are aware of; second, it appeases the irritation of doubt; and, third, it involves the establishment in our nature of a rule of action, or, say for short, a habit' (1878/1923, p. 41). As James put it, beliefs are 'rules for action' (1898/1963, p. 12). Beliefs, as Peirce and James understand them, involve – indeed, *set up* – expectations about how things will be.

Mach had a different terminology and focused on what he called 'judgements'. He characterised them as 'not a matter of belief but naive feelings' and added 'that belief, doubt, unbelief, rest on judgements about agreement or

---

[27] Without this orientation to future experience, Mach's 1883 maxim would have been as liable to criticism by Peirce as 'an arbitrary and indefensible limitation of useful knowledge' as Comte's strictures on verifiable hypotheses; see Peirce (1904). I thank Alex Klein for discussion of this point and the reference.

disagreement between often very complicated sets of judgements' (1905/1976, p. 90, note 14). Practical judgement came first; belief was a higher-order phenomenon and concept. What mattered was, ultimately, how an organism was prepared for new experiences – whether that was conscious or not was not important – and that, for Mach, was a matter of judgement. And to do that, judgement too set up expectations. Thus, he could decree: 'A judgement . . . that we find appropriate to the physical or mental finding to which it relates, we call correct and see knowledge in it. . . . Knowledge is invariably a mental experience directly or indirectly beneficial to us. . . . Knowledge and error flow from the same mental sources, only success can tell one from the other' (1905/ 1976, p. 83). Mach evidently agreed with the thought behind Peirce's 'truth, which is distinguished from falsehood simply by this, that if acted on it will carry us to the point we aim at and not astray' (1877/1923, p. 31).

Conclusion

The precise meaning of pragmatism was already in dispute between the founders of American pragmatism, but here we may appeal to Hilary Putnam's characterisation. Pragmatism rejects scepticism about human knowledge but fully embraces fallibilism; it rejects unbridgeable dichotomies and it accepts the 'primacy of practice' in demanding that human ideation be suitably grounded in experience to be held trustworthy.[28] After the foregoing, I hope readers can agree that Mach's principle of scientific significance – indeed, the entire conception of knowledge underlying it – was deeply pragmatist in spirit. His forthright embrace of the incompleteness of the scientific world view exemplifies his rejection of scepticism and acceptance of fallibilism. His theory of elements in *The Analysis of Sensations* (not discussed here), far from grounding a phenomenalist ontology, rejected the metaphysical dualism of mind and body.[29] And, most clearly, he accepted the primacy of practice. Mach's historicist naturalism can be considered an original form of pragmatism.

References

Banks, Erik C. 2014. *The Realistic Empiricism of Mach, James and Russell: Neutral Monism Reconceived*. Cambridge University Press.
Blackmore, John T. (ed.) 1992. *Ernst Mach – A Deeper Look: Documents and New Perspectives*. Kluwer.

---

[28] See Putnam (1994), and compare also Hookway (2008/2013).
[29] 'It depends on the definite purpose of the observation whether we view [an element] as a sensation or a thing' (Mach, Diary Entry 1880, quoted in Haller 1980/1992, p. 219).

Carnap, Rudolf 1963. 'Intellectual Autobiography', in P. A. Schilpp (ed.), *The Philosophy of Rudolf Carnap*. Open Court, pp. 3–84.

Feigl, Herbert 1949. 'Naturalism and Humanism: An Essay on Some Issues of General Education and a Critique of Current Misconceptions Regarding Scientific Method and The Scientific Outlook in Philosophy,' *American Quarterly* 1: 135–148. Reprinted in H. Feigl and M. Brodbeck (eds.), *Readings in the Philosophy of Science*. Appleton-Century-Crofts, 1953, pp. 8–18.

Feyerabend, Paul K. 1970. 'Philosophy of Science: A Subject with a Great Past', in R. Stuewer (ed.), *Historical and Philosophical Perspectives of Science*. University of Minnesota Press, pp. 172–183.

　1984. 'Mach's Theory of Research and Its Relation to Einstein', *Studies in History and Philosophy of Science* 15: 1–22.

Frank, Philipp 1917. 'Die Bedeutung der physikalischen Erkenntnistheorie Ernst Machs für das Geisteslebens unserer Zeit', *Die Naturwissenschaften* 5: 65–80. Translated as 'The Importance for our Times of Ernst Mach's Philosophy of Science', in Frank's *Modern Science and its Philosophy*. Harvard University Press, 1949, pp. 61–79.

Haller, Rudolf 1980. 'Poetische Phantasie und Sparsamkeit. Ernst Mach als Wissenschaftstheoretiker', in *Festkolloquium am 12. November 1979: 20 Jahre Ernst-Mach-Institut 1959–1979*, Freiburg, pp. 20–48. Translated as 'Poetic Imagination and Economy. Ernst Mach as Theorist of Science', in J. Agassi and R. S. Cohen (eds.), *Scientific Philosophy Today*. D. Reidel, 1981, pp. 71–84, and in John T. Blackmore, (ed.) 1992. *Ernst Mach – A Deeper Look: Documents and New Perspectives*. Kluwer, pp. 215–228.

　1988. 'Grundzüge der Machschen Philosophie', in R. Haller and F. Stadler (eds.), *Ernst Mach. Werk und Wirkung*. Hölder-Pichler-Tempsky, pp. 64–86.

Hiebert, Erwin N. 1970. 'Mach's Philosophical Use of the History of Science', in R. Stuewer (ed.), *Historical and Philosophical Perspectives of Science*. University of Minnesota Press, pp. 184–203.

Hookway, Christopher 2008. 'Pragmatism', *The Stanford Online Encyclopedia of Philosophy*, updated 2013. Available from https://plato.stanford.edu/entries/pragmatism/

Horkheimer, Max and Adorno, Theodor 1947. *Dialektik der Aufklärung*. Querido. Translated as *The Dialectic of Enlightenment*. Stanford University Press, 2007.

Kant, Immanuel 1784. 'Beantwortung der Frage: Was ist Aufklärung?', *Berlinische Monatsschrift* 2: 481–494. Translated as 'An Answer to the Question: "What is Enlightenment?"', in I. Kant, *An Answer to the Question: 'What is Enlightenment?'* Penguin, 2009, pp. 1–11.

James, William 1898. 'Philosophical Conceptions and Practical Results', *The Univeristy Chronicle* (Berkeley, California). Reprinted in W. P. Alston and G. Nakhnikian (eds.) *Readings in Twentieth-Century Philosophy*. The Free Press, 1963, pp. 12–25.

1907. *Pragmatism. A New Name for some Old Ways of Thinking.* Longmans, Green & Co. Reprinted, Prometheus, 1991.

Lampa, Anton 1912. *Ernst Mach.* Verlag Deutsche Arbeit.

Lübbe, Hermann 1972. 'Positivismus und Phänomenologie: Mach und Husserl', in *Bewußtsein in Geschichten. Studien zur Phänomenologie der Subjektivität: Mach-Husserl-Schapp-Wittgenstein.* Rombach, pp. 33–62. Translated as 'Positivism and Phenomenology: Mach and Husserl', in T. Luckmann (ed.), *Phenomenology and Sociology.* Penguin, 1978, pp. 119–141.

Mach, Ernst 1872. *Die Geschichte und die Wurzel des Satzes von der Erhaltung der Arbeit.* Reprinted, J. A. Barth, 1909. Translated as *History and Root of the Principle of the Conservation of Energy.* Open Court, 1911.

1882. *Die ökonomische Natur der physikalischen Forschung.* Translated as 'The Economical Nature of Physical Inquiry', in Ernst Mach, 1898. *Popular Scientific Lectures,* 3rd edn. Open Court. Reprinted 1986, pp. 186–213.

1883. *Die Mechanik in ihrer Entwicklung historisch-kritisch dargestellt,* 9th edn. Brockhaus. Translated as *The Science of Mechanics: A Critical and Historical Account of its Development.* Open Court, 1960.

1884. *Über Umbildung und Anpassung im natutrwissenschaftlichen Denken.* Translated as 'On Transformation and Adaptation in Scientific Thought', in Ernst Mach, 1898. *Popular Scientific Lectures,* 3rd edn. Open Court. Reprinted 1986, pp. 214–235.

1886. *Beiträge zur Analyse der Empfindungen.* Fischer. Translated as *Contributions to the Analysis of Sensations.* Open Court, 1897.

1896. *Die Prinzipien der Wärmelehre,* 2nd edn. J. A. Barth. Translated as *Principles of the Theory of Heat.* D. Reidel, 1986.

1898. *Popular Scientific Lectures,* 3rd edn. Open Court. Reprinted 1986.

1905. *Erkenntnis und Irrtum.* J. A. Barth. Translated as *Knowledge and Error.* D. Reidel, 1976.

1907. *Die Rassenfrage.* Neue Freie Presse. Reprinted in K. D. Heller (ed.), *Ernst Mach: Wegbereiter der modernen Physik.* Springer, 1964, pp. 98–102.

1910a. 'Die Leitgedanken meiner naturwissenschaftlichen Erkenntnislehre und ihre Aufnahme durch die Zeitgenossen,' *Scientia* 7: 225–240 and *Physikalische Zeitschrift* 11: 599–606. Reprinted in L. Mach (ed.), *Die Leitgedanken meiner naturwissenschaftlichen Erkenntnislehre und ihre Aufnahme durch die Zeitgenossen. Sinnliche Elemente und naturwissenschaftliche Begriffe. Zwei Aufsätze von Ernst Mach.* J. A. Barth, pp. 3–18. Excerpts translated as 'The Leading Thoughts of My Scientific Epistemology and Its Acceptance by Contemporaries', in John T. Blackmore (ed.), *Ernst Mach – A Deeper Look: Documents and New Perspectives.* Kluwer, 1992, pp. 133–139.

1910b. *Populär-wissenschaftliche Vorlesungen,* 4th edn. J. A.Barth. Reprint of 5th edn, Xenomoi, 2014.

Peirce, Charles S. 1877. 'The Fixation of Belief', *Popular Science Monthly* 12: 1–16. Reprinted in Charles S. Peirce, *Chance, Love and Logic. Philosophical Essays* (ed. M. R. Cohen). Harcourt, Brace and Co., 1923, pp. 7–31.

1878. 'How to Make Our Ideas Clear', *Popular Science Monthly* 12: 286-303. Reprinted in Charles S. Peirce, *Chance, Love and Logic. Philosophical Essays* (ed. M. R. Cohen). Harcourt, Brace and Co., 1923, pp. 32–60.

1904. 'Comte's Philosophy. Review of *The Philosophy of Auguste* Comte by L. Levy-Bruhl', *The Nation*, 28 April 1904.

Pojman, Paul 2009. 'Ernst Mach', *The Stanford Online Encyclopedia of Philosophy*, updated 2018. Available from https://plato.stanford.edu/entries/ernst-mach/

Putnam, Hilary 1994. 'Pragmatism and Moral Objectivity', in his *Words and Life* (ed. J. Conant). Harvard University Press, pp. 151–181.

Uebel, Thomas 1996. 'On Neurath's Boat', in N. Cartwright, J. Cat, L. Fleck, and T. Uebel, *Otto Neurath: Philosophy Between Science and Politics*. Cambridge University Press, pp. 89–166.

2003. 'History of Philosophy of Science and the Politics of Race and Ethnic Exclusion', in F. Stadler and M. Heidelberger (eds.), *Wissenschaftsphilosophie und Politik/Philosophy of Science and Politics*. Springer, pp. 91–117.

von Mises, Richard 1938. *Ernst Mach und die empiristische Wissenschaftsauffassung. Zu Ernst Machs 100. Geburtstag am 18. Februar 1838.* Van Stockum. Translated as 'Ernst Mach and the Empiricist Conception of Science', in R. S. Cohen and R. J. Seeger (eds.), *Ernst Mach: Physicist and Philosopher*. D. Reidel, 1970, pp. 245–270.

6

On the Philosophical and Scientific Relationship between Ernst Mach and William James

ALEXANDER KLEIN

Recent Interest in Mach and James

Perhaps nobody more clearly embodies the productive and complicated relationship between American Pragmatism and Logical Positivism[1] than two respective forefathers of those movements, William James and Ernst Mach. Positivists themselves had long understood that these men were personal friends and that they engaged one another's work (Feigl 1963/1981, p. 41, 1969/1981, p. 69) for more than a quarter of a century.

There has been a quiet uptick in scholarly attention to the relationship between Mach and James recently, largely fuelled by interest in the historical connections between the big philosophical movements they inspired.[2] Attention has focused on the pragmatist (small 'p') tendencies of three figures who constituted the so-called First Vienna Circle (henceforth FVC; Haller 1991), and who began meeting regularly in 1907: Philipp Frank, Hans Hahn, and Otto Neurath. This group was particularly strongly influenced by Mach, and they also developed explicit sympathies for American Pragmatism.

The FVC was well aware of James's 1907 book *Pragmatism* upon its publication (it appeared in both English and German in the same year that the group began meeting in a Vienna coffee house),[3] and they *later* came to see common ground with the American movement which that book helped inspire.[4] But one striking revelation in recent scholarship is that, nevertheless,

---

[1] Mach's scientific epistemology was of course a major inspiration for Logical Positivism (Stadler 1992, 2015, prologue and ch. 1).

[2] The most important developments in this literature, for my purposes, are to be found in Uebel's recent work (cited throughout the text); also see Ferrari (2017, esp. pp. 22–27), Visser (2001), Hiebert (1976, pp. xiii, xxvi), Stadler (2017) and Blackmore (1972, pp. 126–128). Holton has also treated the Mach/James relationship in material largely repeated in Holton (1992, pp. 33–36, 1993a, pp. 50–51, 1993b, pp. 7–11).

[3] Wilhelm Jerusalem's German translation carried a 1908 imprint, but in fact it was first released in 1907 (Uebel 2017, p. 84).

[4] Uebel dates the acknowledgement of deep commonalities to about 1929 (Uebel 2015, pp. 6–7).

the pragmatist tendencies one finds among members of the FVC are likely to have been 'home-grown' (Uebel 2014, p. 632), largely inspired by Mach and to a lesser extent by his Viennese ally, Wilhelm Jerusalem, and *not* (at least primarily and at least not directly) by James or any other American figure (also see Uebel 2015, p. 2, 2017, pp. 93–94; Stadler 2017, p. 14).

Mach began publicly developing what we might well regard as a pragmatist outlook on science at least as early as his 1872 *Die Geschichte und die Wurzel des Satzes von der Erhaltung der Arbeit* (*History and Root of the Principle of the Conservation of Energy*) (Mach 1872/1911; see Uebel 2014, p. 634), decades before James first employed the word 'pragmatism' in print (in 1898), and five years before Peirce published what are commonly regarded as the founding documents, *avant la lettre*, of the American Pragmatist movement (Peirce 1877, 1878).[5] There can be no question that Mach's scientific methodology had a major impact on the FVC (and on many of their later allies; Stadler 1992, 2015, prologue and ch. 1). In contrast, whatever influence James might have had seems to have been less direct,[6] and the FVC's early engagement with his work typically resulted in an ambivalent response (as in Neurath's 1909 review of *Pragmatism*; Uebel 2015, p. 9), so that it would be a stretch to call James a serious inspiration, at least for the *First* Vienna Circle. And Peirce remained largely unknown in Europe until the early 1930s, so is unlikely to have exerted much early influence either (Uebel 2014, p. 628, Ferrari 2017, p. 19).

This important research (led by Thomas Uebel in particular) has not yet received the attention it deserves. It raises crucial questions for, among others, historians of American Pragmatism. Some Pragmatist historians have recently claimed that Analytic Philosophy can include Peirce and (to a lesser extent) James among its founding fathers (Misak 2013, 2015, 2016; Aikin and Talisse 2017; cf. Klein 2018). The thinking is that, after the 1930s, one can find distinct pragmatist threads in the Analytic tradition – threads that run right through to today. If Quine, Wittgenstein, or Ramsey (to give three oft-cited examples) were influenced in important ways by Peirce or James (or perhaps Dewey), then one can think of those pragmatist

---

[5] Mach's 1872 book appeared in the same year that Peirce and James's fabled 'Metaphysical Club' convened in Cambridge (Menand 2001, pp. 201–232), but this was a private group that Mach would have been unlikely to hear about. James had not yet been hired to a permanent position at Harvard and had yet to publish anything but short reviews, and it would be ten years before James would meet Mach in person.

[6] One must not overstate this point. Ferrari offers evidence that James's *Pragmatism* was probably 'much discussed' in the FVC (Ferrari 2017, pp. 27 and ff.), although one can learn about various reservations they had by consulting Uebel (2014, p. 628, 2015, pp. 4–5, 9, 15–16). One major worry concerned what they perceived as James's psychologism (Uebel 2015, p. 12; cf. Klein 2016).

threads that have long been part of mainstream Analytic philosophy as having a Pragmatist (big 'P') provenance. I am sympathetic with this line. But note that if Uebel is right, then these pragmatist threads may also owe something important to a Machian influence on Logical Positivism (first via the FVC, later perhaps via Carnap).

This chapter pursues a necessary next step for scholars interested in the historical origins of pragmatist thinking in Analytic philosophy: to investigate in more detail the relationship between Mach and James, for if Mach and James came around to similarly pragmatist views about science, it might be that one friend influenced the other. Perhaps Mach co-opted pragmatist moves from his American friend and interlocutor, such that we can preserve the revisionist story that Peirce and James *are* the original fabricators of pragmatist threads in Analytic philosophy.

In the next section, I shall explain why this response is not quite right. For one thing, James and Mach seem rarely to have discussed philosophical issues at much length with one another, and in published work, they cite one another's philosophical views only in passing. Instead, instrumentalism about science seems to have been in the transatlantic air during much of their careers, so that American Pragmatism and Machian Empiricism are better regarded as partaking in, and contributing to, this larger trend in scientific philosophy. To the extent that there is evidence of philosophical influence, the evidence suggests that the influence flowed from Europe to the USA – particularly from Mach to James, who was fluent in German (and extraordinarily cosmopolitan). There is less evidence of a reciprocal philosophical influence from James to Mach.

The next question becomes: if philosophical pragmatism was not the basis for James and Mach's intellectual relationship, then what *was* that basis? It turns out that the majority of references each makes to the other's work concern empirical matters, most particularly the role of the semicircular canals in the perception of bodily orientation and the question of whether there is a distinctive 'feeling of effort' (*Innervationsgefühl*). And given that Mach, Helmholtz, and others had appealed to a supposed *Innervationsgefühl* as a cue in their accounts of spatial perception, this latter issue inevitably crops up as a topic as well. In the third section of this chapter, I shall touch on the debate over the *Innervationsgefühl* in order to emphasise the actual, empirical matters that the two men actually spent the most energy engaging one another on. That debate is interesting in part because it ended in James helping to change Mach's mind.

We remember Mach as a master experimentalist and James as a philosophical renegade, so it is a surprise to learn that the main *philosophical* influence apparently flowed from Mach to James, while the main influence when it comes to matters of *empirical* interest actually flowed quite the other way.

Philosophical Arrows of Influence

James and Mach died in 1910 and 1916, respectively, but institutional and intellectual connections between Pragmatism and Logical Positivism continued, particularly after leading advocates of the latter movement migrated to the USA around the time of World War II. Many Pragmatists and Positivists saw one another as fellow travellers thanks in part to a shared commitment to empiricism, and alliances between figures such as Herbert Feigl, Charles Morris, Ernest Nagel, Otto Neurath, John Dewey, and Philipp Frank were instrumental in helping the refugees settle into new professional positions (Richardson 2003; Reisch 2005; Klein 2007, pp. 389–399). It is therefore not surprising that in the recent literature on James and Mach the focus has tended to be on the shared affinities between the two men for *pragmatism* (which Mach himself acknowledged in a 1907 letter; see Thiele 1966, p. 305) and for *neutral monism*, another shared commitment that at one time influenced the development of both Pragmatism and Analytic Philosophy at large.[7]

However, when one examines surviving letters of James and Mach, along with their published writings, pragmatism and neutral monism are not topics that seem to have animated much discussion between the two. Certainly, James both was happy to regard Mach as a fellow pragmatist traveller[8] and also was deferential (in correspondence) towards Mach's pioneering work on neutral monism.[9] But the book *Pragmatism* does not engage Mach very much

---

[7] Some important treatments on James and Mach on the issue of pragmatism include Stadler (2017), Ferrari (2017), Weinberg (1937), and Perry (1935, vol. II, pp. 579–580), and some important treatments that deal with their shared neutral monism include Banks (2014) and Hatfield (2002). 'Neutral monism' is Russell's term for the view he came to share with Mach and James: viz., that minds and bodies are not two different kinds of things, but both rather are composites built from a single kind of underlying stuff (hence 'monism') that is itself 'neutral' between counting as psychical or physical (Banks 2004, p. 41).

[8] In a letter to F. C. S. Schiller on 16 January 1906, James specifically calls Mach's *Erkenntnis und Irrtum* 'excellent wise stuff, and very pragmatic' (James 1992–2004, 11.147). Interestingly, in a 1 June 1907 letter to Charles A. Strong, James expresses some reservation about whether Mach is really a pragmatist, since the latter's 'pure phenomenism is expressly denied by him to be a *philosophy*[;] it is only a point of view which he calls sufficient for scientific purposes' (James 1992–2004, 11.372). For Mach's own denial that he himself is a philosopher, see Mach (1886/1914, p. 30, note): 'I make no pretensions to the title of philosopher. I only seek to adopt in physics a point of view that need not be changed the moment our glance is carried over into the domain of another science; for, ultimately, all must form one whole.' Also see Mach (1926/1976, pp. XXXI–XXXII).

[9] For James's acknowledgement of Mach's influence on his radical empiricism, see his letter to Mach of 19 November 1902: 'I am now trying to build up before my students a sort of elementary description of the constitution of the world as built up of "pure experiences" (in the plural) related to each other in various ways, which are also definite experiences in

beyond namechecking him; and perhaps more surprisingly, Mach's name does not even appear once in *Essays in Radical Empiricism*, the posthumous collection of essays in which James lays out his own neutral monism.

Nevertheless, by the time James began talking about either pragmatism or neutral monism in public (in James 1898, 1904, respectively), the two men had already been in regular contact for almost two decades, and James had been citing Mach in print for almost three. The first published reference to Mach in James's writing is from an 1875 book review on Wundt's *Grundzüge der physiologischen Psychologie* (James 1987, p. 297).[10] Mach is mentioned only in passing there, but his 1875 *Grundlinien der Lehre von den Bewegungsempfindungen* (*Fundamentals of the Theory of Movement Perception*) is preserved in James's personal library (James was indeed working on movement perception at the time).[11] The two men did not meet until November of 1882 when James was in Prague, and Mach seems not to have noticed James's work before then. A few years later – in the 1886 *Beiträge zur Analyse der Empfindungen* (*Contributions to the Analysis of the Sensations*) – we find Mach's first reference to James (Mach 1886, p. 70, note). Subsequently, the two men engaged each other's works through the years, and also maintained a personal correspondence until they died.

Still, it is really empirical work that forms the basis for the connection between the two men, with philosophical issues typically falling into the background. Tellingly, in the book that James thought 'very pragmatic'[12] – Mach's 1905 *Erkenntnis und Irrtum* (*Knowledge and Error*) – the numerous references to James are all to the latter's psychological work, particularly from *The Principles of Psychology*. Mach also dedicated the first, 1895 edition of his *Popular Scientific Lectures* to James – but again, the basis for the dedication was James's *scientific* popularisation.[13]

James was keen to recruit Mach as an ally of Pragmatism, though Mach's return endorsement was somewhat 'perfunctory', as Perry put it (Perry 1935,

---

their turn. . . . I wish you could hear how frequently your name gets mentioned, and your books referred to' (James 1992–2004, 10.150). In addition, James's annotations in his copy of the fourth (German) edition of Mach's *Analysis of Sensations*, published in the following year, are clearly aimed at probing the overlaps and divergences between the two men on neutral monism (at Houghton, WJ 753.13; see esp. ch. 1).

[10] The first evidence we have of James mentioning Mach in the classroom is from lecture notes for Philosophy 3, which James taught between 1879 and 1885 (James 1988 170).

[11] This volume has moderate marginalia and can be found at the Houghton Library at call number WJ 753.13.4. 'Wm. James / 20 Quincy St. / Cambridge' is inscribed in James's hand in the front flyleaf. James lived at that address from 1866 through late 1889. Perry reports only one Mach volume from James's library having been sold – the 1895 McCormack translation of *Popular Scientific Lectures*; see Houghton (bMS Am 1092.9 (4578)) for Perry's list of books that were sold off.

[12] See Note 8.

[13] A full history of this dedication can be found in Stadler (2017).

vol. II, p. 463). Thus, in a 1911 letter quoted by the late Erik C. Banks, Mach wrote the following to the Danish philosopher Anton Thomsen:

> The center of his [James's] work certainly lies in his excellent Psychology. I cannot quite come to terms with his Pragmatism. 'We cannot give up the concept of God because it promises too much.' That is a rather dangerous argument.[14]

Banks writes that although the two men shared a remarkable intellectual respect for one another, Mach was nevertheless uncomfortable with the way James sought to 'squeeze in "*Spiritualismus und Schwärmerei*" [spiritualism and fanaticism; Banks attributes these words to Mach][15] alongside science'.

At least a general sympathy for James's pragmatism nevertheless did develop among Mach and Mach's followers, and this is not entirely surprising. James had portrayed pragmatism as a scientific 'tendency' in philosophy, a tendency he claimed was already being exemplified by various European scientists and scientifically minded philosophers, including Mach, Sigwart, Ostwald, Pearson, Milhaud,[16] Poincaré, Duhem, and Ruyssen[17] (James 1907/ 1975, pp. 34, 93). Early Logical Positivists would have looked favourably upon many of these figures, particularly Mach.

Here is James sketching a similar story in his 1904 review of F. C. S. Schiller's *Humanism*:

> Thus has arisen the pragmatism of Pearson in England, of Mach in Austria, and of the somewhat more reluctant Poincaré in France, all of whom say that our sciences are but *Denkmittel* [instruments of thought – more on this term, below] – 'true' in no other sense than that of yielding a conceptual shorthand, economical for our descriptions. Thus does

---

[14] This translation is from Banks (2003, p. 143). The original German letter can be found in Blackmore and Hentschel (1985, p. 86).

[15] This is actually a minor misquote. In an earlier letter to Thomsen (4 September 1909), Mach had written that he thinks James's *Principles of Psychology* is 'the best of the current books' on the topic. Mach wrote that James did seem to be 'somewhat prone to fanaticism and *spiritism* [*Schwärmerei und Spiritismus*]', but that nevertheless, 'one does well to retain his other services' (my translation; Blackmore and Hentschel 1985, pp. 62–63). For more on the correspondence between Mach and Thomsen, see Koch (1991).

[16] Gaston Milhaud was trained as a mathematician, and he established and occupied the first chair in the History of Philosophy in Its Relation to the Sciences at the Sorbonne (Chimisso 2008, p. 24). Abel Rey, who later established the Sorbonne's Institut d'Histoire des Sciences, would later occupy Milhaud's chair (Brenner 2005, p. 435). Along with Mach, Poincaré, and Duhem, the young Rey was another key influence on the FVC (Haller 1991, p. 97).

[17] James apparently read and notated Theodore Ruyssen's 1904 *L'Evolution Psychologique du Jugement*. Ruyssen's name only appears in this list starting in the fourth impression of *Pragmatism*; the first three impressions instead listed Gerardus Heymans, a Dutch philosopher and psychologist (see note 34.3 in James 1907/1975, p. 162).

Simmel in Berlin suggest that no human conception whatever is more than an instrument of biological utility; and that if it be successfully that, we may call it true, whatever it resembles or fails to resemble. Bergson, and more particularly his disciples Wilbois, Le Roy, and others in France, have defended a very similar doctrine. Ostwald in Leipzig, with his 'Energetics', belongs to the same school, which has received the most thoroughgoingly philosophical of its expressions here in America, in the publications of Professor Dewey and his pupils in Chicago University, publications of which the volume *Studies in Logical Theory* (1903) forms only the most systematised installment.

(James 1987, p. 551)

From this passage, one can see how so-called scientific philosophers[18] attracted to Mach's economy of thought might appreciate James's pragmatism, here characterised as the view that scientific theory amounts to 'conceptual shorthand, economical for our descriptions', rather than quasi-images meant to 'resemble' their objects. He also portrays this kind of philosophical view as in line with the scientific methodology and philosophy of respected European figures. For young scientific philosophers, James might have seemed a fellow traveller, even if his propensity for offering more of a *Weltanschauung* than a targeted philosophy of science seems to have limited his actual influence (Uebel 2015, pp. 5, 9).

And if we come at their relationship from Mach's perspective, we can also see some deep affinities. Following C. B. Weinberg (1937), we can usefully think of Mach as having two epistemologically basic commitments, both of which James shared, broadly speaking. The first is Mach's commitment to empiricism or positivism, summed up in his slogan: 'where neither confirmation nor refutation is possible, science is not concerned' (Mach 1883/1893, p. 490). In a similar vein, he often portrays science as being motivated by humans' basic, biological need to have their thoughts 'conform to what they have observed' (Mach 1895, p. 224), or as he elsewhere puts it: '[s]cience always takes its origin in the adaptation of thought to some definite field of experience' (Mach 1886/1897, p. 24). Mach's empiricism culminates in his treatment of sensations as the 'fundamental' 'elements of the world' out of which the facts of both physics and psychology alike are to be built (Mach 1886/1897, pp. 10, 25).

For his part, James also advocated a reliance on experience as that to which our theorising must ultimately be responsible, calling his position 'radical empiricism': 'To be radical, an empiricism must neither admit into its constructions any element that is not directly experienced, nor exclude from them

---

[18] For two relevant discussions of the history of so-called scientific philosophy during this era, see Richardson (1997, 2003).

any element that is directly experienced' (James 1912/1976, p. 22). And like Mach, James would also cultivate a form of 'neutral monism' (as Russell would later term the position) in his own empiricist soil (James 1912/1976, p. 81; Banks 2014).

Mach's other core epistemological commitment is to construe science as ultimately aiming at saving mental labour. 'It is the object of science to replace, or save, experiences, by the reproduction and anticipation of facts in thought.... This economical office of science, which fills its whole life, is apparent at first glance' (Mach 1883/1893, p. 481). This is his doctrine of the 'economy of thought', a doctrine that has often been thought to resonate with pragmatism (e.g. Weinberg 1937), for instance because of Mach's related rejection of a copy-theory of truth and his emphasis on the role of interest in enquiry:

> In the reproduction of facts in thought, we never reproduce the facts in full, but only that side of them which is important to us, moved to this directly or indirectly by a practical interest. Our reproductions are invariably abstractions. Here again is an economical tendency.
>
> (Mach 1883/1893, p. 482)

Like Mach, James was interested in the psychology of science, including the evolutionary psychology of science. In fact, even before the (just-quoted) 1883 *Die Mechanik in ihrer Entwickelung historisch-kritisch dargestellt* (*The Science of Mechanics: A Critical and Historical Account of Its Development*), we find James's early *Mind* essay 'The Sentiment of Rationality' also suggesting that we theorise for the purpose of saving energy:

> [A] philosophic conception of nature is thus in no metaphorical sense a labour-saving contrivance. The passion for parsimony, for economy of means in thought, is thus the philosophic passion *par excellence*, and any character or aspect of the world's phenomena which gathers up their diversity into simplicity will gratify that passion, and in the philosopher's mind stand for that essence of things compared with which all their other determinations may by him be overlooked.
>
> (James 1879, p. 320)[19]

There is evidence, though it is extremely limited, that this passage *might* have been directly inspired by Mach. As I have mentioned, Mach had developed similar themes already in his 1872 book on the history of the conservation of energy, and James references this book (in this connection) in the 1907

---

[19] We find James expressing similar views almost three decades later in *Pragmatism*, where confirmed hypotheses are called the 'sovereign triumphs of economy in thought' (James 1907/1975, p. 93; also see pp. 18, 109).

Pragmatism (p. 105);[20] but it is hard to ascertain whether James might have read that work early enough for it to have influenced the formulation of his own instrumentalism in 'The Sentiment of Rationality'.[21] In addition, Mach's 1875 *Grundlinien* articulates a similar view, but it does so only briefly and in passing. The passage in question is unmarked in James's personal copy.[22] We do know that James read this work in the 1870s, but the work is concerned with a technical treatment of bodily-movement perception (as are James's explicit references to that work); the brief reflection on scientific methodology comes up quite in passing.

In any case, Mach first became aware of his friend's statements about the economising aims of science, and the role of interest in enquiry, from this 1879 paper. We know from an 1884 letter that James sent this essay to Mach shortly after their original meeting two years earlier in Prague (Perry 1935, vol. I, p. 588; Thiele 1966, p. 300). And in the 1886 *Beiträge zur Analyse der Empfindungen*, in presenting his own views on the economy of thought, Mach says that James had pointed out in conversation that 'The Sentiment of Rationality' had articulated a kindred view of '*der Begriffe als öconomische Mittel*' ('concepts as economical instruments'; Mach 1886, p. 141, note 83).

Let us now take a closer look at this construal of concepts as instruments – '*Mittel*' in the Mach passage just quoted, or '*Denkmittel*' in the James passage quoted above. James began using the latter, German phrase regularly in 1903 to describe concepts or theories that we retain because they are helpful for organising experience.

One obvious source for James's use of the word is Mach himself, the only figure of the three who James initially mentions as holding that 'our sciences are but *Denkmittel*' who wrote in German. Mach only seldom used this

---

[20] James reproduces an epigram of Lessing that he says 'Mach quotes somewhere', without identifying the source. The quotation appears in Mach (1872, p. 1).

[21] James's personal library at Houghton does not include Mach (1872), and Perry does not list it as having been sold (bMS Am 1092.9 (4578)). By his own account, Mach had come in his teaching to portray science in instrumental terms as early as 1861, and he connected this instrumentalism to an *economy* of thought in 1864 under the influence of his friend, the political economist E. Hermann (Weinberg 1937, p. 3). James does cite a short, technical piece from this time period (Mach 1863) in the *Principles* (at James 1890/1981, p. 413, note), and James's chapter on time repeatedly cites Mach's experiments on aural temporal perception from another short, technical piece of the era (Mach 1865). But the 1872 *Conservation* book seems to be the earliest of Mach's publications in which James might plausibly have encountered Mach's instrumentalist reflections on methodology, and it is difficult to know *when* James read that work.

[22] Mach writes that 'the essence of natural science' is to learn the 'rules governing how ["natural phenomena"] ... reoccur'. Knowledge of these rules is desirable because it 'allows us to avoid fully observing the phenomena each time' (Mach 1875/2001, p. 54).

unusual term. But James might well have had in mind the following usage from his friend's 1883 *Mechanik*:

> The division of labour [*Die Theilung der Arbeit*], the confinement of a researcher to a small domain, the investigation of these domains as a life's work, is the necessary condition [*die nothwendige Bedingung*] for the fruitful development of science. Only with this one-sidedness and limitation can the special economical *instruments of coping* [*Mittel zur Bewältigung*] with the domain achieve the necessary refinement [*Ausbildung*]. At the same time, however, there lies here the danger of overestimating these instruments [*diese Mittel*] – with which we are always occupied, and which are nothing but tools of the trade – of holding them as the actual object of science [*eigentliche Ziel der Wissenschaft*].
>
> 2. In our opinion, such a state of affairs has really been created in physics due to its disproportionately large, formal development as compared with the rest of the natural sciences. Most natural inquirers attribute to *the instruments of thought* [*Den Denkmitteln*] of physics – the concepts mass, force, atom, and so on, whose sole function is economically and orderly to revive experiences – a reality beyond and independent of thought.
>
> (Mach 1883, p. 476, emphasis added, my translation, in consultation with McCormack's translation from Mach 1883/1893, p. 505)

Mach's view is that each special science must narrow its scope to attain precision, and in turn to effect a more fruitful 'division of labour' in science at large. The conceptual 'tools of the trade' that scientists use to achieve this precision he here calls '*Denkmitteln*' – instruments of thought – and cautions not to treat these concepts (such as mass, force, and atom) as corresponding to real, natural objects themselves. Physics is not in the business of investigating some things in the world called 'forces' that act on other natural objects called 'atoms', according to Mach; instead, he thinks physics employs the concepts of 'force' and 'atom' as *Denkmitteln*, tools for constructing theories that efficiently and economically give us predictive control over our future experiences.

This passage in James's copy is indeed highlighted,[23] and it reverberates with James's own writing. Reflecting on scientific methodology in the wake of the publication of *The Principles of Psychology* (1890), James would similarly argue that each special science begins with convenient presuppositions about its subject matter. These presuppositions help narrow the domain of each science so as to encourage precision, James thinks; and when the presuppositions are properly fashioned, they also help 'distribute the labour' between

---

[23] The passage contains sidelining and 'NB', in James's hand (WJ 753.13.6). The flyleaf bears 'Wm. James / 95 Irving St. / Cambridge', which suggests that he only acquired this volume sometime after late 1889, when he and his family moved to that address.

specialised disciplines in 'the most efficient' way (James 1983, p. 273; for a discussion, see Klein 2008).

Even more to the point, Mach's use of '*Denkmittel*' here matches nicely with James's use of that term in the 1904 passage I quoted above – key scientific terms do not denote real natural entities, but rather are to be treated as instruments for coping with experience. Indeed, one respect in which Frank, Hahn, and Neurath's views have all been called 'pragmatic' is that they all shared 'the view that scientific statements and theories are tools and instruments' for gaining predictive control over nature, and must be assessed as such (Uebel 2015, p. 13) – in other words, they shared a commitment to what I will call 'instrumentalism'.

So to sum up, we have Mach articulating an early version of his 'economy of thought'-style instrumentalism in his 1872 *Conservation of Energy* book. We do not see James referencing this work until much later, in his own 1907 *Pragmatism*. But in the meantime, we have James developing an instrumentalism about scientific 'conceptions' similar to Mach's in 1879. James subsequently gave Mach a copy of this article, and Mach acknowledged the overlap in their views in 1886. Meanwhile, we have Mach describing scientific concepts as '*Denkmitteln*' in 1883, in a passage that James read, but probably sometime after 1889. James echoes some themes from Mach's '*Denkmitteln*' passage in an 1892 methodological reflection on psychology, but James does not yet use that word. Starting in 1903, James begins repeatedly using the word '*Denkmittel*' to flesh out his own pragmatic treatment of scientific concepts and to portray pragmatism as allied with a larger European movement in scientific philosophy.

Thus, Mach's instrumentalism largely predated his awareness of James. The only reference I can find in Mach to James's pragmatic (small 'p') ideas is the early reference in a footnote to James's 1879 'The Sentiment of Rationality'. There are no references in Mach's published work to James's neutral monism and no references in James's papers on neutral monism to Mach's published work. And neither pragmatism nor neutral monism constituted important threads in any correspondence between the two. Still, there is a case for a (quasi-)philosophical influence between the two, and it goes from Mach (and Central Europe more generally) to James, and not vice versa, for James likely encountered Mach's instrumentalism early on. James's use of '*Denkmittel*', along with his citing of a host of other European scientific philosophers, suggests that Mach and other Central European sources inform a core aspect of James's pragmatism – especially his instrumentalism about scientific theories.

Of course, James would have encountered instrumentalist tendencies among American friends such as C. S. Peirce and Chauncey Wright as well, so that perhaps it is most accurate to say that this sort of view was just 'in the air' at the time. Still, neither James, Peirce, nor Wright seems to have influenced the development of *European* instrumentalism of the era; the general influence is more likely to have been moving from east to west.

The Empirical Conversation between Mach and James

Earlier, I suggested that empirical and not philosophical topics constituted the more common focus of intellectual exchange between the two friends, and here we see a surprising influence going from James to Mach.

James's first two important, in-print confrontations with Mach appear in 1880. 'The Feeling of Effort' (James 1880) contains a rebuttal of some German views on volition that Mach had accepted (we will return to this), and James's review of a book by Karl Spamer on the physiology of the semicircular canals contains a passing reference to Mach's work on this topic (James 1987, p. 376). The physiological function of the semicircular canals turns out to be another important source of early contact between the two, and I will briefly canvas their work on this latter topic before turning to a disagreement they had over volition.

In 1842, Jean-Pierre Flourens had published studies showing that pigeons with portions of their semicircular canals removed often lose their equilibrium. Friedrich Goltz had later suggested that the semicircular canals are organs of balance (Henn and Young 1975, p. 139).

In his 1875 *Grundlinien*, Mach then offered a mechanistic account of *how* the semicircular canals produce sensations of bodily motion and how these sensations help agents keep their balance (Mach 1875/2001). He ran some experiments that sought to rule out rival hypotheses concerning the possible source of bodily-motion perception, including pressure on the soles of the feet, blood flow in the body at large, and pressure on the head. Mach concluded that the semicircular canals indeed contribute to the sensation of motion, and that they do so by detecting bodily acceleration (other important, contemporaneous sources for this view were Josef Breuer and Alexander Crum Brown).

Now James had conducted some experiments in hopes of supporting the Machian position in this debate. At first, he failed to produce conclusive results, but he reported the work nonetheless in the 1880 review of Spamer (James 1987, p. 375, note). James had prepared twenty-one frogs by destroying pairings of semicircular canals in the same plane (i.e. left anterior and right posterior, or vice versa; or left and right horizontal), and then he had whirled the frogs around on spinning bowls to test their resulting vertigo.

More importantly, he also conducted an extensive study of deaf students in support of the general Mach–Breuer–Brown theory, which he published in 1881. It is these experiments that Mach eventually took notice of. James wrote that if the semicircular canals were organs 'of translation through space, which in its more extreme degrees becomes the feeling of dizziness or vertigo', then one should expect that 'some, at least, of the inmates of deaf and dumb institutions ought to prove insusceptible of experiencing this latter sensation' (James 1983, p. 125).

He travelled to a series of such institutions, using a swing apparatus to try to instil dizziness (James 1983, pp. 125–126). He also distributed questionnaires to gather self-reports of susceptibility to dizziness from other deaf subjects he could not examine directly. Of the 519 deaf subjects about whom he eventually gathered data, James reported that 320 of them were either not susceptible to dizziness at all or were only slightly so susceptible. In contrast, of 200 non-deaf Harvard professors, only one was found to be unsusceptible to dizziness at all (James 1983, p. 128).

James and Mach's personal relationship dates to a month after James published the completed results of his work on dizziness in deaf subjects (James 1882), so this is an apt place to say a bit more about that relationship. To begin with, in a colourful letter to his wife, James describes meeting Mach in Prague on 1 November 1882. Stumpf and Mach had insisted on 'trotting me about, day & night, over the whole length & breadth of Prague'. James describes his conversation with Mach in particular as 'unforgettable':

> I don't think any one ever gave me so strong an impression of pure intellectual genius. He apparently has read everything & thought about everything, and has an absolute simplicity of manner and winningness of smile when his face lights up that are charming.
>
> (James 1992–2004, 5.285–286)

The first surviving correspondence between the two men is a letter from Mach to James dated 'Prague, Jan. 29, 1884' (alluded to above). Mach acknowledges receipt of, but has not yet had time to read, James's 'The Sentiment of Rationality', and he says, 'Your fine experiments on rotary dizziness you will find already taken account of in my new book.'

The book in question (Mach's *Beiträge*) would be published two years later. In a long footnote discussing the Mach–Breuer–Brown hypothesis, Mach calls James's observations 'the most remarkable' (*merkwürdigsten*) of all the empirical evidence available in its favour (Mach 1886, p. 70).

The issue that would eventually command the most sustained discussion between the two, however, was volition. Throughout the nineteenth century, psychologists and philosophers had come to associate volition with the feeling of muscular effort (Scheerer 1989, pp. 41–42). James extensively criticises an account of the physiological basis for this feeling that he claims was pioneered by Johannes Müller. Müller is supposed to have identified volition with the feeling of the *efferent* (outflowing) nerve current – for (James's) Müller, the feeling of effort *just is* the feeling of the efferent nerve 'telling' my muscles to contract.[24]

Wundt and Helmholtz both supported the efferent hypothesis by appealing to observations of patients with various forms of paresis (Wundt 1863, Helmholtz 1867). These patients reportedly experience a feeling of effort when they attempt to move a wholly or partially paralysed limb. Such a feeling cannot arise from the afferent (inflowing) nerve current, the efferent-theorists reason, because the feeling reportedly arises even when the limb is actually moved little or not at all (see esp. Wundt 1863, vol. I, p. 222). Wundt termed this feeling of effort the '*Innervationsgefühl*' (literally, innervation-feeling).

James was a staunch critic of the *Innervationsgefühl*. Here he is registering his dissent in his 1880 essay 'The Feeling of Effort':

> In opposition to this popular view, I maintain that the feeling of muscular energy put forth is a complex afferent sensation coming from the tense muscles, the strained ligaments, squeezed joints, fixed chest, closed glottis, contracted brow, clenched jaws, etc., etc. That there is over and above this another feeling of effort involved, I do not deny; but this latter is purely moral and has nothing to do with the motor discharge.
>
> (James 1983, p. 85)

James claimed that the feeling of effort arises only from *afferent* nerve currents. When I lift a heavy object, for James, the feeling of effort is the feeling *of* my muscles having flexed, not the antecedent feeling of *attempting* to flex.

But how can James explain effort feelings in partially paralysed patients, then? James points to Alfred Vulpian's observations of hemiplegic patients who are asked to try to squeeze a ball with their paralysed hand. Vulpian confirms Wundt's report that such a patient *does* experience a feeling of effort. But such a patient also 'unconsciously performs this action with the sound' hand at the same time. James says he 'repeatedly verified' Vulpian's observations himself (James 1983, p. 92).

James takes Vulpian's observation to support a general conjecture: that *whenever* 'effort' is felt in connection with a totally or partially paralysed body part, that effort is coming from the 'tense muscles, the strained ligaments, squeezed joints, fixed chest, closed glottis, contracted brow, clenched jaws, etc., etc.' that are being moved *elsewhere in the body*, perhaps without the subject's direct awareness.

---

[24] At any rate, this is according to James (1983, p. 84), who cites Müller (1837, vol. II, p. 500). In the cited location, I do not see Müller saying that the *only* feeling of effort comes from efferent nerves. In fact, he elsewhere seems happy to accept the existence of afferent muscular feelings as well (as Scheerer 1989, p. 44 points out; see Müller 1837, vol. II, p. 363).

Mach enters the discussion with more sophisticated, abductive versions of the argument from paresis, with Helmholtz offering similar considerations. He and Helmholtz both appeal to the *Innervationsgefühl* as a cue in spatial perception. They reason as follows. Objects normally appear to move in two cases:[25]

(1) When the retinal image changes relative position but the eyeballs remain stationary or
(2) When the retinal image retains the same relative position but the eyeballs move.

Now, *eyeball* paresis can introduce translocation illusions. For example, when a patient who can only move an eyeball with difficulty is asked visually to track a moving object, the object may appear to move more in the direction of eyeball rotation (Helmholtz 1856–1867/2005, vol. III, pp. 245–246). The efferent theorist says that what *would* explain such illusions is if subjects typically judge themselves to be in the type (2) situation not based on having felt their eyeballs actually moving, but based on feeling themselves *trying* to move their eyeballs – that is, when they have a feeling of the efferent nerve current (the supposed *Innervationsgefühl*).[26] If that is correct, then the translocation illusion associated with eyeball paresis constitutes a confirmatory prediction of the efferent hypothesis.

James has a simple response: when the paralysis (or paresis) is in the *right* eye, one finds that the *left* eyeball continues to move even after the right one is impeded. The feeling of this *left* eyeball motion could provide the misleading (i.e. illusion-inducing) cue (James 1983, pp. 96–97).

Mach's 1886 *Beiträge* does not reference James on the feeling of effort, but it adds a curious new experiment that gives an implicit rejoinder to James's right eye/left eye account. Using himself as a subject, Mach rotated his eyeballs as far to the left as possible, then jammed 'two large lumps of fairly firm putty' (*zwei grosse Klumpen von ziemlich festem Glaserkitt*) on his eyeballs, in the rightmost corner of each eye. He reports getting the same sort of translocation illusion reported by patients with paresis in one eyeball. But the experiment is important because the putty effectively induces eyeball paresis *in both eyes at once*, so James cannot account for the illusion by appealing to the feeling of unimpeded movement in a non-paralysed eye. Mach concludes, perhaps overly dramatically: 'The will to perform movements of the eyes or the innervation of the act, is itself the space-sensation' (Mach 1886, p. 57, 1886/1897, p. 60).

---

[25] This is James's way of framing the issue (James 1983, p. 94), but it goes back at least to Müller (1837, vol. II, p. 363).

[26] An important early version of this argument in Mach is in his 1886 *Beiträge* (Mach 1886, pp. 55–57, 1886/1897, pp. 57–59).

James reported that he did not get the expected illusion when he jammed putty into his own eyes. But he responded that even if the effect could be reliably elicited, 'the conditions are much too complicated for Professor Mach's theoretic conclusions to be safely drawn' (James 1890/1981, p. 1118) because having large foreign objects stuck to your eyeballs produces too many 'peripheral sensations'. These are strong (and unusual) enough that they may cause any number of illusions, and do so 'quite apart from the innervation feelings which Professor Mach supposes to coexist'.

Surprisingly, when Mach expanded the *Analysis of Sensations* for the second, 1900 edition (he now dropped the 'Contributions to . . .' (*Beiträge*) prefix), he came around to James's view. Mach added a completely new chapter on 'Will', in which he now *denied* that there is any feeling associated with the efferent nerve current at all.[27]

Citing both his own experience with a recent stroke along with James's work on the subject, Mach says it now 'seems to me plausible to suppose that this [his own feeling of effort when trying to move his own stroke-paralysed legs] was caused by the energetic innervation of other muscle-groups in addition to the muscles of the paralysed extremities' (Mach 1886/1914, p. 175). The most important consideration seems to be this:

> The hypothesis of specific sensations of innervation is not required for the explanation of the phenomena, and, on the principle of economy, is consequently to be avoided. Finally, sensations of innervation are not directly observed.
>
> (Mach 1886/1914, p. 174)

Considerations of parsimony,[28] ultimately, have pushed Mach over to James's side. He sees that James is able to account for the variety of paresis cases that have been discussed, and to do so *without* postulating any special feeling of innervation. And such a feeling would indeed have to be postulated, Mach now concedes, because it cannot be directly observed.

The final feeling one is left with from following three decades of discussion and debate between Mach and James, through their private correspondence and their professional publications, is that these are two men who admired one another enormously. They shared a broad philosophical outlook that involved turning empiricism in a strongly instrumentalist direction. I have provided

---

[27] He actually retained the earlier essay on spatial perception (complete with the *Innervationsgefühl* employed as a cue) largely unchanged, because 'I do not wish to conceal the method by which I was led to my theory', though he now acknowledges the correctness of James's view (he also cites Münsterberg and Hering as important sources, along with James; Mach 1886/1914, p. 168, note).

[28] James had long argued that parsimony demands that we abandon the *Innervationsgefühl* (James 1983, p. 86).

limited evidence that Mach might have exerted a modest influence on James in this capacity, but philosophical issues only motivated a small minority of their intellectual exchanges; the two friends largely settled into their philosophical views independently, it seems. They are much more likely to cite one another, to engage one another, to argue with one another, and to praise one another on matters of experiment. And here the surprise is that on the most contentious experimental issue that divided them, it was James who ultimately helped changed Mach's mind, not the reverse.

References

Aikin, Scott F., and Talisse, Robert B. 2017. *Pragmatism, Pluralism, and the Nature of Philosophy* Routledge.

Banks, Erik C. 2003. *Ernst Mach's World Elements: A Study in Natural Philosophy.* Kluwer.

2004. 'The Philosophical Roots of Ernst Mach's Economy of Thought', *Synthese* 139: 23–53.

2014. *The Realistic Empiricism of Mach, James, and Russell: Neutral Monism Reconsidered.* Cambridge University Press.

Blackmore, John T. 1972. *Ernst Mach: His Work, Life, and Influence.* University of California Press.

Blackmore, John T. and Hentschel, Klaus (eds.) 1985. *Ernst Mach als Aussenseiter.* W. Braumüller.

Brenner, Anastasios 2005. 'Réconcilier les Sciences et les Lettres: Le Rôle de L'Histoire des Sciences selon Paul Tannery, Gaston Milhaud et Abel Rey', *Revue D'Histoire des Sciences* 58: 433–454.

Chimisso, Cristina 2008. *Writing the History of the Mind: Philosophy and Science in France, 1900 to 1960s.* Ashgate.

Feigl, Herbert 1963/1981. 'The Power of Positivistic Thinking: An Essay on the Quandaries of Transcendence', in R. S. Cohen (ed.), *Inquiries and Provocations: Selected Writings, 1929–1974.* D. Reidel, pp. 38–56.

1969/1981. 'The *Wiener Kreis* in America', in R. S. Cohen (ed.), *Inquiries and Provocations: Selected Writings, 1929–1974.* D. Reidel, pp. 57–94.

Ferrari, Massimo 2017. 'William James and the Vienna Circle', in Sami Pihlström, Friedrich Stadler, and Niels Weidtmann (eds.), *Logical Empiricism and Pragmatism.* Springer, pp. 15–42.

Haller, Rudolf 1991. 'The First Vienna Circle', in T. Uebel (ed.), *Rediscovering the Forgotten Vienna Circle: Austrian Studies on Otto Neurath and the Vienna Circle.* Springer Netherlands, pp. 95–108.

Hatfield, Gary C. 2002. 'Sense-Data and the Philosophy of Mind: Russell, James, and Mach', *Principia: Revista Internacional de Epistemologia* 6: 203–230.

Helmholtz, Hermann von 1856–1867/2005. *Treatise on Physiological Optics*, 3 vols. Trans. James Powell Cooke Southall. Dover.

1867 *Handbuch der Physiologischen Optik, Allgemeine Encyklopädie der Physik*. Leopold Voss.

Henn, V. and Young, L. R. 1975. 'Ernst Mach on the Vestibular Organ 100 Years Ago', *Journal of Otorhinolaryngology and its Related Specialties* 37: 138–148.

Hiebert, Erwin N. 1976. 'Introduction', in Ernst Mach, *Knowledge and Error: Sketches on the Psychology of Enquiry*. Trans. Thomas J. McCormack from the 5th German edn. D. Reidel, 1926/1976, pp. xi–xxx.

Holton, Gerald 1992. 'Ernst Mach and the Fortunes of Positivism in America', *Isis* 83: 27–60.

1993a. 'From the Vienna Circle to Harvard Square: The Americanization of a European World Conception', in F. Stadler (ed.), *Scientific Philosophy – Origins and Developments*. Kluwer, pp. 47–73.

1993b. *Science and Anti-Science*. Harvard University Press.

James, William 1879. 'The Sentiment of Rationality', *Mind* 4: 317–346.

1880. 'The Feeling of Effort', *Anniversary Memoirs of the Boston Society of Natural History* (Boston, MA).

1882. 'The Sense of Dizziness in Deaf–Mutes', *American Journal of Otology* 4: 239–254.

1890/1981. '*The Principles of Psychology*', in F. H. Burkhardt, F. Bowers, and I. K. Skrupskelis (eds.), *The Works of William James*. Harvard University Press.

1898. 'Philosophical Conceptions and Practical Results', *The University Chronicle (University of California)* 1: 287–310.

1904. 'Does "Consciousness" Exist?', *Journal of Philosophy, Psychology, and Scientific Methods* 1: 477–491.

1907/1975. *Pragmatism*, in F. Bowers and I. K. Skrupskelis (eds.), *The Works of William James*. Harvard University Press.

1912/1976. *Essays in Radical Empiricism*, in F. Bowers and I. K. Skrupskelis (eds.), *The Works of William James*. Harvard University Press.

1983. *Essays in Psychology*, in F. H. Burkhardt, F. Bowers, and I. K. Skrupskelis (eds.), *The Works of William James*. Harvard University Press.

1987. *Essays, Comments and Reviews*. in F. H. Burkhardt, F. Bowers, and I. K. Skrupskelis (eds.), *The Works of William James*. Harvard University Press.

1988. Manuscript Lectures, in F. H. Burkhardt, F. Bowers, and I. K. Skrupskelis (eds.), *The Works of William James*. Harvard University Press.

1992–2004. *The Correspondence of William James*, 12 vols. Edited by I. K. Skrupskelis and E. M. Berkeley. University Press of Virginia.

Klein, Alexander 2007. *The Rise of Empiricism: William James, Thomas Hill Green, and the Struggle over Psychology*. Doctoral dissertation, Indiana University.

2008. '*Divide Et Impera!* William James's Pragmatist Tradition in the Philosophy of Science', *Philosophical Topics* 36: 129–166.

2016. 'Was James Psychologistic?', *Journal for the History of Analytical Philosophy* 4: 1–21.

2018. 'In Defense of Wishful Thinking: James, Quine, Emotions, and the Web of Belief', in M. Baghramian and S. Marchetti (eds.), *Pragmatism and the*

European Traditions: Encounters with Analytic Philosophy and Phenomenology before the Great Divide. Routledge, pp. 228–250.

Koch, Carl Henrik 1991. 'The Correspondence of Ernst Mach with a Young Danish Philosopher', *Danish Yearbook of Philosophy* 26: 97–112.

Mach, Ernst 1863. 'Zur Theorie des Gehörorgans', *Sitzungsberichte der kaiserlichen Akademie der Wissenschaften, Mathematisch-naturwissenschaftliche Classe* 48: 283–300.

    1865. 'Untersuchungen über den Zeitsinn des Ohres', *Sitzungsberichte* 51: 133–150.

    1872. *Die Geschichte und die Wurzel des Satzes von der Erhaltung der Arbeit.* J.G. Calve'sche.

    1872/1911. *History and Root of the Principle of the Conservation of Energy.* Transl. by Philip E. B. Jourdain. Open Court.

    1875/2001. *Fundamentals of the Theory of Movement Perception.* Transl. by Laurence R. Young, Volker Henn, and Hansjörg Scherberger. Kluwer Academic/Plenum Publishers.

    1883. *Die Mechanik in ihrer Entwickelung: Historisch-kritisch dargestellt.* F. A. Brockhaus.

    1883/1893. *The Science of Mechanics: A Critical and Historical Account of its Development.* Transl. by Thomas J. McCormack from the 2nd (1888) edn. Open Court.

    1886. *Beiträge zur Analyse der Empfindungen.* Gustav Fischer.

    1886/1897. *Contributions to the Analysis of the Sensations.* Transl. by C. M. Williams from the 1st edn. Open Court.

    1886/1914. *The Analysis of Sensations and the Relation of the Physical to the Psychical.* Transl. by C. M. Williams and S. Waterlow. From the 5th German edn. Open Court.

    1895. *Popular Scientific Lectures.* Transl. by Thomas J. McCormack from the 1st (1896 [*sic*]) edn. Open Court.

    1926/1976. *Knowledge and Error: Sketches on the Psychology of Enquiry.* Transl. by Thomas J. McCormack from the 5th German edn. D. Reidel.

Menand, Louis 2001. *The Metaphysical Club: A Story of Ideas in America.* Farrar, Straus, and Giroux.

Misak, Cheryl J. 2013. *The American Pragmatists.* Oxford University Press.

    2015. 'James on Religious Experience', presented at Conference on Issues in Modern Philosophy: God, New York University, 7 November 2015.

    2016. *Cambridge Pragmatism: From Peirce and James to Ramsey and Wittgenstein.* Oxford University Press.

Müller, Johannes 1837. *Handbuch der Physiologie des Menschen,* 2 vols. Verlag von J. Hölscher.

Peirce, Charles Sanders 1877. 'The Fixation of Belief', *Popular Science Monthly* 12: 1–15.

    1878. 'How to Make Our Ideas Clear', *Popular Science Monthly* 12: 286–302.

Perry, Ralph Barton 1935. *The Thought and Character of William James*, 2 vols. Harvard University Press.

Pihlström, Sami, Stadler, Friedrich, and Weidtmann, Niels (eds.) 2017. *Logical Empiricism and Pragmatism*. Springer.

Reisch, George A. 2005. *How the Cold War Transformed Philosophy of Science: To the Icy Slopes of Logic*. Cambridge University Press.

Richardson, Alan W. 1997. 'Toward a History of Scientific Philosophy', *Perspectives on Science* 5: 418–451.

    2003. 'Logical Empiricism, American Pragmatism, and the Fate of Scientific Philosophy in North America', in G. L. Hardcastle and A. W. Richardson (eds.), *Logical Empiricism in North America*. University of Minnesota Press, pp. 1–24.

Scheerer, Eckart 1989. 'On the Will: An Historical Perspective', in W. A. Hershberger (ed.), *Volitional Action: Conation and Control*. North-Holland, pp. 39–62.

Stadler, Friedrich 1992. 'The "Verein Ernst Mach": What Was It Really?', in J. T. Blackmore (ed.), *Ernst Mach – A Deeper Look: Documents and New Perspectives*. Kluwer, pp. 363–378.

    2015. *The Vienna Circle: Studies in the Origins, Development, and Influence of Logical Empiricism*. Springer.

    2017. 'Ernst Mach and Pragmatism – the Case of Mach's Popular Scientific Lectures (1895)', in Sami Pihlström, Friedrich Stadler, and Niels Weidtmann (eds.), *Logical Empiricism and Pragmatism*. Springer, pp. 3–14.

Thiele, Joachim 1966. 'William James und Ernst Mach', *Philosophia Naturalis* 9: 298–310.

Uebel, Thomas 2014. 'European Pragmatism? Further Thoughts on the German and Austrian Reception of American Pragmatism', in M. C. Galavotti et al. (eds.), *New Directions in the Philosophy of Science*. Springer, pp. 627–643.

    2015. 'American Pragmatism and the Vienna Circle: The Early Years', *Journal for the History of Analytical Philosophy* 3: 1–35.

    2017. 'American Pragmatism, Central-European Pragmatism and the First Vienna Circle', in Sami Pihlström, Friedrich Stadler, and Niels Weidtmann (eds.), *Logical Empiricism and Pragmatism*. Springer, pp. 83–102.

Visser, Henk 2001. 'Wittgenstein's Machist Sources', in J. T. Blackmore, R. Itagaki and S. Tanaka (eds.), *Ernst Mach's Vienna, 1895–1930, or, Phenomenalism as Philosophy of Science*. Kluwer, pp. 139–158.

Weinberg, Carlton Berenda 1937. *Mach's Empirio-Pragmatism in Physical Science*. Albee Press.

Wundt, Wilhelm Max 1863. *Vorlesungen über die Menschen- und Tier-Seele*, 2 vols. Leopold Voss.

7

Ernst Mach and Friedrich Nietzsche

On the Prejudices of Scientists

PIETRO GORI

In the specialised literature on the history of philosophy and the philosophy of science, Ernst Mach and Friedrich Nietzsche are often associated with each other. The reason for this is mainly rhetorical: given that Mach and Nietzsche apparently belong to different domains of our culture, scholars often try to surprise their readers by showing that similarities between these authors' views can in fact be found – and that they are not at all secondary. This happens on both sides. Mach scholars, whose audience has little patience for non-systematic, aphoristic modes of expression such as Nietzsche's and for the philosophical perspectives Nietzsche inspired, refer to Nietzsche in order to argue that Mach participated in a rich and multifaceted cultural movement. On the other hand, Nietzsche scholarship has been especially intrigued by Mach, for his viewpoint may cast light on certain interpretative problems involving questions approached by Nietzsche and inspired by post-Kantianism and modern epistemology (e.g. the problem of subjectivity, the contradictory concept of the 'thing-in-itself', and the actual value of scientific knowledge).

The aim of this chapter is to provide a thorough account of this association. As I will argue, the consistency which can be discovered between Mach and Nietzsche is much more substantial than one may imagine. On the basis of their conception of knowledge and truth, it is possible to outline a complete parallelism between them – and I use this Machian expression intentionally, for I believe that Mach and Nietzsche concerned themselves with the same issues concerning our intellectual relationship with the external world and that it is only their different interests that allow us to interpret these issues as pertaining to ethics (Nietzsche) or to the actual practice of scientific research (Mach). They in fact dealt with the very same questions and indeed pursued a common general aim, namely the elimination of worn-out conceptions from the world of modern culture. But an accurate investigation reveals even more, namely that Mach's and Nietzsche's research interests converge on the problem of realism versus anti-realism, a heated topic in contemporary philosophy

I would like to thank John Preston for his careful reading and valuable remarks which helped in improving my paper.

of science and, more generally, a fundamental philosophical issue. As will be shown, that problem permeates the reflections of both authors, and it is in the light of this particular issue that Mach and Nietzsche can be compared.

Phenomenalism

Mach's unique official pronouncement on Nietzsche is as strong as it is negative. In the revised and extended edition of *The Analysis of Sensations*, at the end of the paragraph in which Mach famously argues that 'the ego must be given up' (Mach 1959, p. 24), he declares that 'the ethical ideal founded on . . . the freer and more enlightened view of life', holding that the I is a mere theoretical product, a view 'which will preclude the disregard of other egos and the overestimation of our own, . . . will be far removed from . . . the ideal of an overweening Nietzschean "superman", who cannot, and I hope will not be tolerated by his fellow-men' (Mach 1959, p. 25). Mach's friend and populariser Hans Kleinpeter[1] had a hard time persuading Mach that his observation was biased by a received view which neglected the most interesting aspects of Nietzsche's thought. The almost exclusive interest that intellectuals had in the highly poetic work *Thus Spoke Zarathustra* at the beginning of the twentieth century meant that they remained stuck on the surface level of what was otherwise a profound and multifaceted thought. But a study of Nietzsche's posthumous writings outlines an image that differs greatly from the one that had been popularised, making it possible to assess the actual philosophical value of Nietzsche's work (cf. Kleinpeter 1912/1913). For Kleinpeter, that value rests on a biologically grounded conception of truth which leads to the sort of epistemological relativism defended by Mach.[2]

Kleinpeter passed on to Mach the idea that Nietzsche developed some interesting theoretical issues (cf. Gori 2011, pp. 294–298), and he worked extensively on this between 1911 and 1912. His 1913 book *Der Phänomenalismus. Eine naturwissenschaftliche Weltanschauung* includes the results of that research, namely the idea that Nietzsche defended the principles of the world-conception outlined by the guiding figures of a *scientific philosophy* (e.g. Mach, Richard Avenarius, John Stallo, and William Clifford). Kleinpeter (1913, p. 95) focuses on Nietzsche's reflections on the merely symbolic value of concepts and words that one finds in his posthumous writings. Apparently, Nietzsche maintained that our knowledge is only a collection of intellectual instruments developed in the evolution of mankind, whose fruitfulness for the preservation of the species – the fact that they allow

---

[1] Kleinpeter has been a minor, albeit interesting figure in the history of scientific philosophy. He was a strong supporter of Mach, whose thesis Kleinpeter discussed in a series of articles published, for example, in *The Monist*.

[2] Kleinpeter was not the first to notice an affinity between Mach and Nietzsche. Before him, for example, Rudolf Eisler (1902) stressed their similarities.

us to manage the world in an operationally efficacious way – has determined our ordinary faith in their ontological consistency. But this is a terrible mistake, for we have no access to 'things in themselves', and therefore we should remain agnostic at least concerning their properties (and perhaps also their actual existence). For Kleinpeter, this biological conception of knowledge lies at the origin of the scientific instrumentalism defended by Mach and by pragmatist thinkers such as William James and F. C. S. Schiller (Kleinpeter 1913, pp. 27, 123, 209, 251; cf. also Kleinpeter 1912, 1912/1913).[3] According to that view, 'truthfulness' is not related to the actual agreement of our world-description with what lies beyond the phenomena; on the contrary, it only pertains to the result of a process through which we test our expectation over a well-defined experience. Insofar as it is not a property of things, truth depends on the observer, their interests, and/or the field of the research they are pursuing. As defended by Mach in particular, for example, scientific truth is the momentary resting point of research, a collection of thought-symbols that allows us to save time and effort in future research.[4]

Kleinpeter's view of a phenomenalist Nietzsche is sustained by the observations on the metaphorical character of truth that are coherently developed throughout Nietzsche's works. As early as 1873, Nietzsche in fact argued that our truths are only illusions and metaphors and that we have forgotten their standing as mere intellectual elaborations of the received sensorial data (Nietzsche 1999, p. 146). In 1878, he conceived of the whole world of representation as 'the outcome of a host of errors and fantasies which have gradually arisen and grown entwined with one another in the course of the overall evolution of the organic being' (Nietzsche 1996, p. 20); accordingly, in 1882, he describes knowledge as a collection of intellectual errors (Nietzsche 2001, p. 110). The list of passages of that sort goes on, but what is important is that Kleinpeter found good reason to argue that 'Nietzsche understood better than any other thinker the philosophical importance of ... Mach's, Stallo's, Clifford's ... attempt to rid science of metaphysics' (Kleinpeter 1913, p. 95). In fact, the idea that our knowledge is nothing more than a collection of operationally fruitful labels and that we are stuck within the boundaries of this intellectually constructed world of appearances calls into question the very dichotomy between the 'apparent' and the 'real' or 'true' world, thus leading us to the fundamental problem of realism versus anti-realism.

When Kleinpeter tells Mach that he believes Nietzsche to be 'much better than his reputation suggests' (letter to Mach dated 22 December 1911, in Gori 2011, p. 295), he therefore means that, in Nietzsche, Mach could have found an ally in his crusade against the naïve realism one encounters in the ordinary scientific conception. Insofar as Nietzsche stresses the importance of appearances as actual limits of our knowledge and criticises philosophers' tendency

---

[3] On this, cf. Gori (2019a, chapter 4.1).
[4] Cf. Mach's principle of the economy of thought (e.g. in Mach 1897, chapter 9).

to 'call real what in fact is merely *conceptual*,[5] he is a true upholder of phenomenalism, as much as Mach is (Kleinpeter 1912/1913, p. 7; 1913, pp. 27, 143). Furthermore, Nietzsche's epistemological concerns led him to a complete rejection of substance concepts of all kinds (e.g. that of the subject). In his letter to Mach, Kleinpeter especially stresses that Nietzsche conceived of the I as a multiplicity of elements, not at all as a substantial unity (Kleinpeter 1913, p. 181; Gori 2011, p. 295). But if Nietzsche defended an anti-metaphysical view of the ego consistent with the Machian one, then Mach should have reconsidered his critical remarks on the ethical ideal of the overweening 'superman', given that his argument is based on the idea that this figure is an expression of the selfish 'overestimation of our own' which follows from the ordinary conception of the ego – which it is not, in fact.[6]

As far as we know, Kleinpeter's attempt to make Mach change his mind on Nietzsche was destined to fail.[7] But that was just the first step in a body of work that deserved further exploration, the value of which can be assessed on the broader plane of the history of philosophy. The question that Kleinpeter left to the interpreters is this: is it viable to compare Mach and Nietzsche? On what basis can we put them into dialogue? And, most importantly, is this enquiry fruitful for Mach studies, or for Nietzsche studies, or perhaps for both? In the following sections, I will try to answer these questions. As I will show, by following the path of the theoretical issues that Mach and Nietzsche both addressed, an intriguing general picture emerges.

Enlightenment

A few years later than Kleinpeter, Philipp Frank endorsed his idea of 'the striking agreement of Mach's view with those of a thinker for whom he cannot have had any great sympathy: Friedrich Nietzsche' (Frank 1970, p. 232). For him, Mach and Nietzsche had similar views on the nature and limits of knowledge. They both criticised the contradictory conception of the 'in itself' and, most importantly, they were both 'enlightenment philosopher[s] of the end of the nineteenth century' insofar as they 'protest[ed] against the

---

[5] On this, *Human, All Too Human* § 11 (Nietzsche 2005a, p. 16) and the 1888 posthumous fragment 14[153] (Nietzsche 1980, vol. 13, p. 336) are especially interesting; see also Gori (2017).

[6] Mach's view of the Nietzschean Superman was popular at his time, as shown by the early reception of Nietzsche's *Thus Spoke Zarathustra* – the one which determined the appreciation of Nietzsche by the worst expression of German twentieth-century politics. In recent times, Nietzsche scholarship has worked hard to correct this and other misinterpretations of Nietzsche's thought through the reconstruction of his philosophical context, which has been especially revealing.

[7] Cf. Kleinpeter's letter to Elisabeth Förster-Nietzsche dated 9 November 1912 (*Goethe-Schiller Archiv*, Weimar) and Gori (2011, p. 293).

misused concepts of [their] time' (Frank 1970, pp. 231–232). What did Frank mean by this?

In the 'Age of Enlightenment', Frank (1970, p. 229) identifies 'a struggle against the misuse of auxiliary concepts'; that is, an attempt to show that concepts have a realm of validity outside of which they cannot be used properly. It is characteristic of the history of science that every period has its auxiliary concepts and every subsequent period misuses them; therefore, 'in every period a new enlightenment is required in order to abolish this misuse.... To this work Mach dedicated himself' (Frank 1970, pp. 229, 231). The attempt to enlighten the mind of his fellow scientists rests at the core of Mach's epistemology and in fact defines his own criticism. In the Preface to the first edition of *The Science of Mechanics*, Mach declares that the aim of the volume is 'to clear up ideas, expose the real significance of the matter, and get rid of metaphysical obscurities' (Mach 1919, p. ix). This 'enlightening or anti-metaphysical intention' (in the original German text, Mach speaks of '*eine aufklärende oder antimetaphysische Tendenz*') clarifies the goal of the entire book, which is to provide a *historico-critical* investigation into the principles of mechanics. From the viewpoint of the history of philosophy, this is of some interest, for the outlined conceptual cluster represents an ideal development of the philosophical methodology. The name that Mach gives to his own approach clearly evokes Kantian *criticism*, one of the most important products of philosophical *enlightenment*, but Mach's version claims to be original, especially since it aims to get rid of the metaphysical obscurities that Kant had not been able to abandon. Mach's criticism is in fact a *historical enquiry* which attempts to cast light on the nature of ordinary scientific concepts by tracing their origins through the obscure paths of past cultures and civilisations.

The importance of history as an anti-metaphysical tool is stressed in Mach's 1871 lecture on the *Conservation of Energy*. For Mach, 'we are accustomed to call concepts metaphysical, if we have forgotten how we reached them' (i.e. if we isolate them from their historical development; Mach 1911, p. 17). For him, concepts are dynamic entities whose actual meaning depends on the cultural framework to which they pertain. Consequently, 'science is unfinished, variable' (Mach 1911, p. 17); it is not a discovery of *true* features of reality, but rather a creative enterprise. Science creates our conceptual system, constructing the entire intellectual edifice through which we manage the world. Accordingly, in *The Analysis of Sensations*, Mach defends the instrumental view that 'the biological task of science is to provide the fully developed human individual with as perfect a means of orientating himself as possible' (Mach 1959, p. 37). Furthermore, in *Knowledge and Error*, he observes that the 'laws of nature are a product of our mental need to find our way about in nature, so that we do not stand estranged and baffled in front of natural processes' (Mach 1976, p. 34). Instead of being an expression of the actual development of natural events, these laws must be considered only as the most recent 'attempt at orientation' produced by our 'current state of culture'. These are but sketches of a general picture that Mach defended from his early years onwards, stimulated by the urge to contrast the 'naïve realism ... of the average man' with higher cultural products, such as the scientific world-description, on which that naïve realism casts its shadows (Mach 1959, p. 37). The 'constant correction of ordinary thought' that determines 'progress in scientific thought' is made possible by the growth of civilisation (Mach 1976, p. 2), and history plays an important role in that development, from a philosophical point of view. The historical approach is in fact crucial, for it reveals the inner nature of science as much as the inconsistency of the ordinary metaphysical commitment that rests at its core. Most importantly, this *anti-metaphysical criticism* allows us to redefine the boundaries of scientific knowledge – that is, to deal with fundamental epistemological questions such as 'In what sense are scientific concepts meaningful?' or 'To what extent are the general results [of science] 'true'?' (Einstein 1992, p. 154).

Is it possible to find a similar approach in Nietzsche? In fact, it is. Firstly, Mach's 1871 definition of 'metaphysical concepts' is consistent with Nietzsche's early observations on truth, namely the famous idea that 'truths are illusions of which we have forgotten that they are illusions', stated in the 1873 unpublished essay 'On Truth and Lying in a Non-Moral Sense' (Nietzsche 1999, p. 146).[8] Nietzsche's approach to that issue has much in common with Mach's argument. For Nietzsche, our 'truths' are the result of an intellectual activity of schematisation and categorisation which worked well in terms of human adaptation. Given its great practical value, and considering that we cannot leave it aside, we have ultimately become familiar with that world-description. As a result, we now pretend 'that with words [we are] expressing the supreme knowledge of things' (Nietzsche 1996, p. 16), whereas in truth we are only dealing with fruitful designations. Secondly, in the opening section of *Human, All Too Human*, Nietzsche introduces 'historical philosophy' as the proper methodology to cast light on metaphysical obscurities. For him, 'lack of historical sense is the family failing of all philosophers' who embrace a metaphysical viewpoint, 'assuming for the more highly valued thing a miraculous source in the very kernel and being of the "thing in itself"'[9]

---

[8] I defended this idea in Gori (2014) and Gori (2019a, chapter 5.2).

[9] These observations are a direct development of Nietzsche's 1873 reflections. In 'On Truth and Lying', he in fact argues that 'when different languages are set alongside one another it becomes clear that, where words are concerned, what matters is never truth, never the full and adequate expression; otherwise there would not be so many languages. The "thing-in-itself" (which would be, precisely, pure truth, truth without consequences) is impossible for even the creator of language to grasp, and indeed this is not at all desirable. He designates only the relations of things to human beings, and in order to express them he avails himself of the boldest metaphors' (Nietzsche 1999, p. 144).

(Nietzsche 1996, pp. 12–13). One of the fundamental problems of traditional philosophy thus resides in the illusory belief in a fascinating realm of eternal 'truth'. But 'there are *no eternal facts*, just as there are no absolute truths. Consequently, what is needed from now on is *historical philosophizing*, and with it the virtue of modesty' (Nietzsche 1996, p. 13, emphasis in original).

Apparently, Nietzsche is as much an anti-realist and instrumentalist as Mach. They both consider knowledge a creative intellectual production of signs and labels for the aim of orientation; they both argue that language is imbued with metaphysical commitment; most importantly, both Mach and Nietzsche defend a critical approach to these issues based on the historical method, which they consider the only method that allows us to shed light on ordinary world-description and to overcome its naïve realism. Thus, it is possible to view Mach and Nietzsche as Enlightenment philosophers, as Frank did, and from that viewpoint one can argue that their contribution to the growth of philosophical studies rests especially in the way they dealt with an issue as crucial as the dualism between realism and anti-realism.

Interpretative Problems

The similarities between Mach's and Nietzsche's epistemological concerns, and the semantic consistency between their reflections on the topic, has intrigued Nietzsche scholarship, especially since work on Nietzsche's private library and his sources has revealed his great interest in the outcomes of modern science.[10] This interest seems to be inconsistent with a critical attitude towards the ordinary scientific world-description reiterated in Nietzsche's writings, such as when he argues that 'physics is only an interpretation and exegesis of the world (to suit us, if I may say so!) and *not* a world-explanation' (Nietzsche 2002, p. 15), or that scientific concepts are 'conventional fictions for the purposes of designation and communication – *not* explanation' (Nietzsche 2002, p. 21, emphasis in original). These observations especially conflict with passages where Nietzsche defends an empiricist epistemology and demonstrates his commitment to an ontology based on our sensual experience. In plain language, there seems to be a contradiction within Nietzsche's view that science provides a world-image created only to make the world manageable for us – one that, consequently, does not straightforwardly report the world as it is – and his idea that our 'senses do not lie' (Nietzsche 2005, p. 167) and that

---

[10] Over the past decades, studies on this theme have been published in Babich and Cohen (1999), Brobjer and Moore (2004), Heit et al. (2012), and Heit and Heller (2014).

we can accept 'sensualism at least as a regulative hypothesis, if not as a heuristic principle' (Nietzsche 2002, p. 16).[11]

These interpretative problems have been addressed by Nadeem Hussain in two seminal papers inspired by the presence of Mach's 1886 *Beiträge zur Analyse der Empfindungen* in Nietzsche's private library.[12] Hussain explores in particular the problematic issue of Nietzsche's commitment to a 'falsification thesis' that claims that our language is only about our own representations and that any attempt to grasp the actual features of the world existing beyond them is destined to fail. Consequently, the truth-value of our world-description is put into question, and the fundamental problem of epistemological relativism or scepticism arises. For Hussain, a Machian approach to the matter can shed light on Nietzsche's most difficult passages. In fact, he believes that 'Mach provides for us a basis on which we can interpret much of what Nietzsche says about scientific theories and the role of the senses in a way that would be compatible, at least by Mach's own lights, with some kind of falsification thesis' (Hussain 2004a, p. 350).[13]

The falsification thesis was originally explored by Maudemarie Clark in her seminal and controversial *Nietzsche on Truth and Philosophy* (Clark 1990). In that book, Clark argued that Nietzsche's use of the notion of 'truth' is problematic given his late view of the role of sense organs and his critical approach to the Kantian 'thing-in-itself', which Nietzsche ultimately rejects as involving a *contradictio in adjecto* (Nietzsche 2002, p. 16). For Clark, the only consistent use of the notion of truth is a matter of correspondence with the actual features of reality; if one has no access to the world as it is independently of us, nothing can be claimed about our actual knowledge – not even that it is a falsification of the original datum. Therefore, together with the thing-in-itself, Nietzsche had to give up his falsification thesis, which Clark claims he did.[14]

---

[11] Kleinpeter (1913, p. 69) observes that the idea that 'the senses do not deceive' can be found in Goethe as much as in Mach (e.g. 1959, p. 10). It is, however, in fact a Kantian observation (Kant 1998, p. 384; A 293/B 350).

[12] Apparently, Nietzsche was one of the few who were interested in that book when it first appeared (cf. Baatz 1997, p. 85). A thorough account of Nietzsche's personal copy of Mach's 1886 essay is provided in Gori (forthcoming).

[13] As Robert Cohen points out, Mach insisted on the merely descriptive power of scientific knowledge in a way that can be compared with Nietzsche's critical remarks. For Cohen (1970, p. 132), Mach 'seems to conclude [that] science is not an attempt to understand the world as it is but only to describe the world as we experience it; and epistemology, to be scientific, must likewise be not an attempt to understand the phenomenon of science but only a description of it'. Cohen does not emphasise the consistency between Mach and Nietzsche on this point, but he invokes Nietzsche later in his paper (p. 135).

[14] Green (2002) and Anderson (1996) disagree with her, for example. Apparently, Clark does not accept that Nietzsche attempted a revaluation of the notion of truth, as suggested by other scholars (cf. e.g. Clark 1990, p. 33). I am inclined to believe that, in his late writings, Nietzsche called into question the traditional notion of truth as correspondence

As Brian Leiter puts it (Leiter 2002, p. 14), Clark's interpretation is based on the fact that 'it is impossible to reconcile' the falsification thesis with 'Nietzsche's explicit *empiricism* – his view that "all evidence of truth comes only from the senses"'. But Hussain argues that Nietzsche might have defended the sort of sensualism one finds in Mach and, consequently, that it is possible 'to interpret him as simultaneously rejecting the thing-in-itself, accepting a falsification thesis and defending empiricism' (Hussain 2004a, p. 357).

The core of Hussain's argument rests on his interpretation of Mach's 'sensualism': his neutral monism, or, as Hussain calls it, 'Machian positivism' (Hussain 2004a, p. 328). For Hussain, Mach defended an original kind of empiricism, which can be seen as a development of the phenomenalism endorsed by neo-Kantian thinkers such as Friedrich A. Lange, Afrikan Spir, and Gustav Teichmüller (three key figures in Nietzsche's philosophical education).[15] Within this framework, it was maintained that sensory evidence does not provide direct access to reality. On the contrary, we are physiologically structured for falsification insofar as both our sense organs and our intellect modify the original datum which we receive from the external world, the very existence of which cannot be denied, although one can say nothing about it. For Hussain (2004a, p. 334), this conception was 'part of a standard story about how physiology and the materialistic world view undermine themselves'. That story was '"in the air" early enough to influence Nietzsche' (Hussain 2004b, p. 120), who studied the work of several neo- and post-Kantian authors with great interest. Furthermore, Hussain (2004b, pp. 120–121) stresses that Mach's 'particular brand of empiricism, and the more specific monistic claim it involves, is at least one natural development of certain shared conceptions of the role of sensory evidence and the nature of scientific theories that were widespread at the time'. As is well known, Mach's conception of 'sensation' is in fact built upon a rejection of the Kantian thing-in-itself as a non-concept (cf. Mach 1959, p. 30) – a notion which is meaningless to us insofar as we 'know' only what falls within the boundaries of our bodily experience. That is why Mach suggested that we talk of elements *or* sensations, thus creating the interpretative difficulties we are accustomed to dealing with today.[16] He believed that no proper knowledge is possible without the mediation of our body; therefore, we can only experience the elements as

---

and, instead of getting rid of that very notion, argued that it must be reconceived from a new viewpoint. As I tried to show in Gori (2019a), this approach can be compared to William James's pragmatist view of truth – which, incidentally, was inspired by Mach (cf. e.g. Hiebert 1976; Holton 1992).

[15] Hussain (2004a, p. 347). On Nietzsche's engagement with these authors, see, for example, Stack (1983), Green (2002), and Emden (2014).

[16] On the interpretative problems related to Machian elements, see Banks (2003).

related to that body itself – that is, as sensations (Mach 1959, p. 12). But these sensations do not provide access to the actual features of reality; they do not provide us with superior knowledge of things. We can trust them only because they are the very origin of our world-description and the actual limits of meaningful speech, for any discourse which does not deal with sensations and the relation between sensory elements is inconsistent (Mach 1992, p. 119). For Mach, in fact, everything that goes beyond the limits of the realm of phenomena is merely a metaphysical assumption which can have no value in high-level cultural products such as science.

Hussain argues that Mach's untraditional brand of empiricism, which attempts to overcome the metaphysical problems implied by positivism and neo-Kantianism, shows us that 'one can take science and the senses quite seriously without straightforwardly accepting "the common sense picture of the world of relatively enduring middle-sized objects or the scientific world-view"' to which Clark (1990, p. 108) refers (Hussain 2004a, p. 358). Moreover, that view can help Nietzsche scholarship to understand 'what Nietzsche might mean by sensualism' in his late writings (Hussain 2004a, p. 336) and ultimately to see that Nietzsche's commitment to the truthfulness of the sensorial image can be consistent with the anti-metaphysical view he reiterates and even radicalises in the late period. The only way to make sense of Nietzsche's observations seems to be to accept that he developed the physiological conception debated at the time in the same way that Mach did, namely by assuming an anti-realist view which is moderate insofar as it takes care of the instrumental (i.e. fictional) character of our world-description alone, while the existence of an external world acting upon us is not rejected.[17] In order to affirm that the features of the world of which we have knowledge are human products, the phenomenalist account must admit that there is, in fact, an external world acting on us. But – as noted above – insofar as it lies beyond the boundaries of that knowledge, anything we pretend to say about it is ill-founded and therefore meaningless to us.[18]

---

[17] I defended this interpretation of Mach's and Nietzsche's anti-realism in Gori (2018) and Gori (2019a), respectively.

[18] This only apparently means falling back to the Kantian picture Mach and Nietzsche claimed to disavow, for the question of the 'thing-in-itself' has to be addressed on the pure theoretical level. The existence of an external realm acting on our sensorial apparatus is mandatory if one wants to be a phenomenalist and not an idealist. But this does not mean that one must also be a realist about the properties *we* attribute to things. These are a pure human product, whereas the world remains completely unknowable – therefore of no use – to us. As Nietzsche argues, 'thing-in-itself' is a *contradiction in adjecto* (Nietzsche 2002, p. 16) precisely because it is not possible to even conceive a world-feature independently from our knowing activity (i.e. to know *is* to create world-properties, properties which only make sense for us observers and world-interpreters). Thus, the intellectual path that Nietzsche famously draws in the section 'How the "True

As Hussain conclusively remarks (2004a, p. 358), the crucial point is the way in which Mach and Nietzsche approached 'the distinctions between appearance and reality and the internal and external world'. Both of them, in fact, assumed that this distinction is fictitious insofar as it is a product of the ordinary viewpoint, which has faith that our linguistic activity reproduces the external world adequately, thus allowing us to 'possess knowledge of the world' (Nietzsche 1996, p. 16). But, as Nietzsche famously argued in 'Twilight of the Idols' – in a way that, as will be shown, strikingly resembles Mach's observations on the topic – once the 'apparent world' is interpreted as the actual limit of human knowledge, the 'real' or 'true' world becomes 'an idea that is of no further use, ... a superfluous idea, *consequently* a refuted idea' (Nietzsche 2005, p. 171, emphasis in original). But along with the true world, we also get rid of the apparent one, Nietzsche remarks, for the latter is a pure intellectual notion created in contraposition to the realm of human designations, which common sense mistakes as the criterion of reality.[19] This obliteration is the 'trademark' of Mach's and Nietzsche's anti-metaphysical viewpoint, and, consequently, that on which one must focus if one is to develop a productive comparison between the two authors.

A Broader Viewpoint

Further studies on Nietzsche's engagement with modern science have been carried out since Hussain presented his initial case, and the Machian issues one encounters in Nietzsche have been thoroughly explored. On the one hand, these studies reveal that Nietzsche and Mach could not have influenced each

World" Finally Became a Fable' of 'Twilight of the Idols' (Nietzsche 2005b, p. 171) can be seen as an invitation to assume the sort of 'critical standpoint' Hans Vaihinger outlines in his 1911 book *The Philosophy of As-If*, which admits the instrumental value of our concepts conceived as intellectual aids for orientation, rather than the form of metaphysical anti-realism which is ordinarily attributed to Nietzsche, with all of the problems it raises (for more on this, see Gori 2019b). I think the same goes for Mach. For him, too, it would be inconsistent to argue against the very existence of an external world. As I will try to argue in the next section, a possible solution to this riddle is to uphold a sort of epistemological agnosticism or moderate realism which limits the realm of the 'knowable' (i.e. meaningful to our intellectual products).

[19] As Hussain aptly observes (2004b, p. 117), Mach and Nietzsche agree that 'language, and conscious reasoning that must occur in language, misleads'. They both insist on the instrumental value of concepts, on their mere symbolic nature and economical fruitfulness (cf. e.g. Mach 1897, p. 192; Nietzsche 1980, vol. 13, p. 336, 2003, p. 9, 2005a, p. 16). Furthermore, for Hussain (2004b, p. 117), 'within a Machian reading an interpretation of the world, and thus a perspective on the world, is a theory of the world that sets up names for particular clusters of sensory elements and the relations they stand in. Such interpretations in general will involve falsification, since grammar misleads us, perhaps necessarily, to think that our theory refers to objects and picks out explanatory causal relations.'

other directly (Gori, forthcoming). On the other hand, however, Hussain's observations have for the most part been confirmed. Namely, it is today widely accepted that Nietzsche's view of truth and knowledge is indebted to neo-Kantianism, and that he also developed and ultimately overcame that tradition in an original way. Furthermore, the framework that Mach shares with Nietzsche has been defined as a debate on the physiological limitations of human knowledge and the apparent impossibility of solving certain funda-mental 'world riddles'.[20] Against the early attempt to make either a positivist or an anti-positivist out of Nietzsche (he apparently endorses both views at different stages of his philosophical activity), a new image of Nietzsche as a post-Kantian and post-positivist has been defended, and it can be argued that Nietzsche was an important representative of a post-empiricist philosophy of science which claims that

> 'in natural science data is not detachable from theory; . . . facts have to be reconstructed in the light of interpretation; . . . theories are not models externally compared to nature in a hypothetico-deductive schema; . . . what counts as facts are constituted by what the theory says about their interrelations with one another; . . . the language of natural science is irreducibly metaphorical and inexact; . . . [and that] meanings in natural science are determined by theory; they are understood by theoretical coherence rather than by correspondence with facts'
>
> (Hesse 1972, p. 280).

Broadly speaking, it is possible to say that, like Mach, Nietzsche dealt with the outcomes of the Kantian legacy from a non-sceptical point of view, thus developing a view which is comparable to other approaches still debated within the current philosophy of science, such as the pragmatist, fictionalist, constructivist, and coherence theories of truth.[21] That is, both Mach and Nietzsche embraced the account of an active and creative mind which modern research reported, without interest in finding any realm of fixed truths or meanings beyond the phenomena. As shown above, this view implies that we cannot surpass the boundaries of our humanly categorised world and, there-fore, that nothing meaningful can be said of what lies beyond it.[22]

This is what a thorough investigation of Nietzsche's concept of 'perspecti-vism' reveals. In *The Gay Science* § 354, Nietzsche intertwines that notion with 'phenomenalism', defined as the view that 'due to the nature of *animal consciousness*, the world of which we became conscious is merely a

---

[20] Emile Du Bois-Reymond's 'Ignorabimus!' makes the case for the nineteenth-century problematisation of the scientific world-explanation and shows us that epistemological relativism was the fundamental problem of that time (cf. Bayertz et al. 2007). As I tried to argue in Gori (2019a), Nietzsche's 'death of God' can be profitably interpreted in the light of that issue.

[21] On this see, for example, Remhof (2015).

[22] This conception is explored, for example, in Pihlström (2008).

surface- and sign-world',[23] and ultimately declares his agnosticism about 'the opposition between subject and object', a distinction that Nietzsche leaves 'to those epistemologists who have got tangled up in the snares of grammar (of folk metaphysics)' (Nietzsche 2001, pp. 213–214, emphasis in original). As Nietzsche continues: 'Even less am I concerned with the opposition between "thing in itself" and appearance: for we "know" far too little to even be entitled to *make* that distinction. We simply have no organ for *knowledge*, for "truth".' This observation is the final step of a long yet consistent reflection on human knowledge that Nietzsche developed from 1873 onwards, a reflection which led him to the idea that the popular notion of 'knowledge' must be reconceived, for we erroneously believe it to provide proper access to reality.[24] On this, Nietzsche seems to agree with Mach, who in *Knowledge and Error* attempts to correct the ordinary conception that puts too much trust in imagination, due to its fruitfulness as a means of conceptual completion (Mach 1976, p. 1). For Mach, 'the imagination rounds off incomplete findings in the way that is most familiar to it, thus occasionally falsifying them', leading ordinary thought 'to the opposition between illusion and reality, between appearance and object' (Mach 1976, p. 7). This is of course problematic, for we misconceive the only fact to which we have access – the testimony of our sense organs – and create an illusory realm of non-existing entities. More precisely, Mach argues that 'once this opposition has emerged, it tends to invade philosophy as well, and is not easily dislodged. The weird and unknowable "thing-in-itself" behind appearances is the ordinary object's unmistakable twin, having lost all other significance' (Mach 1976, p. 7; cf. also Mach 1959, p. 6).

Insofar as they are expressed by an expert in the field of experimental psychology, Mach's remarks are more concrete and precise than Nietzsche's, and the same can be said of his epistemological agnosticism, which coherently follows from the monistic conception endorsed in *The Analysis of Sensations*. In that book, too, Mach criticised the 'common and popular way of thinking and speaking [which] contrasts "appearance" with "reality"', arguing that 'to speak of "appearance" may have a practical meaning, but cannot have a scientific meaning. Similarly, the question which is often asked, whether the world is real or whether we merely dream it, is devoid of all scientific meaning'

---

[23] It can be argued that '*Phänomenalismus*' in Nietzsche identifies precisely the post-Kantian world-picture outlined in works such as Friedrich Lange's 1866 work *History of Materialism*, which Nietzsche carefully studied. On this, see, for example, Stegmaier (2012, p. 262 ff.).

[24] Quite significantly, in the 1886–1887 posthumous fragment 7[60], where Nietzsche contrasts positivism with perspectivism, we read that 'inasmuch as the word "knowledge" has any meaning at all, the world is knowable' (Nietzsche 2003, p. 139). For a discussion of that note in the light of the phenomenalist framework, see Gori (2019a, chapter 2).

(Mach 1959, p. 11). Accordingly, in *Knowledge and Error*, Mach argues that once we get rid of the illusory dualism between appearance and reality, the more economic conclusion (i.e. the inference to the best explanation) is to avoid any metaphysical commitment that cannot be demonstrated; therefore, agnosticism seems to be preferable to anti-realism.[25] Mach indeed believes that 'after misconstruing the boundary between the internal and external and thereby imposing the stamp of illusion on the ego's entire content', we have no 'further need for an unknowable something outside the confines that the ego can never transcend' (Mach 1959, p. 7). Thus, the anti-metaphysical and monistic view which Mach advocates allows us to abandon the very distinction between appearance and reality, as Nietzsche argues. That is, if explored from the point of view of the 'complete parallelism of the psychical and physical' (Mach 1959, p. 60), and if, consequently, we take care of the *functional relations* between the elements only, 'the question as to illusion and reality loses its sense' (Mach 1976, p. 7).

As noted above, this obliteration is crucial for both Mach and Nietzsche. Their criticism of common-sense realism and its faith in the existence of substance concepts such as the I, the body, etc., is in fact based on the assumption that the distinction between two separate realms of reality is artificial and illusory. Nietzsche is quite clear on this in 'Twilight of the Idols', for example, where he blames language for having induced us to believe in the existence of a causal agent called 'I'. In doing this, language proved to be a product of 'psychology in its most rudimentary form', for it introduced a 'fetishistic mindset' which is essentially metaphysical and of which we can rid ourselves only insofar as we give up our 'faith in grammar' (Nietzsche 2005, pp. 169–170; cf. the passage from *The Gay Science* § 354 quoted above). The same critical attitude towards the notion of subjectivity can be found in *Beyond Good and Evil*.[26] There, Nietzsche reflects on the 'atomistic' (i.e. 'metaphysical') need which spurs on philosophers as much as scientists (Nietzsche 2002, p. 14). For him, it is because of this need that we ordinarily believe in the existence of both spiritual and material atoms – that is, in substance entities imagined as a psychical unity or 'residual earth [i.e. a particle of matter] out of which the effects are produced' (Nietzsche 2002, p. 18). With his remarks, Nietzsche aims to dismantle an old article of faith and, consequently, to rid philosophy of its inherited fundamental prejudices.[27]

---

[25] On Mach's commitment to or against metaphysical realism, I disagree with Banks (2004). I tried to make my case for this in Gori (2018).

[26] On Nietzsche's criticism of subjectivity in the light of Lange's and Mach's approaches to scientific psychology, cf. Gori (2015).

[27] The first chapter of *Beyond Good and Evil* is in fact titled 'On the Prejudices of Philosophers', and the whole book is conceived as a 'Prelude to the Philosophy of Future'.

Therefore, Nietzsche aimed to enlighten the minds of his fellow philosophers just as Mach attempted to clear up the scientific ideas of his time. In *The Analysis of Sensations*, Mach argues against 'a very widespread prejudice', based on the monistic conception that the illusory substance concepts I and body, matter and soul, 'disintegrate into elements' (Mach 1959, pp. 5, 310).[28] This prejudice is precisely the idea that 'there is [a] rift between the psychical and the physical, inside and outside' – that there are '"sensation[s]" to which an external "thing", different from sensation, correspond[s]' (Mach 1959, p. 310). But there is nothing of that sort for Mach, for whom the world we experience is constituted by 'one kind of elements, out of which this supposed inside and outside are formed – elements which are themselves inside or outside, according to the aspect in which, for the time being, they are viewed' (Mach 1959, p. 310). In claiming this, Mach contrasts the traditional view uncritically accepted by his colleagues and defends a sort of pragmatically orientated sensualism without metaphysics, according to which 'the boundary-line between the physical and the psychical is solely practical and conventional' (Mach 1959, p. 311) and theories and concepts have a mere instrumental value. Within this picture, Mach stresses the purely logical value of complex symbols such as atoms and molecules, arguing that 'if ordinary "matter" must be regarded merely as a highly natural, unconsciously constructed mental symbol for a relatively stable complex of sensation elements, much more must this be the case with the artificial hypothetical atoms and molecules of physics and chemistry' (Mach 1959, p. 311). On this, too, Mach's viewpoint is strikingly similar to Nietzsche's, for they both defend a pragmatic anti-realism which aims to revaluate the value of traditional concepts, once their truth value – their value as an adequate reproduction of reality – is put into question.[29] Indeed, the passage from *The Analysis of Sensations* continues as follows: 'The value of these implements [i.e. atoms and molecules] for their special, limited purposes is not one whit destroyed. As before, they remain economical ways of symbolizing experience. But we have as little right to expect from them, as from the symbols of algebra, more than we have put into them, and certainly not more enlightenment and revelation than from experience itself. We are on our guard now, even in the province of physics, against overestimating the value of our symbols' (Mach 1959, p. 311).

Within this picture, it is necessary to redefine the very function of both philosophy and science. The task of the latter ceases to be merely

---

[28] Richard von Mises (1970, p. 256) argued that 'Mach's analysis of knowledge provides an approach to' the 'dilemma of the world of appearances of our senses and the "true" world of science', to the problem of the thing-in-itself and the general issue of realism. The effect of that analysis 'is liberating because it clears away accumulated debris and it provides release by opening our eyes to an unprejudiced view'.

[29] This thesis is advocated in Gori (2019a).

representative and descriptive for Nietzsche and Mach, who instead focus on the role it plays as a means of orientation. Mach is especially clear on this in *The Analysis of Sensations*, where he argues that science aims to provide us 'with as perfect means of orientating himself as possible' (Mach 1959, p. 37), and in *Knowledge and Error*, where the laws of nature are described as 'a product of our *psychological* need to find our way about in nature', the most recent 'attempt at orientation' produced by our 'current state of culture' (Mach 1976, p. 354, emphasis in original, my translation).[30] In Nietzsche, the question of orientation is subtler but also more substantial, for it is related to the fundamental problem of relativism (the 'death of God') with which he is so deeply concerned and which in *The Gay Science* (§ 125) he in fact presents as a lack of reference points for mankind. Future philosophy as a critical enterprise which undermines the value of the old 'truths' is expected to deal with that relativism and to outline the path towards a valuational system of a different kind.

It is precisely in the light of this that we can compare Mach and Nietzsche and conclude that they belong to the same movement in the history of Western thought. Relativism is a fundamental issue of modernity, an issue that involves the 'hard sciences' as much as any other expression of philosophical thought. It is 'a stream in the philosophy of the past two hundred years that began as a trickle' and has ultimately 'swelled into a roaring torrent' (Bernstein 1983, p. 13). Relativism imbues heated topics of contemporary philosophy of science such as the realism versus anti-realism debate and the problem of the objectivity of our world-description – two issues that Mach and Nietzsche wrestled with in an original way. As I have tried to show, their contributions to cultural studies can be evaluated on the basis of their attempt to protect us from the destructive power of that torrent, but at the same time to take advantage of that power itself, in order to allow the highest manifestations of European civilisation to rid themselves of metaphysical obscurities.

References

Anderson, R. Lanier 1996. 'Overcoming Charity: The Case of Clark's Nietzsche on Truth and Philosophy', *Nietzsche-Studien* 25: 307–341.
Baatz, Ursula 1997. 'Ernst Mach and the World of Sensations', in S. E. Bronner and P. F. Wagner (eds.), *Vienna: The World of Yesterday, 1889–1914*. Humanities Press, pp. 82–92.
Babich, Babette and Cohen, Robert S. 1999. *Nietzsche and the Sciences*, 2 vol. Kluwer.

---

[30] Cf. also *Knowledge and Error* (Mach 1976, pp. 2–4) and, for more on this, Gori (2019c).

Banks, Erik C. 2003. *Ernst Mach's World Elements: A Study in Natural Philosophy.* Kluwer.

2004. 'The Philosophical Roots of Ernst Mach's Economy of Thought', *Synthese* 139: 23–53.

2014. *The Realistic Empiricism of Mach, James, and Russell: Neutral Monism Reconceived.* Cambridge University Press.

Bayertz, Kurt, Gerhard, Myriam, and Jaeschke, Walter (eds.) 2007. *Weltanschauung, Philosophie und Naturwissenschaft im 19. Jahrhundert. Vol. 3: Der Ignorabimus-streit.* Meiner.

Bernstein, Richard 1983. *Beyond Objectivism and Relativism. Science, Hermeneutics, and Praxis.* University of Pennsylvania Press.

Brobjer, Thomas and Moore, Gregory (eds.) 2004. *Nietzsche and Science.* Ashgate.

Clark, Maudemarie 1990. *Nietzsche on Truth and Philosophy.* Cambridge University Press.

Cohen, Robert S. 1970. 'Ernst Mach: Physics, Perception and the Philosophy of Science', in Robert S. Cohen, and Raymond J. Seeger (eds.), *Ernst Mach: Physicist and Philosopher.* D. Reidel, pp. 126–164.

Einstein, Albert 1992. 'Ernst Mach', in J. T. Blackmore (ed.), *Ernst Mach – A Deeper Look: Documents and New Perspectives.* Kluwer, pp. 154-159.

Eisler, Rudolf 1902. *Nietzsches Erkenntnistheorie und Metaphysik.* Haake.

Emden, Christian 2014. *Nietzsche's Naturalism: Philosophy and the Life Sciences in the Nineteenth Century.* Cambridge University Press.

Frank, Philipp 1970. 'The Importance of Ernst Mach's Philosophy of Science for Our Times', in Robert S. Cohen, and Raymond J. Seeger (eds.), *Ernst Mach: Physicist and Philosopher.* D. Reidel, pp. 219–234.

Gori, Pietro 2011. 'Drei Briefe von Hans Kleinpeter an Ernst Mach über Nietzsche', *Nietzsche-Studien* 40: 290–298.

2014. 'Nietzsche and Mechanism. On the Use of History for Science', in Helmut Heit and Lisa Heller (eds.), *Handbuch Nietzsche und die Wissenschaften.* De Gruyter, pp. 119–137.

2015. 'Psychology without a Soul, Philosophy without an I. Nietzsche and 19th Century Psychophysics (Fechner, Lange, Mach)', in J. Constâncio, M. J. Mayer Branco, and B. Ryan (eds.), *Nietzsche and the Problem of Subjectivity.* De Gruyter, pp. 166–195.

2017. 'On Nietzsche's Criticism towards Common Sense Realism in Human, All Too Human I, 11', *Philosophical Readings* 9: 207–213.

2018. 'Ernst Mach and Pragmatic Realism', *Revista Portuguesa de Filosofia* 74: 151–172.

2019a. *Nietzsche's Pragmatism. A Study on Perspectival Thought*, Eng. trans. De Gruyter.

2019b. 'Nietzsche's Fictional Realism: A Historico-Theoretical Approach', *Estetica. Studi e Ricerche* 9: 167–181.

2019c. 'What Does It Mean to Orient Oneself in Science? On Ernst Mach's Pragmatic Epistemology', in F. Stadler (ed.), *Ernst Mach – Life, Work, Influence* (Vienna Circle Institute Yearbook). Springer, pp. 525–536.

forthcoming. 'Ernst Mach. Beiträge zur Analyse der Empfindungen (1886)', *Studia Nietzscheana* / *nietzschesource.org*

Green, Michael S. 2002. *Nietzsche and the Transcendental Tradition*. University of Illinois Press.

Heit, Helmut, Abel, Günter, and Brusotti, Marco (eds.) 2012. *Nietzsches Wissenschaftsphilosophie*. De Gruyter.

Heit, Helmut and Heller, Lisa (eds.) 2014. *Handbuch Nietzsche und die Wissenschaften*. De Gruyter.

Hesse, Mary B. 1972. 'In Defence of Objectivity', *Proceedings of the British Academy* 57: 275–292.

Hiebert, Erwin N. 1976. 'Introduction', in Ernst Mach, *Knowledge and Error. Sketches on the Psychology of Enquiry*. D. Reidel, pp. xi–xxxviii.

Holton, Gerald 1992. 'Ernst Mach and the Fortunes of Positivism in America', *Isis* 83: 27–60.

Hussain, Nadeem 2004a. 'Nietzsche's Positivism', *European Journal of Philosophy* 12: 326–368.

2004b. 'Reading Nietzsche through Ernst Mach', in Thomas Brobjer and Gregory Moore (eds.), *Nietzsche and Science*. Ashgate, pp. 111–129.

Kant, Immanuel 1998. *Critique of Pure Reason*, Eng. trans. Cambridge University Press.

Kleinpeter, Hans 1912. 'Der Pragmatismus im Lichte der Machschen Erkenntnislehre', *Wissenschaftliche Rundschau* 20: 405–407.

1912/1913. 'Die Erkenntnislehre Friedrich Nietzsches', *Wissenschaftliche Rundschau* 3: 5–9.

1913. *Der Phänomenalismus. Eine naturwissenschaftliche Weltanschauung*. J. A. Barth.

Leiter, Brian 2002. *Routledge Philosophy Guidebook to Nietzsche on Morality*. Routledge.

Mach, Ernst 1897. *Popular Scientific Lectures*. Open Court.

1911. *History and Root of the Principle of the Conservation of Energy*. Open Court.

1919. *The Science of Mechanics: A Critical and Historical Account of its Development*. Open Court.

1959. *The Analysis of Sensations*. Dover.

1976. *Knowledge and Error. Sketches on the Psychology of Enquiry*. D. Reidel.

1992. 'Sensory Elements and Scientific Concepts', in J. T. Blackmore (ed.), *Ernst Mach - A Deeper Look: Documents and New Perspectives*. Kluwer, pp. 118–126.

Nietzsche, Friedrich 1980. *Sämtliche Werke. Kritische Studienausgabe in 15 Bänden*. Eds. G. Colli and M. Montinari. DTV/De Gruyter.

1996. *Human, All Too Human*. Cambridge University Press.

1999. *The Birth of Tragedy and Other Writings*, Cambridge University Press.

2001. *The Gay Science*. Cambridge University Press.

2002. *Beyond Good and Evil*. Cambridge University Press.

2003. *Writings from the Late Notebooks*. Cambridge University Press.

2005. *The Anti-Christ, Ecce Homo, Twilight of the Idols, and Other Writings*. Cambridge University Press.

Remhof, Justin 2015. 'Nietzsche's Conception of Truth: Correspondence, Coherence, or Pragmatist?', *Journal of Nietzsche Studies* 46: 229–238.

Pihlström, Sami 2008. *Pragmatist Metaphysics*. Continuum.

Stack, George 1983. *Nietzsche and Lange*. De Gruyter.

Stegmaier, Werner 2012. *Nietzsches Befreiung der Philosophie*. De Gruyter.

von Mises, Richard 1970. 'Ernst Mach and the Empiricist Conception of Science', in Robert S. Cohen, and Raymond J. Seeger (eds.), *Ernst Mach: Physicist and Philosopher*. D. Reidel, pp. 245–270.

8

Abstraction, Pragmatism, and History in Mach's Economy of Science

LYDIA PATTON

Introduction

Ernst Mach's appeal to the 'economy of science' has been interpreted as an overarching principle of minimization (Stein 1967, 1977), promoting the increasing simplification of scientific knowledge via principles that increase calculating power without adding substantively to the knowledge embedded in empirical facts. There is a growing literature (Gori, Banks, Pojman, Wulz, Serra, and Maia) arguing for a more robust understanding of Mach's 'economy of science'. Machian 'economy' appeals to the continuity between scientific experiences and concepts, but also to the increasing complexity of scientific concepts, building on connections between what Mach called world-elements or sensation elements (Banks 2003). Mach's account emphasises not only continuities between experiences that allow for simplification, but also areas of divergence that promote the branching of scientific concepts and methods. I emphasise the roles of abstraction, pragmatism, and history in Mach's economy of science, and I argue that these elements allow Mach to investigate the productive tension between creative and conservative moments in the history of science.[1]

Mach's World-Elements

Ernst Mach is one of the nineteenth-century scientists, like Hermann von Helmholtz, whose work is not easy to fit into contemporary disciplinary categories. He is not merely a physicist, physiologist of perception, or philosopher of mind. Certainly, Mach considered himself to be a physicist, but he

Deep gratitude is due: to John Preston, for the invitation to contribute and for his editorial guidance and suggestions for revision; to Richard Staley, for far-reaching and perceptive comments that shaped the final revision of the chapter; and to Thomas Uebel, for excellent comments that helped to define and to clarify the project of the chapter and for sharing his own very illuminating chapter from this volume. This chapter is dedicated to my friend Erik C. Banks.

[1] While this line of thought is of course inspired by Kuhn, I am not suggesting (here) that it inspired him.

would have wished his work in physiology and psychology, among other areas, to be recognised. Erik C. Banks (2003) reads him as fitting into the tradition of 'natural philosophy', which included physics, mathematics, what we would now call philosophy of mind, theology, and the life sciences.

Still, Mach can be placed in the nineteenth-century tradition of 'Erkenntnistheorie' or theory of knowledge, in which physiology, physics, and the life sciences worked in interaction. That tradition was increasingly responsive to work in sense physiology and what is now called biology. But it was also affected by debates over historical methods and over philosophical approaches, including Kantianism, materialism, and empiricism (Patton 2004, chapters 1 and 2).

The account that follows emphasises Mach's approach to a central concept in his scientific reasoning: the economy of science. There is a productive tension in Mach's work between the creative moment in the history of science, in which scientists can introduce principles or concepts that make new connections between experiences, and the minimisation project that Mach took equally seriously, of explaining experiences and reasoning economically using accounts of our physical and physiological interaction with the environment. Both the creative and the minimisation projects turn our attention to Mach's account of the distinction between the subjective and objective in epistemology. I will argue that Mach's overarching concern, in accounting for the role of the scientist in the economy of science, is with the *pragmatic history* of the experiencing and creative knower, rather than with exclusive, reductive phenomenological or biological explanations.

Debates over Mach's work in the 1960s, 1970s, and 1980s centred on a number of key issues: Mach's dispute with Ludwig Boltzmann over the 'reality' of atoms and his remarks about the epistemological foundations of Newtonian space and time (including the well-known 'bucket' and 'two spheres' thought experiments).

John Blackmore's copious publications in the 1970s and 1980s followed on V. I. Lenin's earlier reading of Mach as a phenomenalist (Lenin 1909/1972). Ivor Grattan-Guinness (1974) observes in a review of Blackmore's *Ernst Mach* that Blackmore opposes 'representationalism', which 'asserts that the real world is unattainable and represented to us only through mental appearances', to Mach's supposed 'presentationalism', which 'advocates that the external world *is* the appearances we receive of it' (Grattan-Guinness 1974, p. 76, emphasis in original).

The latter seems to be Blackmore's sense of 'phenomenalism': that the external world of sense is not a representation of something beyond sensed particulars, but rather the appearance just *is* the reality. Blackmore supports the 'representationalist' view, that the appearances are indications or signs of an underlying reality:

> To clarify and occasionally criticise Mach's philosophical ideas especially those of a phenomenalistic or Buddhistic drift, I have often contrasted his

point of view with what I call 'common sense'. I mean the representatio-
nalistic view of Galileo, Boyle, Locke, and Newton – ideas widely accepted
and used by practical people today and the only epistemology compatible
with a reasonable understanding of the process of perception as accepted
by most scientists in the field.

(Blackmore 1972, pp. x–xi)

Along with other critics, Seaman (1975) is sceptical of Blackmore's claim that
science and common sense converge on representationalism, and he questions
Blackmore's assertions that Mach is a phenomenalist and that his positions are
in conflict with any 'reasonable understanding of the process of perception' by
either 'scientists in the field' or 'practical people'.

   Erik Banks' first book, *Ernst Mach's World Elements* (2003), presents a
reading of Mach that does not interpret him as a phenomenalist, and thus
avoids Blackmore's criticisms. Banks' reasoning is subtle. He builds a reading
of Mach on two fronts that are relevant here:

(1) A detailed account of Mach's understanding of the process of perception,
    which leads to
(2) A detailed account of Mach's 'neutral monism', which Banks argues is
    neither equivalent to nor reducible to phenomenalism.

Banks' second book, *The Realistic Empiricism of Mach, James, and Russell*
(2014), follows up on the second point in detail, presenting Mach as an early
supporter of neutral monism along with James and Russell. Scott Edgar
(2013) explains how Mach's neutral monism allows him to deny coherently
the existence of things in themselves, a problem also faced by F. A. Lange,
who found a less consistent response. A key point that Edgar raises is that
Mach's denial of things in themselves reaches to denying an absolute *subject*
as well:

> In rejecting the view that there are any permanent substances or things in
> themselves underlying the sensation-like elements of experience, he
> rejects the idea of a permanent subject or mind that contains sensations,
> just as he rejects the idea of permanent bodies underneath our sensations
> of objects in the world. Thus for Mach, these sensation-like elements are
> strictly speaking neither mental nor physical considered in themselves.
> Rather, he thinks we consider an element a physical property when we
> regard it as part of a complex that represents a 'body' and we consider it a
> sensation when we regard it as part of a complex that represents an 'ego'.

(Edgar 2013, p. 113, note 11)

The upshot of new readings of Mach from scholars including Banks, Edgar,
Alix Hui (2013), and Richard Staley is to reject the idea that Mach was a
naïve phenomenalist about perception and to defend a more complex
picture of Mach as a neutral monist. To Mach, perception presents us with

'world-elements' that are neither exclusively objective nor exclusively subject-ive. Any attempt to reduce one set of elements to another, or to show that we can make inferences from 'subjective' perceptions to a realm of 'real', 'object-ive' underlying objects, is ruled out. This is not because Mach is committed to phenomenalism – it is because he is committed neither to phenomenalism nor to representationalism.

Mach explains that we use world-elements to construct a world by analysing our sensations, but we do not know a priori whether those sensations are 'mere' phenomena (i.e. merely a response of our sensorium to stimuli) or whether they are indications of external objects. Banks points out that, to Mach, most of our sensations are both. As I would gloss it, sensations include information about the 'object' of perception – which is not mind-independent, but rather what is objective *in* perception – and information about the 'subject' of perception, 'direct' information about which is not accessible independently of perception either.[2]

The object-directedness of perception, for Mach, is not something we can *know about* immediately. This is not the position that we must construct a priori concepts of objects in order for perception to be object-directed: Mach rejects that position definitively in the introduction to *The Analysis of Sensations*. Rather, Mach defends the position that perception is not given to us as object-directed: we cannot know immedi-ately what in our perceptions, as we experience them, is objective and what is subjective.

To Mach, the physiology of perception can establish one perspective on the contents of perception. That perspective allows us to distinguish those parts of perception that are *due to* the process of perception and those parts that are, for instance, introduced by measurement processes..

If sensation is to contribute to the depth of scientific knowledge, it must be able to be managed as part of the economy of science. For Mach, this means first of all that experienced sensation must be analysable, in the classic, chemical sense that sensation can be divided into distinct parts of recognisable types. Mach uses techniques from physics and psychology, especially the theory of manifolds, to make this explicit (Mach 1903).

Objectivity and History in Mach and Helmholtz

Mach's contemporary Helmholtz also thinks that whether sensations allow for inferences about subjective or objective content is a scientific matter that should be investigated empirically, not a determination that can be made on

---

[2] See Banks (2003) for detailed historical accounts.

the basis of conceptual analysis alone.[3] But Helmholtz argued that we must assume an a priori causal law that describes the stable relationships between external objects, conceived as the sources of our sensations and representations, and the physiological constitution and perceptual system of the subject (Hatfield 2018; Patton 2018).

Mach does not base his system on an a priori law of causality as Helmholtz does. Helmholtz's reliance on causal reasoning and on induction is his way of securing the stability and justification of inference from perceptual experience. Helmholtz argues that perceptions are signs, and not copies, of the external objects that are their sources, a position Brian Tracz has dubbed 'Ignorance' (Tracz 2018, p. 64): 'In Helmholtz's terms, "our images of the things in our representation are not similar to their objects" [Helmholtz 1867, § 2.3, pp. 590]. And since Helmholtz maintains that we can only obtain knowledge of objects through perception, this fundamental dissimilarity leaves us ignorant, in some respect, of the things we represent' (Tracz 2018, p. 64).

But Helmholtz maintains that we can find what is true in our representations by finding the causal laws that describe the actual relationships between perceptual signs and what they depict (De Kock 2014, 2015):

> Each law of nature states that on preconditions that are similar in certain respects consequences that are similar in certain other respects always occur. Since similars in our sensible world are indicated by similar signs, the nomological sequence of similar effects following similar causes corresponds to an equally regular sequence in the realm of our sensations.
>
> (Helmholtz 1878, p. 13)

Helmholtz argues that inferences from previous experience can influence present experience as it occurs. Thus, the *history* of our perceptual experience becomes important for him:

> ... the remembered images from earlier experiences work together with present sensations to bring forth an intuitive image, which intrudes upon our power of perception with compelling force, without what is given through memory and what is given through present perception being separated in consciousness.
>
> (Helmholtz 1867, § 26, pp. 436–437, my translation)

Helmholtz distinguishes between the inductive inferences and prior perceptual images that can be operative in occurrent perception and the causal laws that he sees as underlying veridical inferences from perception.

While Mach also emphasises the role of history in the growth of scientific knowledge, he does not make this distinction. For Mach, the laws of nature

---

[3] See Patton (2018), including references to other work.

and other 'economical' scientific methods replace pre-theoretical 'instinctive knowledge'. But scientific methods and principles arise out of instinctive knowledge. The section following will explain Mach's concept of instinctive knowledge and its role in his account of the development of basic laws of mechanics: to obviate reliance on a priori elements.[4]

Instinctive Knowledge and the Laws of Nature

Mach, like Helmholtz, argues that prior experience affects the way we behave in ordinary life and in scientific research. Mach's position can be found in *The Development of Mechanics*.[5] There, Mach argues that previous perceptual experience impinges upon current experience. While Helmholtz focuses on the *lack* of resemblance between our perceptions and the external objects that are their sources, Mach emphasises that observations are indications of 'the processes of nature':

> How does instinctive knowledge originate and what are its contents? Everything which we observe in nature imprints itself uncomprehended and unanalysed in our percepts and ideas, which, then, in their turn, mimic the processes of nature in their most general and most striking features. In these accumulated experiences we possess a treasure-store which is ever close at hand and of which only the smallest portion is embodied in clear articulate thought.
>
> (Mach 1919, p. 28)

A common feature of Mach's and Helmholtz's work is the focus on *series* or *sequences* of observations, as opposed to *singular* perceptions. This feature distinguishes their work from that of philosophers who emphasised the role of singular observations in 'confirming' hypotheses about external objects or processes. Mach does identify singular aspects of perceptions, but only in the context of series or sequences. These sequences have patterns, features that 'mimic the processes of nature in their most general and most striking features'.

One of the earliest chapters of *The Development of Mechanics* focuses on the role of 'instinctive knowledge' in Simon Stevin's derivation of the principle of the inclined plane, 'a veritable acme' of Stevin's career.[6] Stevin, a Flemish mathematician and engineer, performed thought experiments on the 'wreath

---

[4] I am grateful to Thomas Uebel for clarification of this passage.

[5] Mach (1919, first edition published in 1883). The title chosen for the English translation of the book was unfortunately, and wrongly, *The Science of Mechanics*.

[6] Dijksterhuis (2012/1970, p. 52). Serra et al. (2018) discuss Mach's analysis of instinctive knowledge in detail, including previous analyses by Pojman, Feyerabend, and Buchler (§ 2).

Figure 8.1   Left: Stevin's inclined plane with endless chain (Mach 1919, pp. 24–25). Right: Stevin's plane with the symmetrical portion of the chain removed (Mach 1919, pp. 24–25)

of spheres' or 'clootcrans' (figures 19 and 20 in Stevin 1586, p. 41, discussed in Mach 1919, pp. 24–25) (see Figure 8.1).

As Mach summarises his reasoning,

> He imagines a triangular prism with horizontally placed edges, a cross-section of which ABC is represented in Fig. 19. For the sake of illustration we will say that AB=2BC; also that AC is horizontal. Over this prism Stevinus[7] lays an endless string on which 14 balls of equal weight are strung and tied at equal distances apart. We can advantageously replace this string by an endless uniform chain or cord. The chain will either be in equilibrium or it will not. If we assume the latter to be the case, the chain, since the conditions of the event are not altered by its motion, must, when once actually in motion, continue to move for ever, that is, it must present a perpetual motion, which Stevinus deems absurd. Consequently only the first case is conceivable. The chain remains in equilibrium. The symmetrical portion ADC may, therefore, without disturbing the equilibrium, be

[7] Mach calls Stevin 'Stevinus'.

removed [figure 20]. The portion AB of the chain consequently balances the portion BC. Hence: on inclined planes of equal heights equal weights act in the inverse proportion of the lengths of the planes.

(Mach 1919, pp. 24–25)

Dijksterhuis summarises Stevin's theorem as follows:

> Proposition XIX. Given a triangle, whose plane is at right angles to the horizon, with its base parallel thereto, while on each of the other sides there shall be rolling spheres, of equal weight to one another: as the right side of the triangle is to the left side, so is the apparent weight of the spheres on the left side to the apparent weight of the spheres on the right side.
>
> (1970/2012, p. 52, trans. corrected)

'Apparent weight' is a translation of the term '*Stallwicht*', 'meaning the component of an acting force which is actually exerting an influence' (Dijksterhuis 1970/2012, p. 52). Dijksterhuis notes:

> The admiration we owe to Stevin's wreath of spheres need not blind us, however, to the fact that the intuition on which it rests is not equalled in value by the logical force of the reasoning. The salient point of the demonstration obviously consists in the impossibility of a perpetual motion. Now the whole contrivance is conceived in the ideal realm of rational mechanics, where all disturbing influences, such as friction and air resistance, are believed not to exist. But in this realm a perpetual motion is by no means impossible: a simple pendulum, which has been pulled to one side and then released, forms an example of it. The wreath of spheres would indeed perform a perpetual motion if it were given an initial velocity.
>
> (Dijksterhuis 1970/2012, p. 54)

In *The Development of Mechanics*, though, Mach explains Stevin's reasoning quite differently. Dijksterhuis focuses on the question of whether 'the ideal realm of rational mechanics' allows for the possibility of a perpetual motion. Mach argues, instead, that Stevin is appealing to the 'instinctive knowledge' of the people Stevin wishes to persuade with his derivation of the theorem. In particular, Mach argues that Stevin is appealing to our lack of experience of a perpetual motion of the kind required to refute the law of the inclined plane:

> He feels at once, and we with him, that we have never observed anything like a motion of the kind referred to, that a thing of such a character does not exist . . . we accept the conclusion drawn from it respecting the law of equilibrium on the inclined plane without the thought of an objection, although the law if presented as the simple result of experiment . . . would appear dubious.
>
> (Mach 1919, p. 26)

Mach concludes that Stevin places 'instinctive knowledge' above simple observations made in experiment. The *basis* of the instinctive knowledge in question is a sequence of observations, and so we might wonder how Mach makes this distinction. Mach goes on to say:

> That Stevin ascribes to instinctive knowledge of this sort a higher authority than to simple, manifest, direct observation might excite in us astonishment if we did not ourselves possess the same inclination ... We feel clearly, that we ourselves have contributed nothing to the creation of instinctive knowledge, that we have added nothing to it arbitrarily, but that it exists in absolute independence of our participation. Our mistrust of our own subjective interpretation of the facts is thus dissipated.
>
> (Mach 1919, p. 26)

There are two salient points about Machian 'instinctive knowledge' that should be emphasised. First, it is not a 'subjective interpretation', but rather based on autonomous, independent facts. That does not mean that it has no subjective component, in the sense discussed in the sections above. Rather, it means that our *interpretation* of the observations involved does not rest on grounds exclusive to the subject, but rather on grounds independent of the subject's participation.

Second, Mach's instinctive knowledge rests not on *single* observations, but on *sequences* of observations that express relations between *groups* of facts. In another context, Mach writes along the same lines of the more formal knowledge of physics, represented in equations[8]:

> [W]hat we have found is that between the [elements] of a group of facts ... a number of equations exists. The simple fact of change brings it about that the number of these equations must be smaller than the number of the [elements]. If the former be smaller by one than the latter, then one portion of the [elements] is uniquely determined by the other portion.
>
> (Mach 1986, pp. 180–181)

As Banks emphasises, the method of differences is fundamental to Mach's method in science:

> The tension between the elements and their ordering into this general manifold of space, time, and matter is a general problem in Mach's philosophy of nature. The divide falls between his Heraclitean view that the elements are transitory unique events, arising and vanishing and possessing always an individual existence, and his view that space, time, and matter, however unreal they may be on a fundamental level, represent for Mach economical permanencies that must be acknowledged as a task

---

[8] I am grateful to Richard Staley for clarification of this passage.

of science ... Mach said that he considered the real facts of nature to be the existence of 'differences' or inequalities ... Mach's elements are the differences of state in the world and, by a careful tracking of their effects on one another, the determinations of the rates and magnitudes of those effects, Mach thought one could deduce the existence of independent potential sources and relations of intensity from this raw data by finding orderings in it.

(Banks 2003, p. 239)

Mach's account is based on the following principles:

- The world is constructed from functional connections between world-elements.
- Different functional connections between those world-elements can bring different states of affairs into view. These functional connections may draw on previous experience.
- Features of experience, such as its *continuity* and whether we can provide a coherent *history* of our experience in giving explanations, support efficient orderings of the data of experience.
- Anti-reductionism: there is no one correct or accurate state of affairs to which a complex of world-elements must be reduced. There can be multiple accurate expressions of the content of a complex of world-elements.

These principles form the core of Mach's economy of science. In the section that follows, I will present a dynamical account of this economy and provide an example of how it works.

The Dynamics of the Economy of Science

Following on the earlier work of Paul Feyerabend (1987, 1999), Paul Pojman (2000, 2011), and Banks (2004), there has been a recent resurgence of interest in Mach's economy of science, including work by Pietro Gori (2018, 2019), Isabel Serra and Elisa Maia (2018), Monika Wulz (2015), and myself (2018). The more recent scholarship focuses on Mach's economy as a subject of interest in itself. Some earlier work (Stein 1967, 1977) was more interested in the question of why Mach seemingly dismissed Newton's methods. On the other hand, work on Mach's empirically founded research by Don Howard, especially, has made much of the differences between Mach and the neo-Kantians and other apriorist thinkers.

In *The Development of Mechanics*, Mach uses the words 'economy' and 'economical' in a number of ways, all of which contribute to his picture of science as a system. I have gathered his uses into Table 8.1 (Patton 2018).

The *methods* on the left side of Table 8.1, which we will discuss in a moment, contribute to the economy of science. The *features* of science on

Table 8.1 *Typology of the economy of science*

Methods that promote economy	The effects of these methods
'Instinctive knowledge'	Transparency
The method of differences	Empirical fruitfulness
The law of continuity of experience	The ability to disregard details
Relations of the whole	Computational power
	Minimisation
	Completion of experience

the right side of Table 8.1 are desirable results of the employment of those methods. One common interpretation of Mach on the economy of science is to read him as arguing in a simplistic way that 'economy' adds up to minimisation, or increased computational power, alone – and certainly, Mach does say that economy involves minimisation. But that statement is made in the context of *other* statements that require the economy of science to include *every aspect* of science in its methods, to focus on the *completeness* and *continuity* of the experience that lies at the basis of science, and to conceive of every scientific result within the whole of knowledge (*Erkenntnis*).

Here, it will be instructive to examine another of Mach's discussions of 'instinctive knowledge': his critique of Archimedes on the law of the lever, discussed early in the twentieth century by Giovanni Vailati, and recently by Paolo Palmieri (2008) and Maarten van Dyck (2009).

As Mach observes, Archimedes begins from a form of 'instinctive' knowledge. He starts with two propositions that Archimedes takes to be self-evident:

a. Magnitudes of equal weight acting at equal distances (from their point of support) are at equilibrium.
b. Magnitudes of equal weight acting at unequal distances (from their point of support) are not in equilibrium, but the one at the greater distance sinks. (Mach 1919, pp. 8–9)

As Mach notes, the first proposition can be supported by a form of instinctive knowledge. If a 'spectator' looks at an experimental set-up with two masses of equal weight and at equal distances from the fulcrum of a balance, we can perceive, 'determined by the symmetry of our own body', that they will be in equilibrium, and this will be 'a highly imperative *instinctive* perception' (Mach 1919, p. 10, emphasis in original). Thus far, Archimedes' derivations are similar to Stevin's, examined above. Just as we instinctively perceive that no perpetual motion will occur in Stevin's wreath of circles, we also perceive that two weights will not move if they are equal and placed at equal distances from the fulcrum of a balance (assuming a proper set-up).

Mach goes on to say that Archimedes attempts to *demonstrate* the law of the lever from a process of reasoning that ultimately fails. As van Dyck puts it,

> Both the spirit and content of Mach's criticism are best captured by his well-known rhetorical question about how it would be possible to derive 'by speculative methods' the inverse proportionality of weight and lever-arm 'from the mere assumption of the equilibrium of equal weights at equal distances'. Mach's answer is ... that Archimedes' rational argument ... presupposes what needs to be proved. He locates the exact step in the argument where this happens, in what Palmieri calls the equilibrium-preserving assumption. This assumption basically states that equilibrium is not disturbed if one of the equilibrating weights is replaced with two weights placed symmetrically around the original weight, each having half the weight of the original body.
>
> (van Dyck 2009, pp. 315–316)

Mach argues that 'Archimedes pursued exactly the same tendency as Stevinus, only with much less good fortune' (Mach 1919, p. 27). This 'tendency' was to formalise the reasoning from 'instinctive knowledge' that pre-theoretical reasoners tended to apply. The difference between Archimedes and Stevin, to Mach, is that Stevin appeals to instinctive knowledge in his proof of a proposition that is properly supported by that knowledge. Archimedes, on the other hand, appeals to a general proposition – *not* instinctive knowledge – in his attempt to support another general proposition. Thus, Archimedes tries to prove that a proposition (the law of the lever) is applicable generally by appealing to a general principle, and not by showing that it is reducible to a form of instinctive knowledge that rules out error. As Mach puts it:

> [W]hen Archimedes substitutes for a large weight a series of symmetrically arranged pairs of small weights, which weights extend beyond the point of support, he employs in this very act the doctrine of the centre of gravity in its more general form, which is itself nothing else than the doctrine of the lever in its more general form.
>
> (Mach 1919, p. 14)

Stevin appealed to the impossibility of a perpetual motion in his derivation of the law of the inclined plane. But this is a virtuous appeal, according to Mach's account, because it appeals only to experiences that every subject on earth has had, and thus his 'procedure is no error. If an error were contained in it, we should all share it' (Mach 1919, p. 27). If Archimedes had simply rested on the instinctive demonstration of the principles of equilibrium (a and b above), he would not have made an error.

Instead, Archimedes tried to find a *deductive* proof: in it, Archimedes appeals not to instinctive or empirical knowledge, but to another, equivalent general principle. Archimedes assumes, rather than appealing to, 'the assumption that the equilibrium-disturbing effect of a weight P at the distance L from the axis of rotation is measured by the product P•L' (Mach 1919, p. 14). Mach remarks:

> It is characteristic, that he will not trust on his own authority, perhaps even on that of others, the easily presented observation of the import of the product P•L, but searches after a further verification of it. Now as a matter of fact we shall not, at least at this stage of our progress, attain to any comprehension whatever of the lever unless we directly *discern* in the phenomena the product P•L as the factor decisive of the disturbance of equilibrium. In so far as Archimedes, in his Grecian mania for demonstration, strives to get around this, his deduction is defective. But regarding the import of P•L as given, the Archimedean deductions still retain considerable value, in so far as the modes of conception of different cases are supported the one on the other, in so far as it is shown that one simple case contains all others, in so far as the same mode of conception is established for all cases.
>
> (Mach 1919, pp. 18–19)

Mach's objection to Archimedes is not merely that Archimedes' reasoning is circular. It is that Archimedes' reasoning is *deductive*, and not a demonstration from instinctive knowledge. Mach disagrees with Archimedes on the *possible ways of demonstrating* the law of the lever: Mach is effectively arguing not that Archimedes has begged the question, but that the knowledge that had been gained by Archimedes' time about the law of the lever and about the centre of gravity were effectively the same, and that both arise from a form of 'instinctive knowledge', resting in part on the symmetry of our own bodies, to which Archimedes did not want to appeal.

The law of the lever and the principle of the centre of gravity formulate instinctive knowledge so that we are able to observe, 'in so far as the same mode of conception is established for all cases', that all cases of measurements of equilibrium have similar features. This form of reasoning, for Mach, does not support a form of abstraction to a priori principles that support deductive reasoning. Rather, it supports the use of these principles to promote the *completeness* and *continuity* of experience and observation, so that our experience can be more informative. All this, in turn, promotes the economy of science: 'The function of science, as we take it, is to replace experience' (Mach 1919, pp. 489–490). Elsewhere in *The Development of Mechanics*, Mach makes very illuminating remarks on this score:

> The most important result of our reflections is that *precisely the apparently simplest mechanical principles are of a very complicated character, that these principles are founded on uncompleted experiences, nay on experiences that never can be fully completed, that practically, indeed, they are sufficiently secured, in view of the tolerable stability of our*

environment, to serve as the foundation of mathematical deduction, but that they can by no means themselves be regarded as mathematically established truths but only as principles that not only admit of constant control by experience but also require it.

(Mach 1919, pp. 237–238, emphasis in original)

Scientific principles are ways of gathering experiences, and finding orderings in them, to show that some can be represented as functionally equivalent to others (see the discussion above of the method of differences). The laws and principles of mechanics are formal statements of previous experiences that have revealed functional relationships between experiences. Those relationships are not complete. They can be used 'as the foundation of mathematical deduction', but they can be used only within the experimental method, which provides '*constant control by experience*'.

These principles are not axioms or rules that have universal validity or applicability, nor can their truth be finally established, since they are founded on experiences that in principle can never be completed. However, they can be productive in the sense that they allow for the goal of the completeness of science, within a reasonable human lifespan:

> We must admit … that there is no result of science which in point of principle could not have been arrived at wholly without methods. But, as a matter of fact, within the short span of a human life and with man's limited powers of memory, any stock of knowledge worthy of the name is unattainable except by the greatest mental economy. Science itself, therefore, may be regarded as a minimal problem, consisting of the completest possible presentment of facts with the least possible expenditure of thought.

(Mach 1919, pp. 489–490)

The above quotation is emblematic of one reading of Mach on economy: that the principles of the economy of science are only intended to make the stunning breadth of scientific experience and knowledge available in more concise form.

But, as should be evident from the above, that is not the only aim of Mach's economy of science. Archimedes' derivation of the law of the lever is 'economical' in this sense, but Archimedes' understanding of it does not contribute to Machian economy. As Archimedes uses the principle, it does not promote the completeness or coherence of our experience, but rather introduces an element that Archimedes believes to be outside our experience, a speculative principle. Mach argues that, instead, we should take the principle to be formulated in a way that shows how it arises from experience and indeed from instinctive knowledge.[9]

---

[9] Richard Staley's comments were extremely helpful in clarifying the reading in this passage.

The challenge for readings of Mach's economy of science is to explain how economical reasoning can be productive if it consists only in the formulation of knowledge that already existed. One way to explain this is the usual interpretation of 'biological-economical': people find the formal principles or rules describing experience and scientific knowledge that are the most efficient, given our environment and our physiology.

Machian economy is not merely a way to promote efficiency, but promotes distinct epistemic goals as well, including the completeness and continuity of experience and knowledge and the ability to relate facts in one area of science to facts in another. Interpretations of Machian economy should be developed in a way that demonstrates how to work effectively in science using economic principles and methods.

One such way draws on the work of Herbert Breger, who argues that 'mathematical progress' can take place 'by formally recognizing a know-how which existed before' (Breger 2000, p. 224). Breger takes the approach of looking at textbooks and handbooks from previous mathematical traditions (what Ludwik Fleck calls 'vademecum science' and Thomas Kuhn calls the 'context of pedagogy'). Without making a systematic study, he analyses a number of important cases, including Christiaan Huygens, Felix Klein's *Erlangen* programme, and a number of textbooks on calculus and group theory.

Breger notes a trend towards increasing 'abstraction' when moving from the nineteenth to the twentieth century. Klein himself, on Breger's reading, did not see much point in abstraction: 'Later on, in his *Geschichte der Mathematik im 19. Jahrhundert*, Felix Klein mentions the three well-known axioms (associativity, existence of an inverse element, existence of a neutral element), but just for the sake of completeness; reference to them is considered to be an unnecessarily abstract approach' (Klein 1926, pp. 335–336; Breger 2000, p. 224). On Breger's reading, Klein's work on group theory grew from *working with* groups concretely. The more abstract statements about axioms, permutations, transformation rules, and so on are useful only insofar as they allow us to work more effectively with those groups.

Breger's paper does not reach any particular systematic conclusion, nor is it meant to do so. It is meant as a historical intervention to explain that the contemporary structures of abstraction in mathematics were constructed, accompanying a shift in the way mathematics is taught and practiced. Klein emphasised working directly with groups and group actions, and on the acquisition of something like what Mach calls 'instinctive knowledge'. On Breger's account, those who later formalised Klein's programme replaced that instinctive knowledge with abstract, speculative principles that worked more seamlessly with the mathematical methods that took over the field during that time.

The key point is that, for Mach as for Klein, abstract structures or principles are of use insofar as they allow us to grasp the instinctive or tacit knowledge (to use Michael Polanyi's later term) available to us in mathematical and physical practice. Going beyond what Breger argues, employing similar reasoning, Mach argues that mathematical reasoning allows us to apply that instinctive knowledge by analogy across domains (see, especially, Mach 1903).[10]

We could use a number of accounts of the nature of mathematics to describe the role of these principles or structures:

(1) *Instrumentalism or fictionalism.* We construct structures to respond effectively when working with the domain under study. In this sense, the abstract elements and processes are *constructed* but are not intended to capture any *real* structural elements of the groups under study.

(2) *Structuralism or structural realism.* The structures that are constructed allow for processes and 'objects' to emerge that turn out to be informative about the phenomena we are investigating. The structures themselves turn out to be 'real', in the weaker sense that they are part of a framework for referring to real things and processes, or in the stronger sense that the structures can be proven to correspond to real features of things or processes (including dynamical features). Thus, even though the structures are *constructed*, they nonetheless can be shown to be *real*.

(3) *Abstractionism.* A contemporary extension of Fregean logicism, according to which 'abstraction principles play a crucial role in the proper foundation of arithmetic, analysis, and possibly other areas of mathematics'.[11]

As Gary Hatfield (2018) has noted, Helmholtz's view seems to be consistent with a form of structuralism. I would argue against any such view as being consistent with Mach's account, which points to an interesting distinction between Helmholtz and Mach.

Mach also does not seem to be a thoroughgoing fictionalist, which is where I differ from some readings of Mach's economy of science. Mach does *not* argue that we may construct any principles we like as long as they promote the more efficient or concise formulation of scientific data or facts. To say so is to ignore any number of passages in Mach, including those emphasised by Staley (2013 and this volume), in which Mach draws substantially from his research on physiology to back up his physical reasoning, even the reasoning that influenced Einstein's theory of general relativity. A typical passage is

---

[10] Thanks are due to Richard Staley for pressing for a clarification of this passage.

[11] Ebert and Rossberg (2016, p. 5). Also see other chapters in this volume and in Cook (2007).

found in the 1903 essay 'Space and Geometry from the Point of View of Physical Inquiry':

> Our notions of space are rooted in our physiological constitution. Geometric concepts are the product of the idealisation of physical experiences of space. Systems of geometry, finally, originate in the logical classification of the conceptual materials so gathered. All three factors have left their indubitable traces in modern geometry. Epistemological inquiries regarding space and geometry accordingly concern the physiologist, the psychologist, the physicist, the mathematician, the philosopher, and the logician alike, and they can be gradually carried to their definitive solution only by the consideration of the widely disparate points of view which are here offered.
>
> (Mach 1903, p. 1)

This passage makes two points key regarding Mach on economy. First, Mach sees the 'economy' of science as partly being like a *managed* economy in a society: different sectors of that society may engage in science, and their 'widely disparate points of view' all contribute to the solution of scientific problems (see Schabas 2005).

The second point is a significant and underappreciated element in Mach. To him, logic and mathematics are creative and open-ended, not determinative and reductive:

> Supposing a mathematician to have modified tentatively the simplest and most immediate assumptions of our geometrical experience, and supposing his attempt to have been productive of fresh insight, certainly nothing is more natural than that these researches should be further prosecuted, in a purely mathematical interest. Analogues of the geometry we are familiar with, are constructed on broader and more general assumptions for any number of dimensions, with no pretention to being regarded as more than intellectual scientific experiments and with no idea of being applied to reality.
>
> (Mach 1903, p. 27)

While Mach criticises Archimedes for introducing a deductive method when he should have appealed to instinctive knowledge, he speaks approvingly in this essay of multiple systems of geometry, and of 'idealising' and 'schematising' concepts:

> Geometry, accordingly, consists of the application of mathematics to experiences concerning space. Like mathematical physics, it can become an exact deductive science only on the condition of its representing the objects of experience by means of schematising and idealising concepts. Just as mechanics can assert the constancy of masses or reduce the interactions between bodies to simple accelerations only within the limits

of errors of observation, so likewise the existence of straight lines, planes, the amount of the angle-sum, etc., can be maintained only on a similar restriction. But just as physics sometimes finds itself constrained to replace its ideal assumptions by other more general ones, viz., to put in the place of a constant acceleration of falling bodies one dependent on the distance, instead of a constant quantity of heat a variable quantity, – so a similar procedure is permissible in geometry, when it is demanded by the facts or is necessary temporarily for scientific elucidation.

(Mach 1903, pp. 20–21)

It may seem as if Mach is contradicting himself. He has criticised Archimedes for replacing 'instinctive knowledge' with a speculative principle, and now Mach is allowing for 'idealisation' and of speculation in geometry and even in mechanics.

However, in my view, Mach's account is consistent. Mach allows for idealisations and schematised representations of experiments *if* those methods are used in a way that opens up new avenues for showing how experiences might be related, produced, and demonstrated – even, including reasoning about atomism, as long as the abstract, fictional nature of the methods is recognised.[12] A fundamental principle of the Machian economy of science is the principle of the continuity of experience, which I earlier described as anti-reductionism: that world-elements can be related to one another in multiple, open-ended ways, which reveal distinct connections between experiences.

Mach's method is to find the connections between observations and experiences that promote scientific knowledge. To do so requires abstaining from the use of merely speculative hypotheses, and replacing them with idealised and schematic instruments that can be used productively to bring new facts and relationships into view. As a result, Mach's economy of science can be seen, with some justification, as a kind of Machian abstractionism, to distinguish it from Fregean abstractionism. The outlines of Machian abstractionism are sketched above and in Patton (2018), and they remain to be filled out by future research into Mach's work on the economy of science.

Conclusion: Abstraction, Pragmatism, and History as Economical Methods

While accounts that focus on the biological and psychological aspects of Mach's economy of science are well taken, Mach's epistemological methods cannot be captured in their entirety without an account of his pragmatism, historicism, and abstractionism.

---

[12] Thanks are due to Richard Staley for a fruitful amplification of this claim.

María De Paz (2018) reads Mach as a practice-based thinker along the lines of the description of mathematical practice in Ferreirós (2016). José Ferreirós argues that a practice-based epistemology in mathematics focuses on the mathematical agent (the reasoner), and thus is *cognitive*; moreover, an account of mathematical practice is also *pragmatist* and *historical*. De Paz argues, with reason, that this tripartite focus describes Mach's epistemological methods as well.

Ideally, my approach complements De Paz's, contributing an emphasis on the productive tension between the creative moment in Mach and his biological-economical account. Accounts of Mach's methods that emphasise biology focus on ways to characterise the cognitive activities of a scientific reasoner as containing a subjective component, but also as determined by interaction with the environment. For instance, Isabel Serra and Elisa Maia (2018) read Mach as giving the Kantian a priori a biological slant with his account of instinctive knowledge, following on Pojman's (2000, 2011) earlier reading.

My account pulls apart the two strands of instinctive knowledge and abstraction and considers each as making a separate contribution to Machian economy. It is not opposed to biological readings, but does oppose biological *reductionism*. My reading puts particular emphasis on abstraction as part of the history of science.

Mach does not think that an account of our role as embodied, biologically determined subjects of experience exhausts the contribution of the subjective to epistemology or to economy. The problems which scientists take to be important, the history of approaches to those problems, the principles and laws they adopt, and the methods of abstraction they use are all part of the methods of economical reasoning in science. Knowing how to use effective abstraction principles for one's purposes, for instance, can make one's scientific reasoning much more economical.

I take the account above to be friendly to the pragmatic reading in Pietro Gori (2018, 2019) and in Thomas Uebel's chapter for this volume, which emphasise the pragmatism embodied by Mach's reasoning about the economy of science. As Uebel (this volume) observes, 'historical studies for Mach were far from merely antiquarian. Rather, they intended a "critical epistemological enlightenment of the foundations" of scientific disciplines or sub-disciplines (1896/1986, p. 1, trans. amended) and so were essential to sustaining further progress'.

I am very sympathetic to this approach, and I would add merely my emphasis on the role of abstraction principles arising from instinctive knowledge in creative scientific reasoning. Mach's method involves the following claims, adding one to the earlier stock:

> Features of experience, such as its *continuity*, and whether we can provide a coherent *history* of our experience in giving explanations, support efficient orderings of the data of experience.

Anti-reductionism: there is no one correct or accurate state of affairs to which a complex of world-elements must be reduced. There can be multiple accurate expressions of the content of a complex of world-elements.

Abstractionism: the use of principles and laws to make connections between different experiences, to allow access to distinct presentations of world-elements, and thus to promote the continuity and connectedness of experience supports the economy of science. These principles arise from instinctive knowledge, and history describes how one led to the other.

The new element of 'abstraction' can be used historically to enable the 'critical epistemological enlightenment of the foundations' of scientific disciplines that Uebel emphasises. A history of scientific reasoning ought not be restricted to a history of phenomenological experience, but rather should include an analysis of the principles that were used to introduce orderings into that experience in order to achieve certain goals. Machian abstractionism, pragmatism, and historicism work together in the economy of science.

References

Banks, Erik C. 2003. *Ernst Mach's World Elements: A Study in Natural Philosophy.* Kluwer.

2004. 'The Philosophical Roots of Mach's Economy of Thought', *Synthese* 139: 23–53.

2014. *The Realistic Empiricism of Mach, James, and Russell: Neutral Monism Reconceived.* Cambridge University Press.

Blackmore, John T. 1972. *Ernst Mach: His Work, Life, and Influence.* University of California Press.

Breger, Herbert 2000. 'Tacit Knowledge and Mathematical Progress', in E. Grosholz and H. Breger (eds.), *The Growth of Mathematical Knowledge.* Kluwer, pp. 221–230.

Cook, Roy T. (ed.) 2007. *The Arché Papers on the Mathematics of Abstraction.* Springer.

De Kock, Liesbet 2014. 'Voluntarism in Early Psychology', *History of Psychology* 17: 105–128.

2015. 'Hermann von Helmholtz's Empirico-transcendentalism Reconsidered', *Science in Context* 27: 709–744.

De Paz, María 2018. 'Reconsidering Mach in the Light of the Interplay of Practices', *Revista Portuguesa de Filosofia* 74: 219–246.

Dijksterhuis, E. J. 1970/2012. *Simon Stevin: Science in the Netherlands around 1600.* Springer.

Ebert, Philip and Rossberg, Marcus 2016. 'Introduction', in P. Ebert and M. Rossberg (eds.), *Abstractionism.* Oxford University Press, pp. 3–34.

Edgar, Scott 2013. 'The Limits of Experience and Explanation: F. A. Lange and Ernst Mach on Things in Themselves', *British Journal for the History of Philosophy* 21: 100–121.

Ferreirós, José 2016. *Mathematical Knowledge and the Interplay of Practices.* Princeton University Press.

Feyerabend, Paul K. 1987. *Farewell to Reason.* Verso.

  1999. 'Philosophy of Science: A Subject with a Great Past', in P. K. Feyerabend, *Knowledge, Science and Relativism: Philosophical Papers, Volume 3* (ed. J. M. Preston). Cambridge University Press, pp. 127–137.

Gori, Pietro 2018. 'Ernst Mach and Pragmatic Realism', *Revista Portuguesa de Filosofia* 74: 151–172.

  2019. 'What Does It Mean to Orient Oneself in Science? On Ernst Mach's Pragmatic Epistemology', in F. Stadler (ed.), *Ernst Mach – Life, Work, Influence* (Vienna Circle Institute Yearbook). Springer, pp. 525–536.

Grattan-Guinness, Ivor 1974. Review of John T. Blackmore, *Ernst Mach: His Work, Life, and Influence. British Journal for the History of Science* 7: 75–79.

Hatfield, Gary 1990. *The Natural and the Normative.* MIT Press.

  2018. 'Helmholtz and Philosophy', *Journal for the History of Analytical Philosophy* 6: 12–41.

Helmholtz, Hermann von 1867. *Handbuch der physiologischen Optik.* Leopold Voss.

  1878. *Die Thatsachen in der Wahrnehmung.* Hirschwald.

Hui, Alexandra 2013. *The Psychophysical Ear: Musical Experiments, Experimental Sounds, 1840–1910.* MIT Press.

Klein, Felix 1926 *Vorlesungen über die Entwicklung der Mathematik im 19. Jahrhundert.* Springer.

Lenin, Vladimir I. 1909/1972 *Materialism and Empirio-Criticism. Critical Comments on a Reactionary Philosophy,* in *Lenin. Collected Works, Volume 14.* Progress Publishers, pp. 17–362.

Mach, Ernst 1903. 'Space and Geometry from the Point of View of Physical Inquiry', *The Monist* 14: 1–32.

  1905. *Erkenntnis und Irrtum.* J. A. Barth.

  1919. *The Science of Mechanics: A Critical and Historical Account of Its Development,* 2nd English edition, translated from the 1901 3rd German edition by Thomas J. McCormack. Open Court.

  1986. *Popular Scientific Lectures,* trans. Thomas J. McCormack. Open Court.

Palmieri, Paolo 2008. 'The Empirical Basis of Equilibrium: Mach, Vailati, and the Lever', *Studies in History and Philosophy of Science* 39: 42–53.

Patton, Lydia 2004. Hermann Cohen's History and Philosophy of Science. Dissertation, McGill University. Available from http://digitool.library .mcgill.ca/R/?func=dbin-jump-full&object_id=85027&local_base=GEN01-MCG02

  2018. 'Helmholtz's Physiological Psychology', in S. Lapointe (ed.), *Philosophy of Mind in the Nineteenth Century.* Routledge, pp. 96–116.

2019. 'New Water in Old Buckets: Hypothetical and Counterfactual Reasoning in Mach's Economy of Science', in F. Stadler (ed.), *Ernst Mach – Life, Work, Influence* (Vienna Circle Institute Yearbook). Springer, pp. 333–348.

Pojman, Paul 2000. *Ernst Mach's Biological Theory of Knowledge*. Dissertation, Indiana University.

2011. 'The Influence of Biology and Psychology upon Physics: Ernst Mach Revisited', *Perspectives on Science* 19: 121–135.

Schabas, Margaret 2005. *The Natural Origins of Economics*. University of Chicago Press.

Seaman, Francis 1975. Review of John T. Blackmore, *Ernst Mach: His Work, Life, and Influence*. *Journal of the History of Philosophy* 13: 273–276.

Serra, Isabel and Maia, Elisa 2018. 'Looking for Routes in Mach's Epistemology', *Revista Portuguesa de Filosofia* 74: 173–196.

Staley, Richard 2013. 'Ernst Mach on Bodies and Buckets', *Physics Today* 66: 42–47.

2019. 'Revisiting Einstein's Happiest Thought', in F. Stadler (ed.), *Ernst Mach – Life, Work, Influence* (Vienna Circle Institute Yearbook). Springer, pp. 349–366.

Stein, Howard 1967. 'Newtonian Space–Time', *Texas Quarterly* 10: 174–200.

1977. 'Some Philosophical Prehistory of General Relativity', in J. Earman, C. Glymour, and J. Stachel (eds.), *Foundations of Space–Time Theories*. University of Minnesota Press, pp. 3–49.

Stevin, Simon 1586. *De beghinselen der weeghconst*. Druckerye van Christoffel Plantijn.

Tracz, Brian 2018. 'Helmholtz on Perceptual Properties', *Journal for the History of Analytical Philosophy* 100: 64–78.

van Dyck, Maarten 2009. 'On the Epistemological Foundations of the Law of the Lever', *Studies in History and Philosophy of Science* 40: 315–318.

Wulz, Monika 2015. 'Gedankenexperimente im ökonomischen Überschuss. Wissenschaft und Ökonomie bei Ernst Mach', *Berichte zur Wissenschaftsgeschichte* 38: 59–76.

9

Holding the Hand of History

Mach on the History of Science, the Analysis of Sensations, and the Economy of Thought

LUCA GUZZARDI

Introduction

Ernst Mach's historical attitude in analysing the epistemological structure of natural sciences can hardly be overlooked by anyone who approaches his work. Examples can easily be traced throughout his activity. Many of his contributions to epistemology are even explicitly presented as, or entirely devoted to, historical reconstructions of some definite concept or discipline, such as *Die Geschichte und die Wurzel des Satzes von der Erhaltung der Arbeit* (1872), *Die Mechanik in ihrer Entwickelung historisch-kritisch dargestellt* (1883), *Die Prinzipien der Wärmelehre. Historisch-kritisch entwickelt* (1896), and *Die Prinzipien der Physikalischen Optik. Historisch und erkenntnispsychologisch entwickelt* (1921), to quote the most representative in that connection.[1]

Although each of these works deals with different subjects (apart from some inevitable overlapping or repetition), all of them share a similar concept that emerges from the label *historisch-kritisch* and cognate expressions, frequently used in their titles or prefaces (e.g. this is the case with the 1909 preface to Mach 1872/1911, which qualifies such work as 'the first attempt to apply my *erkenntniskritischen* point of view'; p. iii). In their recent introduction to Mach's *Prinzipien der Wärmelehre*, Michael Heidelberger and Wolfgang Reiter have proposed the following characterisation of the historical-critical approach:

I am grateful to Pietro Gori and John Preston for their stimulating criticisms and observations, as well as for having read an early draft of this chapter and having shared with me their comments.

[1] For the English translations, see respectively Mach (1872/1911, 1883/1960, 1896/1986, 1921/1926). Very pragmatically, I do not hesitate to quote from these translations, for they are well in use in Mach studies. Where words are particularly important, or where I do not find the translation entirely convincing, I compare the passages with the original text and refer to the German phrases (although without giving the original pagination, for reasons of brevity). For this purpose, I mainly use the *Ernst-Mach-Studienausgabe* (Xenomoi, 2012–2016), comprising so far five volumes, and the original second edition of *Die Geschichte und die Wurzel* ... (J. A. Barth, 1909).

Broadly understood, 'historical-critical' means to take into consideration the history of a certain discipline while assessing it. However, through that idea Mach expresses much more than this: the historical development of a science provides a reservoir of 'modes of thinking' for human beings in dealing with experience. These can be resumed at any time and become effective again, even in our present sciences ... The resulting critical assessment lets the current development of a certain science appear in an essentially contingent light, and thus it prevents dogmatic solidification.

(Mach 2016, p. XIII)

As with many other traits of Mach's epistemology, this strong commitment to history is a 'many-footed beast'.[2] Erwin Hiebert (1970) pointed to five 'dominant lessons' regarding his approach to the history of science that, I think, are commonly assumed by Mach scholarship: (1) scientific 'concepts, laws, and theories are in constant flux'; (2) everything in science is subject to a perennial 'revision and reformulation' – hence there is no reason to consider our recent views as absolutely correct, although they are more adequate to experience, under certain respects, than previous conceptions; (3) what at first appears as a necessary development and a result of alleged a priori mechanisms reveals itself, after closer examination, as historically contingent and dependent on cultural and even individual inclinations; (4) past achievements are not past once and for all – rather, our 'modern' views are the result of a tradition inherited from that past, which is as ancient as the very first 'instinctual' beginnings of *Homo sapiens*; and (5) since there is no sharp boundary between science and 'instinctual knowledge', the products of science not only reflect a historical-cultural contingency, but also a biological contingency: they demonstrate human cognitive capacities as they appear throughout the history of culture – they 'disclose the mode of cognitive organization of experience'.[3]

---

[2] I borrow this expression from a paper by Erik C. Banks (2004, p. 24), who applied it to the principle of economy of thought. Erik devoted much of his research to exploring Mach's philosophy in a particular turn – neutral monism. He was one of the best and most generous researchers I have ever known; I learned very much from his studies and from our dialogues, which I regret were too few. This chapter is dedicated to his memory as a kind of word of gratitude that I would have liked to say personally to him.

[3] For my 'Hiebert's list' I took inspiration from the original one (Hiebert 1970, pp. 188–189, 197–198), which I tried to rearrange logically and to integrate with other parts of his argument: 'The dominant lessons that shine through Mach's deliberations are these: Science has an incredibly rich history. Its ideas are alive. Its concepts, laws, and theories are in constant flux. They are perennially under revision and reformulation. The conceptual products of science, always incomplete, take on a form at any particular time which reflects the historical circumstances and the focus of attention of the particular investigator – now physicist, now physiologist, now psychologist. Strictly speaking, scientific constructs therefore disclose the mode of cognitive organization of experience ... The history of science, Mach believed, demonstrates that there are no sharp boundary lines separating the scientific from the nonscientific or prescientific experience of man.'

Hiebert's list, complete or not, represented a noticeable advancement in the understanding of, as emphasised by the title of his paper, 'Mach's philosophical use of history of science'. Two important features of Mach's conception, however, remain out of this catalogue: his theory of neutral elements (on which are based both psychical and physical events) and his principle of thought economy, of which Hiebert only says that it may 'help us to understand how the history of science was pressed into the service of the philosophy of science' (Hiebert 1970, p. 201).

The present chapter aims to study what kind of relation these two recurring themes of Mach's conception have to his historical approach. It takes its cue from a question originally raised by Ronald Giere's influential review of the fifth volume of the Minnesota Studies in the Philosophy of Science, entitled *Historical and Philosophical Perspectives of Science*, which included Hiebert's article, along with other papers on the relation between history and philosophy of science (Giere 1973). Giere took for granted that Mach provides a significant case study in this connection; he argued, however, that Hiebert did not clarify how 'Mach's philosophical views, e.g. on the epistemological role of sensations or the economy of thought, have any "logical" or "conceptual" relations to either his physical or historical investigations'; finally, he wondered, if that was the case, 'what precisely these relations might be' (Giere 1973, p. 283). As will become clear in the following sections, I think Giere was wrong about what he called, somewhat inexactly, 'the epistemological role of sensations'; still, it seems to me that he captured a real gap in Mach studies which goes well beyond Hiebert's interpretation. And filling this gap may explain the role of history – not only its 'philosophical use', as if it were a tool that can be employed or put aside – in Mach's thought.

Mach on the Utility of History for Life

In 1909, Mach republished his oldest historical-critical investigation, *Die Geschichte und die Wurzel des Satzes von der Erhaltung der Arbeit*, originally a lecture given in Prague in November 1871 and issued some months later, in 1872, as a sixty-page offprint. As it transpires from his correspondence as well as from the new preface and additions, the reissue was part of a strategy of self-defence from the fierce attack that Max Planck launched against Machian epistemology in his 1908 Leyden lecture, *Die Einheit des physikalischen Weltbildes* (see Guzzardi 2002, pp. 419–420; Mach 2005, pp. 28–31). So Mach (1872/1911, p. 9) begins the new preface by remarking that this essay was his 'first attempt to give an adequate exposition of [his own] epistemological standpoint [*erkenntnistheoretischen Standpunktes*]', based on the principle of economy – which was, in turn, the object of Planck's most bitter criticism.[4]

In the original 1872 essay, such an epistemological point of view was expressed in historical fashion, and the introduction explains why. It opens with an epigram from Gotthold Ephraim Lessing, employed to illustrate allusively how people tend to take as self-evident things that are, in truth, historical events and have happened for some reason – or resulted from some conditions – that one can describe and reconstruct a posteriori.[5] A similar tension between illusory obviousness and the complexity of the real, according to Mach (1872/1911, p. 16), occurs when, in school years, 'propositions which have often cost several thousand years' labour of thought are represented to us as self-evident'. There is, however, one antidote to such a deceptive way of learning: 'historical studies'. Otherwise, the price for the lack of a sense of history can be high, for it promotes a reifying, metaphysical attitude, as far as 'we are accustomed to call concepts metaphysical, if we have forgotten how we reached them'. Contrariwise, 'one can never lose one's footing, or come into collision with facts, if one always keeps in view the path by which one has come' (Mach 1872/1911, p. 17).[6]

In accordance with this view, Mach (1872/1911, pp. 16–17) points out that the essay contains 'considerations which, if I except my reading of Kant and Herbart, ... have arisen quite independently of the influence of others [and] are based upon some historical studies'. They pertain to 'some facts belonging both to natural science and to history', and may also serve to instantiate 'the value of the historical method in teaching' – which is characterised as follows:

---

[4] Planck's criticism came to a climax when he stated that 'should Mach's principle of economy actually be shoved into the center of epistemology, it might well disturb the thought processes of leading minds, lame the flight of their imagination, and by such means interfere with the progress of science in disastrous ways' (Planck 1909, pp. 36–37; for the English translation, see Blackmore 1992, pp. 131–132). Mach's correspondence with friends leaves no doubt that he was profoundly touched by this kind of acrimony (Wolters 1987, pp. 19, 32–33).

[5] Lessing's original epigram ('Hänschen Schlau', from his *Sinngedichte*, collected in Lessing 1970, p. 23) reads as follows: '"*Es ist doch sonderbar bestellt,*" / Sprach Hänschen Schlau zu Vetter Fritzen, / "*Dass nur die Reichen in der Welt / Das meiste Geld besitzen*"' ('"Sure, it is remarkably arranged" / Told Little Smart Hans to Cousin Fritz / "That only to the rich in the world / most of the wealth belongs"').

[6] Relatedly, Mach also complains that the unhistorical attitude of his colleagues promoted a campaign against him: 'The reason why, in discussion of these thoughts with able colleagues of mine, I could not, as a rule, come to agreement, and why my colleagues always tended to seek the ground of such "strange" ["*sonderbar*" in the German text: note that he is using an expression employed by Lessing in the above-quoted epigram] views in some confusion of mine, was, without doubt, that historical studies are not so generally cultivated as they should be' (Mach 1872/1911, p. 16).

If from history one learned nothing else than the variability of views, it would be invaluable. Of science, more than anything else, Heraclitus's words are true: 'One cannot go up the same stream twice'. Attempts to fix the fair moment by means of textbooks have always failed. Let us, then, early get used to the fact that science is unfinished, variable. Whoever knows only one view or one form of a view does not believe that another has ever stood in its place, or that another will ever succeed it; he neither doubts nor tests ... We believe, rather, that it can do us no harm to know the point of view of another eminent nation, so that we can, on occasion, put ourselves in a different position from that in which we have been brought up.

(Mach 1872/1911, pp. 17–18)

The introduction closes, after some lines, with Mach's famous motto: 'Let us not let go the guiding hand of history. History has made all; history can alter all.'

This picture of history as an ever-changing stream with an unpredictable course is well in line with the first three lessons in Hiebert's list. With the image of Heraclitus' river, Mach defends the idea that scientific notions and practices are in constant flux (Hiebert's lesson 1); by saying that one cannot fix them once and for all in textbooks, he indirectly argues for our scientific notions being subject to 'revisions and reformulations' (Hiebert's lesson 2); Lessing's epigram and the related complaint about those who are satisfied with their own view and know nothing of others warn against dogmatism, intimating that things that appear obvious and necessary are, in truth, the result of a contingent development (Hiebert's lesson 3). As to lessons 4 and 5, which will build the pillars of Mach's naturalised epistemology, they will be explored in detail in later works, from *The Science of Mechanics* to the *Principles of the Theory of Heat* and *Knowledge and Error*; but their seeds are already appreciable in the 1872 essay on the principle of the conservation of work. In particular, here Mach traces the idea that it is impossible to create work out of nothing, or 'the principle of excluded *perpetuum mobile*', back to Simon Stevin, Galileo Galilei, Christiaan Huygens, etc., on the grounds that it is an incarnation of the principle of causality. This, in turn, is a sort of regulative, pre-mechanical principle that is applied everywhere the attempt to investigate nature is made (Guzzardi 2014, Wegener, this volume).

I think, however, that the most important aspect of Mach's paean to history is that it helps us to understand – adapting Giere's phrase – the conceptual relation of his philosophical views on sensations to the historical investigations he carried out. This is the argument of the next section.

Was Ist Metaphysik? The History of Science and *The Analysis of Sensations*

The Analysis of Sensations (Mach 1886/1959) is perhaps the less historical of Mach's epistemological books. Examples from the history of science appear only sparingly therein and, when taken into consideration, they play but a marginal role. The book famously opens with a chapter titled 'Anti-metaphysical Premises' ('*Antimetaphysische Vorbemerkungen*'), where Mach presents his ontology made out of neutral elements, in themselves neither physical nor psychical, which constitute, however, both the physical and the psychical. This conception would sometimes be regarded as a kind of phenomenalism, sometimes being called 'neutral monism', associating Mach with thinkers such as William James and Bertrand Russell.[7]

In contrast with a widespread view, textual evidences show that, in *The Analysis of Sensations*, Mach does not give a literal, crystal-clear definition of what should be understood as *metaphysics*, *metaphysical*, and related expressions. Most of the time, Mach uses these terms somewhat generically, in a manner that seems to imply something that is not grounded in experience. For example, Mach (1886/1959, p. 61) sees the difference between his conception and Gustav Theodor Fechner's psycho-physical parallelism in the fact that his own 'has by no means a metaphysical substrate [*Untergrund*] but corresponds only to the generalised expression of experiences' (note that in the current English translation *Untergrund* is rendered as 'background'). A further example is Mach's (1886/1959, p. 171) characterisation of will, which should not be understood as 'any special psychical or metaphysical agent', but must be 'explained by means of the physical forces of the organism alone'.

A more general formulation can be found in the 1902 preface to the fourth edition of *The Analysis of Sensations*: the idea that 'science ought to be confined to the compendious representation of the actual [*Tatsächlichen*], consequently leads to the elimination of all superfluous assumptions which cannot be controlled by experience, and, above all, of all assumptions that are *metaphysical* (in Kant's sense)' (Mach 1886/1959, p. xii, emphasis in original). The reference to Kant is, to a certain extent, clarifying. Autobiographical fragments partly contained in Mach's works and partly in unpublished notes show that Kant, particularly through the *Prolegomena to Any Future Metaphysics*, had a strong influence on the young philosopher–scientist (Blackmore 1972, pp. 10–11, 289; Banks 2003, pp. 24–25). In the *Prolegomena*, recalled in a footnote to *The Analysis of Sensations* (Mach 1886/1959, p. 30), Mach could read that the sources of metaphysical cognition 'cannot be empirical. The principles of such cognition . . . must therefore never be taken from experience' (Kant 2004, p. 15). There seems to be no reason why, in his mature years, Mach should reject the characterisation he learned

---

[7] The most important supporter of the neutral monist interpretation of Mach in recent times is Banks (2003, 2014). For an overview of the neutral monist conception, see Stubenberg (2018). For a comparative appraisal of the two readings, see Preston (this volume).

about in his earliest philosophical excursions (Blackmore 1972, p. 33) – but of course, he attached to it a pejorative connotation that was alien to Kant.

As is well known, the Vienna Circle and its epigones, on whom greatly depends the received view of Mach's epistemology, stuck to assertions of this kind for their preferred image of him as the champion of an 'anti-metaphysical' phenomenalism, having sense-data (sensations) as its proper ontology. After all, this seemed quite consistent with Mach's aversion to non-observable entities such as Isaac Newton's absolute space and time, as well as atoms and forces, which he wished to be eventually eliminated from science. This phenomenalist–sensualist interpretation of Mach's epistemology may underlie Giere's reference to 'the epistemological role of sensations' and its relation, if any, with his 'physical ... investigations'. In any case, a harsh criticism of this image of Mach, which has more in common with the search for a noble 'legendary father' than with an objective historical reconstruction, can already be found in Paul Feyerabend (1970, 1984). (Of course, this does not exhaust all the possibilities of a *phenomenalist* reading of Mach; see Preston, this volume.)

More recently, the legend has been challenged in many ways. Even if the influence of Mach on the adherents of the Viennese movement is beyond question (Stadler 1997, 2007, this volume), significant differences must be taken into account as well. Some scholars have emphasised that Logical Positivists were not committed to an updated version of empiricism, but rather aimed to develop a common formal structure for scientific theories based on the analysis of language (Richardson 1997; Friedman 1999; Mormann 2007; Banks 2013). Others have highlighted that Mach was particularly committed to his elements being neutral (i.e. neither physical nor psychical in themselves). Feyerabend (1984, pp. 10–11) argued that Mach's theory of the elements was 'not a necessary boundary condition of research', but a particular *theory*, formally analogous with the atomic hypothesis; moreover, that 'elements are sensations "but only insofar" as we consider their dependence on a particular complex of elements, the human body; "they are at the same time physical objects, namely insofar as we consider other fundamental dependencies"'. And according to Banks (2003, p. 46), the theory of the elements can even be considered as part of 'a strategy for the future development of science'; as he wrote elsewhere, it was – in a strong, philosophically traditional sense – an example of a 'good metaphysics, even if Mach would have rejected the term', which stands at odds with the Vienna Circle's ideas (Banks 2013, pp. 58–64; see also Banks 2014, this volume).

To the criticisms only briefly sketched here, I would add that the neopositivists overlooked one essential feature of Mach's anti-metaphysical discourse, which is not explicitly included in *The Analysis of Sensations*, but can be understood in light of the above-quoted introduction to the essay on the principle of the conservation of work. As I mentioned, here metaphysical concepts are described as those that have gone through an unconscious process of intellectual solidification, which hinders us from tracing them back to their origin; perhaps they are available for use or we even continue to make use of them as obvious, but 'we have forgotten how we reached them' (Mach 1872/1911, p. 17). On the other hand, history can preserve us from this kind of dogmatic solidification of concepts, for one can remain 'in view of the path by which one has come', and this prevents one from 'losing one's footing, or coming into collision with facts'. In other words, the history of science can have, and in fact has in the 1872 essay on the principle of the conservation of work, an anti-metaphysical function.

Now, what Mach calls analysis in *The Analysis of Sensations* (i.e. the resolution of complex concepts such as matter, bodies, the ego, etc., into their composite elements) has an analogous function, even if it proceeds via very different methods. Still, it possesses the same virtue of reminding us that the concepts we use emerge from certain physical–psychological relations among more elementary materials, which are subject, in turn, to further analysis. It is not my aim to proceed to a more or less complete exposition of Mach's conception in this regard, nor do I intend to assess its benefits and flaws. I just want to emphasise that there is a double parallel here with the idea of history Mach advanced in the essay on the principle of the conservation of work. On the one hand, like the 'historical studies' in that context, the kind of analysis undertaken in *The Analysis of Sensations* reveals a sort of Heraclitean flux underlying apparently permanent compounds – concepts such as matter, the body, the ego, etc. On the other hand, it also helps to explain (and to suggest) where they may find their legitimate use and how to polish from them possible metaphysical residues.

Let us take as an example Mach's famous analysis of the *Ich*, the ego. For the common understanding, this is something unchangeable. However, its apparent stability dissolves as soon as, after closer inspection, the ego manifests itself as a '*relatively* permanent ... complex of memories, moods, and feelings, joined to a particular body – the human body' (Mach 1886/1959, p. 3, translation amended, emphasis in original). The elements that form the ego, Mach continues, are transient events that do not repeat themselves identically but only gradually change, and this ensures our ability to recognise ourselves, as well as others, as time passes and leaves its marks on us. Of course, there is nothing wrong with considering the ego such a '*practical* unity, put together for purposes of provisional, orientating consideration' (Mach 1886/1959, p. 28, emphasis in original); we have reason to act accordingly in our everyday lives, since 'the supposed unities "body" and "ego" are only makeshifts, designed for *provisional* orientation and for definite *practical* ends' (Mach 1886/1959, p. 13, emphasis in original). There is nothing metaphysical here, insofar as we can explain this tendency as a development of our natural needs (in modern terms, as far as we can explain this by means of a naturalised epistemological approach). What is metaphysical is rather to consider the ego as an unanalysed concept we can *obviously* use for research purposes – what is metaphysical is the solidified version of the ego in philosophy and psychology (in particular, according to Mach, in Kant's philosophy: see Banks 2013, p. 65); in other words, the belief that the ego really has an unchanging existence.

With this, I am not arguing that the historical and the analytical approaches are really the same thing or that one approach is (logically, conceptually, historically, or in whatever manner) dependent on another – for example, that Mach's historical-critical mode of examining the emergence and development of scientific notions and practices derives from his analytical attitude in perceptual psychology, or vice versa. Following a hint in the above-mentioned footnote of *The Analysis of Sensations*,[8] I rather think that the two approaches originated independently, but possibly from the same need, in the early stages of Mach's career; then they started to interact and reinforce one another, probably already in the early 1860s, as he was working on the *Compendium der Physik für Mediciner* (1863) and while dealing with the *Vorträge über Psychophysik*.[9]

In other words, for Mach, the historical-critical point of view and the analysis developed in his physiological–psychological studies on sensations were the two barrels of the same anti-metaphysical gun.

---

[8] Footnote 1 of p. 30 (in the current English version) gives a well-known and often-quoted autobiographical sketch of Mach's early interest in (and dismissal of) Kantian philosophy, which is brought into connection with his reflections on 'the world with my ego [appearing . . .] as one coherent mass of sensations, only more strongly coherent in the ego'. This memory also includes a frequently overlooked passage: 'Only by alternate studies in physics and in the physiology of the senses, and by historical-physical investigations (since about 1863), and after having endeavoured in vain to settle the conflict by a physico-psychological monadology (in my lectures on psychophysics . . .), have I attained to any considerable stability in my views' (Mach 1886/1959, p. 30). On this, see also Banks (2003, esp. pp. 23–26), who dates the end of the 'intense intellectual "struggle" during the 1860s, that led Mach to his theory of elements', to late 1871, when he was finishing the essay on the principle of the conservation of work.

[9] The *Vorträge über Psychophysik* (*Lectures on Psychophysics*) were, for Mach, a kind of intellectual laboratory where he attempted a first analytical, elements-based point of view (later partly dismissed or reformulated) on sensations; but he also laid the groundwork for his future critique of the ego, and not only that (see Banks 2003, pp. 19–20, 31–32). On the other hand, the *Compendium* closes with a remark that will be crucial in later historical works, such as the essay on the history of the principle of work and *The Science of Mechanics*: the mechanical conception of the world, Mach maintains, might have exhausted its possibilities and perhaps, in the near future, the whole edifice of science will experience a 'global re-organization' [*gänzlichen Umgestaltung*], possibly based on new foundations from a more profound understanding of electrical and magnetic phenomena (Mach 1863, pp. 271–272; see also Mach 1872/1911, p. 86).

Mach on the Economy of Thought and the History of Science

Let us return to Hiebert's assessment of Mach's epistemology. The other important point mentioned by him (Hiebert 1970, p. 201), but – according to Giere (1973, p. 183) – not adequately examined, is the relation, if any, of Mach's doctrine of the economy of thought (undoubtedly, one of the best-known features of his conception) to his historical-critical approach. In this and the following section, I attempt to fill such a gap and show that the relation, which I claim exists, completes the picture I have tried to sketch so far.

Starting with his 1882 lecture *Über die ökonomische Natur der physikalischen Forschung*, where he first exploited material that was later reshaped and expanded in the various editions of *The Science of Mechanics*, Mach tended to present his theory of science in the light of the economy of thought, and this, in turn, as a result of human evolution. The most concise and clearest formulation of this view probably appears in Mach's final reply of 1910 to Planck's attack, written as a kind of intellectual testament, *Die Leitgedanken meiner naturwissenschaftlichen Erkenntnislehre und ihre Aufnahme durch die Zeitgenossen*:

> Any abstract conceptual expression, summarizing the properties of facts, any substitution of a number table with a formula or a composition rule, i.e. its law, any explanation of a new fact through others that are already known, can be conceived as an economical achievement. The broader and more detailed the analysis of the scientific methods – the systematic, organizing, simplifying, logical-mathematical construction – the more vividly one recognises the scientific doing [*Tun*] as economical.... Darwin's ideas ... were already operating in my lectures at Graz in 1864–1867 and found expression through the conception of competition among scientific ideas as struggle for existence, as the survival of the fittest. Such a view is not at odds with the economical conception, but it supplements and can be combined with this, resulting in a biological-economical description. Expressed as briefly as possible, the following appears as the task of scientific knowledge: *the adaptation of thoughts to facts and to each other*. Every favourable biological process is a process of self-preservation; as such it is likewise an adaptive process, and more economical [*ökonomischer*] than a process which would be disadvantageous to an individual. All favourable knowing processes are special cases or parts of convenient [*günstiger*] biological processes.
>
> (Mach 1910, pp. 225–226, my translation, emphasis in original)

The opening paragraph of chapter 10 of *Knowledge and Error*, entitled 'Adaptation of thoughts to facts and to each other', may serve as a specification of what Mach meant by connecting biology and thought economy into a unified pattern of explanation. Here, Mach (1905/1976, p. 120) remarks that 'ideas [*Vorstellungen*] gradually adapt to facts by picturing them with sufficient accuracy to meet biological need'. At the earliest stages of development, we arguably have very simple answers to basic needs, so that 'the accuracy goes no further than required by immediate interests and circumstances'; but even at this primordial level, results might not be univocal, for interests and circumstances 'vary from case to case, [therefore] the adaptive results do not quite match [*stimmen die Anpassungsergebnisse verschiedener Fälle nicht genau untereinander überein*]'. So far so good for the biological part of the explanation; but why should this adapting process be conceived as having an economical nature? Mach's answer is that an important economical feature, involving the least possible expenditure of resources, is simply embedded in biology, for 'biological interest further leads to mutual corrections of the pictures to adjust the deviations in the best and most profitable way'. Shortly after, he also emphasises that 'economizing, harmonizing and organizing of thoughts are felt as a biological need far beyond the demand for logical consistency' (Mach 1905/1976, p. 128).

What occurs in science for Mach is only more complicated but not essentially different from this very natural (in modern terms, naturalised) process of knowing.[10] Let us come back to the *Leitgedanken*. History of science, he continues here, is important because it proves that science and technology have developed from their 'instinctual', need-based beginnings up to the highest conceptual formulations, in accordance with the guiding principle of thought economy (in the sense specified above). In particular, Mach (1910, p. 228) claims that in his 'historical studies on the science of mechanics [Mach 1883/1960] and the theory of heat [Mach 1896/1986], the biological-economical conception of the process of knowledge has made ... the understanding of the scientific development immensely easier'. Based on case studies that are easily traceable in the above-quoted works, he argues that nothing but the economy of thought has guided scientists towards the generalisation of many laws of statics into the principle of virtual displacement and that of virtual work, summarising many laws of statics (Mach 1883/1960, pp. 49–77), or replacing Johannes Kepler's three laws with the unique gravitational law of Newton (Mach 1883/1960, pp. 187–189), or again dividing the notion of heat into two concepts – temperature and quantity of heat – and then introducing more general notions such as energy and entropy (Mach 1896/1986, pp. 146–281).

Clearly, I do not intend to discuss whether Mach was right or not in claims of this kind; as important as it might be, this is an empirical question that historians of science can debate in light of more thorough information and

---

[10] 'These processes at first occur quite unintentionally and without clear consciousness. When we become fully conscious what we find within us is already a fairly complete world picture. Later, however, we gradually go over to continuing the processes with clear deliberation, and as soon as this occurs, enquiry sets in' (Mach 1905/1976, p. 120).

careful examination. Rather, I want to emphasise that, if my reconstruction so far is right, his stance towards economy of thought seems to highlight a tension with his earlier conception of history as a Heraclitean river that neither has an assigned end or *telos* nor follows a predetermined path.

In the 1872 essay, Mach had proclaimed that 'history has made all; history can alter all'. Now, with the principle of the economy of thought he seems to imply that the history of science shows the operating of a standard of progress – perhaps even a teleological principle – in the development of knowledge, for a scientific discipline is more advanced the more comprehensive and economical its laws are (i.e. the fewer the number of basic elements it needs). So, Mach (1883/1960, p. 485) acknowledges a difference in the developmental degree of, say, the highly mathematised parts of physics and what in his times were called *beschreibende Naturwissenschaften* (descriptive natural sciences), such as botany, zoology, meteorology, geography, etc.: 'In the details of science, its economical character is still more apparent. The so-called descriptive sciences must chiefly remain content with reconstructing individual facts. In sciences that are more highly developed, rules for the reconstruction of great numbers of facts may be embodied in a *single* expression.' Thus, in science as a whole there *is* a direction, which is determined by the economy of thought: science can be considered as a 'minimal problem' insofar as it strives towards 'the completest possible presentment of facts with the least possible expenditure of thought' (Mach 1883/1960, p. 490).

Thought Economy and History Reconciled

I do not think that this is a real tension, though, and not because the polemics against Planck might have forced Mach to emphasise, and maybe overestimate, the meaning of the principle of thought economy in his epistemology. Rather, it seems to me that the economy of thought and the history of science have emerged and remained, throughout the development of his intellectual path, on different levels.

Thought economy already played a crucial role in the 1872 essay on the principle of the conservation of work, which – according to the 1909 preface – recommended 'an arrangement ... of facts' consistent with it (*'eine denkökonomische Ordnung des Tatsächlichen'*; Mach 1872/1911, p. 9). There are primarily two epistemic tasks, according to Mach (1872/1911, p. 55), that are economical in nature: first, the 'collection of as many facts as possible in a synoptical form' (e.g. in a law); and second, 'to resolve the more complicated facts into as few and as simple ones as possible', a process that Mach calls *erklären* (i.e. to explain). By iterating the latter procedure, sooner or later a point is reached where the facts cannot be further resolved. As a consequence, from this notion of explanation it follows that the 'simplest facts, to which we reduce the more complicated ones, are always unintelligible in themselves, that is to say, they are not further resolvable'.

Now, Mach continues, although every researcher strives to pursue the most economical way to explain phenomena, not everyone would necessarily agree at what simplest, not-further-intelligible facts one has to stop:

> It is only, on the one hand, an economical question, and, on the other, a question of taste, at what unintelligibilities we stop. People usually deceive themselves in thinking that they have reduced the unintelligible to the intelligible. Understanding consists in analysis alone; and people usually reduce uncommon unintelligibilities to common ones. They always get, finally, to propositions of the form: if A is, B is, therefore to propositions which must follow from intuition, and, therefore, are not further intelligible. What facts one will allow to rank as fundamental facts, at which one rests, depends on custom [*Gewohnheit*] and on history.
>
> (Mach 1872/1911, pp. 55–56)

Globally, the economy of thought is ubiquitous and guides human behaviour. But locally, it can be realised in many different ways, depending on individual inclinations and material conditions. What does appear economical *to me* will not necessarily do so *to you*, because our 'histories' (the material conditions that have made of us what we actually are) might be as different as possible. In other words, the economy of thought is not a *feature of history*, but to be sure history, most of all the history of science, reveals it as a *feature of human cognitive resources*, a result of human evolution which is partly shared with other living beings (see also Mach 1905/1976, pp. 51–64). For this reason, Mach maintains that, as we have seen in the previous section, there is no fundamental difference, no rigid boundary, between instinctual, ordinary thinking and the most advanced and abstract forms of scientific theorising – which is a *leitmotiv* of *Knowledge and Error*.

If understood in this manner, the respective roles of history and thought economy were not things about which Mach changed his mind.[11] A couple of examples from his later works may be illuminating in this regard. In *Principles of the Theory of Heat* (1896), he devotes the entire chapter XXVI to 'The economy of science'. Among a variety of historical cases, he comments on

---

[11] I do not intend to discuss here what is the real origin of the doctrine of the economy of thought in Mach's conception. Banks (2004) argued that it is 'philosophically rooted' in neutral monism, his more general philosophical view. To be honest, I do not find the very notion of 'philosophical roots' clear, and I think that a documentary investigation would shed more light on the problem. Unfortunately, as Banks also remarked, Mach's notebooks preserved at the Archives of the Deutsches Museum in Munich only begin with 1871; but useful indications might emerge from his correspondence, as well as through a careful analysis of his relationship with Emanuel Herrmann, which he often quotes as his source regarding the idea of thought economy.

Kepler's approximate law of refraction. He observes that, by the time it was formulated in the *Paralipomena to Witelo* (1604), Kepler 'had everything in hand for setting up the dioptrics of Carl Friedrich Gauss. Nevertheless, even after Kepler's time, many did not do this.' According to Mach, only Gauss succeeded in finding a good dioptrical theory, for he dramatically changed the approach: he 'set up the two principal planes and the two principal foci once for all, and did not concern himself any more about the separate refracting surfaces. Thus it is not correct here to say that with the given means only one final result can be attained in one way' (Mach 1896/1986, p. 361).

It seems to me that here Mach is suggesting something like this: even if we concede that, say, both Kepler and Gauss apparently started from the same theoretical premises, and even if they applied the same 'economical' thought processes, there are, indeed, a myriad of other historical conditions that may make, and in fact made, their epistemic pathways different, so that one cannot say a priori, as if science were a purely deductive logical system, what will happen next. Such conditions are somehow included in the 'facts' to which thoughts must adapt. Let us suppose that two scientists have two different explanations ('complexes of representations' might be an expression more in line with the Machian language) for a single complex of facts; even in this case, they might agree or not depending on several boundary conditions, such as conscious or unconscious inclinations, different background knowledge, etc. Similarly, suppose that one scientist gives a different explanation of the same fact on different occasions; they can do this because the conditions were different in one or another respect. Again, it is matter of relatively changed conditions whether one theory – which is nothing but a 'mutual adaptation of thoughts' (Mach 1905/1976, p. 120) – conflicts with another, and which of them will eventually prevail: adaptation of thoughts to facts and to one another 'can ... proceed in *different* ways. Of two conflicting thoughts, the one that seems less important and reliable at the time must tolerate modification to the benefit of the other' (Mach 1905/1976, p. 127, emphasis in original).[12]

---

[12] The generally good English translation of *Erkenntnis und Irrtum* is misleading here, therefore I have changed it. The passage in German reads: '*Die Anpassung kann also verschiedener Weise erfolgen. Jener Gedanke unter zwei widerstreitenden, den man zurzeit für weniger wichtig und vertrauenswürdig hält, muss sich die Modifikation zu Gunsten des anderen gefallen lassen.*' This is the current English translation: 'Adaptation can thus proceed in different ways: of two conflicting thoughts, the one that seems less important and reliable at the time must suffer modification by the other.' A minor inaccuracy is that the word *verschiedener* ('different') is italicised in the original text but not in the translation. A more important concern is punctuation: in the English text, the colon, meaning that what follows is a clarification of the previous sentence, is arbitrarily introduced – instead, in the original text, the sentences are separated by a full stop, suggesting that there is juxtaposition, not clarification. Finally, the phrase 'must suffer modification by the other' can ambiguously be interpreted as if the one theory should be modified by the other, whereas this is certainly not the meaning of the German text.

My second example to substantiate the stability of Mach's views regarding the relation of economy of thought to history comes from chapter XVII of *Knowledge and Error*, entitled 'Pathways of Enquiry'. Here, Mach begins by briefly recalling the history of some 'scientific astronomic ideas': the theories of celestial motions, the method of epicycles, the fate of geocentric systems, the development of the heliocentric system, and so on, up to Newton's gravitational theory. 'This development unmistakably shows', Mach comments, 'the increasingly accurate mental reconstruction of astronomic facts' (1905/1976, p. 213). However, he adds two important remarks: first, 'the process is not finished and may well never be [*Der Prozeß ist nicht abgeschlossen und wohl auch nicht abschließbar*]'. Second, even if in hindsight this can be recognised as an economical feat (indeed, 'we see the mental reconstruction or description becoming ever simpler and more economical'), the long span of time needed for the most important achievements of astronomy suggests that these 'do not rest on inferences of the moment obtainable by means of some formula'.

For Mach (1905/1976, p. 222), this can be explained in light of the fact that the pathways of enquiry are far less linear than they retrospectively appear. Scientific developments of this sort 'begin with very primitive ideas in the depths of prehistory, but are by no means concluded today … Knowledge is gained on very tortuous paths and the single steps, though conditioned by prior ones, are partly determined by purely accidental physical and mental circumstances as well'. Even if the same 'economical' thought processes characterise different researchers through the ages, such circumstances are largely independent of us and do not repeat exactly. So Mach concludes: 'Schematizing the cognitive stages may perhaps benefit further enquiry when similar situations recur, but there can be no widely effective instructions for enquiry by formula. Nevertheless, it remains always correct that we aim at adapting thoughts to facts and to each other' (Mach 1905/1976, p. 223).

In both examples I extrapolated, from *The Principles of the Theory of Heat* and *Knowledge and Error*, respectively, Mach clearly adopts an economical (or biological-economical) point of view and uses it for interpreting the historical case at hand. This is consistent with the above-quoted claim from the *Leitgedanken* according to which such a perspective made easier for him the understanding of scientific development in many fields. But it is also evident, I think, that in neither example is he implying a standard of progress or a teleological principle that guides history. Rather, Mach suggests that the one single cognitive process – namely, the economy of thought – may result in a plurality of contingent views, for the principle only requires that thoughts adapt to *ever-changing* conditions. They change endlessly because they are part of the Heraclitean flux of history, in accordance with his claim in the 1872 essay on the principle of the conservation of work.

Concluding Remarks

In this chapter, I defended a popular thesis in Mach studies, namely that there is no hiatus but consonance between his approach to history (primarily the history of science) and his epistemology. They are not simply different, juxtaposed parts of his work, but two complementary, intertwined expressions of a unique conception. Indeed, his approach may be considered an early example of what we now call an integrated history and philosophy of science, taking for granted that crucial to this is something more than merely the idea of a 'history of science into which some philosophy of science may enter . . . [or a] philosophy of science into which some history of science may enter'.[13] The way I argued for this thesis, however, is probably not as popular, for my focus was not on the contents of the various stories Mach reviews, which would unveil the epistemological use of history, in accordance with Hiebert (1970) and Feyerabend (1984).

Of course, for Mach, history is *also* a 'reservoir of "modes of thinking"' (using the above-quoted phrase of Heidelberger and Reiter). In this regard, Feyerabend cites a telling passage from *Knowledge and Error*: the most noticeable personalities of the past – scientists such as Copernicus, Stevin, Galileo, Gilbert, Kepler, or Newton, to mention some heroes in the Machian pantheon – 'furnish examples of the greatest successes of scientific research that teach us without pomp what were the leading motives of enquiry'. And, since such examples make 'familiar to us in the simplest way' the methods employed (e.g. 'the methods of physical and thought experiments, the principles of simplicity and continuity and so on'), history may even be viewed as a collection of inspiring and partially applicable models of behaviour that can guide researchers in future investigations into unexplored domains (Mach 1905/1976, p. 165; see Feyerabend 1984, p. 16).

However, more than considering this aspect of the *use* of history for philosophical purposes, I was interested in emphasising a *structural* level of Mach's discourse on history. I highlighted that both 'historical studies' as defended in the chapter on the principle of the conservation of work and the deconstruction of experience into its constituting 'elements' conducted in *The Analysis of Sensations* have the same anti-metaphysical function (but, of course, neither of them has *only* an anti-metaphysical function). Although they fulfil the anti-metaphysical task differently (and have emerged independently in Mach's intellectual development), both historical reconstructions and experience deconstruction remind us that the concepts we arrived at and are currently using are not given forever. Historical analyses and the analysis of

---

[13] The quote comes from Springer's advertisement page for Stadler (2017; available from www.springer.com/gp/book/9783319532578). I have not been able to trace this characterization of the subject in the book itself.

sensations epitomise different types of enquiry, but they both express the same prophylactic attitude against the dogmatic solidification of practical, and provisional, employment of concepts into their unalterable versions. These versions *are* metaphysical because – paraphrasing Mach (1872/1911, p. 17) – they have lost their footing in the ever-changing reality from which they emerge, thereby coming into collision with facts.

A qualifying point for this double-sided approach is that, in both cases, the relevant constituent materials are transient events. This can be extrapolated, respectively, from Mach's theory of the elements as advanced in the 'Anti-metaphysical premises' to *The Analysis of Sensations* (Banks 2003, this volume), as well as from the image of Heraclitus' river of history in Mach (1872/1911). Of course, there is a reason for this. Mach maintains that the entirety of nature is a Heraclitean, ever-changing flux of elements and qualities. Both the historical facts and the element compounds are concatenations of *natural* events, and 'nature has but an individual existence; nature simply *is*' (Mach 1883/1960, p. 483, emphasis in original; note that the long English sentence divided by a semicolon renders the very concise, meaningful expression: '*Natur ist nur* einmal *da*'). Why are things there 'just once' and do not repeat themselves? For Mach, there is no answer to this – this is simply the way things are (one might say, this is an unintelligible – i.e. no further resolvable – simplest fact in the sense of Mach 1872/1911, p. 55). Things would repeat themselves only if the same conditions would exactly repeat (for 'in nature only that and so much happens as can happen, and ... this can happen in only *one* way'; Mach 1896/1986, p. 360, emphasis in original). But Mach would have said that this is just abstraction; in truth, the same conditions never repeat, because all is endowed with individual existence only.[14] Therefore, there is neither contradiction nor tension between Mach's Heraclitean view of nature and his commitment as a physicist to the idea of phenomenal repetitiveness, as far as equal conditions are involved. (On this, I cannot agree with Banks 2004, p. 25; see also Banks 2003, p. 239.)

Insofar as history is a natural fact, a chain of unrepeatable, contingent events that have produced a determined result, it does not follow a predetermined path. However, directions may locally appear *within* history, as a result of contingent factors. This explains why some historical events that have taken place (e.g. in the development of one or another scientific discipline) may be economically driven (in the sense of the economy of thought), without the whole of history being so. In the contingency of historical events, humans have been determined by evolution to act economically (one could say, to act

---

[14] 'Recurrences of like cases in which *A* is always connected with *B*, that is, like results under like circumstances, that is again, the essence of the connection of cause and effect, exist but in the abstraction which we perform for the purpose of mentally reproducing the facts' (Mach 1883/1960, p. 483).

economically *is* their own contingency). Case by case, they have been determined by given circumstances to economically react by adapting their behaviour to their needs. That is to say, different humans, being determined by different, unrepeatable circumstances to a certain result of their course of action, may choose different adaptive strategies; 'side by side and following one another, [they] will effect adaptation in different ways. One will overlook this, another that' (Mach 1896/1986, pp. 360–361).

And so, what is the conceptual relation of the principle of the economy of thought to Mach's historical approach, to put it as Giere (1973, p. 283) might? To be sure, the economy of thought is not the guiding hand of the guiding hand of history, which would make it something like a metaphysical substance, a solidified version of a natural fact. Rather, it is one *content* of the contingent history of (human) evolution, a cognitive capacity that manifests itself historically. And contingent as it is, we cannot even be sure whether it can be saved or, like the ego in *The Analysis of Sensations*, will ultimately be '*unrettbar*'.

References

Banks, Erik C. 2003. *Ernst Mach's World Elements: A Study in Natural Philosophy.* Kluwer.

2004. 'The Philosophical Roots of Ernst Mach's Economy of Thought', *Synthese* 139, 23–53.

2013. 'Metaphysics for Positivists: Mach versus the Vienna Circle', *Discipline filosofiche* 23: 57–77.

2014. *The Realistic Empiricism of Mach, James, and Russell: Neutral Monism Reconceived.* Cambridge University Press.

Blackmore, John T. 1972. *Ernst Mach: His Work, Life, and Influence.* University of California Press.

Blackmore John T. (ed.) 1992. *Ernst Mach – A Deeper Look: Documents and New Perspectives.* Kluwer.

Feyerabend, Paul K. 1970. 'Philosophy of Science: A Subject with a Great Past', *Minnesota Studies in the Philosophy of Science* 5: 172–183.

1984. 'Mach's Theory of Research and Its Relation to Einstein', *Studies in History and Philosophy of Science* 15: 1–22.

Friedman, Michael 1999. *Reconsidering Logical Positivism.* Cambridge University Press.

Giere, Ronald N. 1973. 'History and Philosophy of Science: Intimate Relationship or Marriage of Convenience?', *British Journal for the Philosophy of Science* 24: 282–297.

Guzzardi, Luca 2002. 'Teorie stravaganti di un redattore invadente. Un positivista italiano nel dibattito scientifico europeo', *Intersezioni* 3: 419–440.

2014. 'Energy, Metaphysics, and Space: Ernst Mach's Interpretation of Energy Conservation as the Principle of Causality', *Science & Education* 23: 1269–1291.

Hiebert, Erwin N. 1970. 'Mach's Philosophical Use of the History of Science', *Minnesota Studies in the Philosophy of Science* 5: 184–203.

Kant, Immanuel 2004 *Prolegomena zu einer jeden künftigen Metaphysik, die als Wissenschaft wird auftreten können.* Johann Friedrich Hartknoch, 1783. Edited and translated by Gary Hatfield as *Prolegomena to Any Future Metaphysics That Will Be Able to Come Forward as Science.* Cambridge University Press.

Lessing, Gotthold Ephraim 1970. *Werke, vol I.* Hanser.

Mach, Ernst 1863. *Compendium der Physik für Mediciner.* Braumüller.

　　1872/1911. *Die Geschichte und die Wurzel des Satzes von der Erhaltung der Arbeit.* J. A. Barth). Translated by P. E. B. Jourdain as *History and Root of the Principle of the Conservation of Energy.* Open Court, 1911.

　　1883/1960. *The Science of Mechanics: A Critical and Historical Account of its Development.* Open Court.

　　1886/1886/1959. *Beiträge zur Analyse der Empfindungen.* Gustav Fischer, 1886. Further German editions in 1900, 1902, 1903, 1906, 1911, 1918, and 1922. Translated by C. M. Williams and S. Waterlow as *The Analysis of Sensations and the Relation of the Physical to the Psychical.* Open Court, 1914 (reprinted Dover Publications, 1886/1959).

　　1896/1986. *Die Prinzipien der Wärmelehre, historisch-kritisch entwickelt.* J. A. Barth, 1896. Further German editions in 1900 and 1919. Translated as *Principles of the Theory of Heat, Historically and Critically Elucidated,* ed. B. F. McGuinness. D. Reidel, 1986.

　　1905/1976. *Erkenntnis und Irrtum. Skizzen zur Psychologie der Forschung.* J. A. Barth, 1905. Further German editions in 1906, 1917, 1920, and 1926. Translated by T. J. McCormack and P. Foulkes as *Knowledge and Error: Sketches on the Psychology of Enquiry.* D. Reidel, 1976.

　　1910. 'Die Leitgedanken meiner naturwissenschaftlichen Erkenntnislehre und ihre Aufnahme durch die Zeitgenossen', *Scientia* 4: 225–240.

　　1921/1926. *Die Prinzipien der Physikalischen Optik. Historisch und erkenntnispsychologisch entwickelt.* J. A. Barth, 1921. Translated by John S. Anderson and A. F. A. Young as *The Principles of Physical Optics: An Historical and Philosophical Treatment.* Dover Publications, 1926.

　　2005. *Scienza tra storia e critica,* ed. L. Guzzardi. Polimetrica.

　　2016. *Die Prinzipien der Wärmelehre: historisch-kritisch entwickelt,* Neudruck der 2. Auflage (1900). Xenomoi.

Mormann, Thomas 2007. 'The Structure of Scientific Theories in Logical Empiricism', in A. Richardson and T. Uebel (eds.), *The Cambridge Companion to Logical Empiricism.* Cambridge University Press, pp. 136–162.

Planck, Max 1909. *Die Einheit des Physikalischen Weltbildes: Vortrag gehalten am 9. Dezember 1908 in der naturwissenschaftlichen Fakultät des Studentenkorps an der Universität Leiden.* Hirzel.

Richardson, Alan W. 1997. *Carnap's Construction of the World: The Aufbau and the Emergence of Logical Empiricism.* Cambridge University Press.

Stadler, Friedrich 1997. *Studien zum Wiener Kreis: Ursprung, Entwicklung, und Wirkung des logischen Empirismus im Kontext.* Suhrkamp.

2007. 'The Vienna Circle: Context, Profile, and Development', in A. Richardson and T. Uebel (eds.), *The Cambridge Companion to Logical Empiricism.* Cambridge University Press, pp. 13–40.

(ed.) 2017. *Integrated History and Philosophy of Science: Problems, Perspectives, and Case Studies.* Springer.

Stubenberg, Leopold 2018. 'Neutral Monism', *The Stanford Online Encyclopedia of Philosophy,* updated autumn 2018. Available from https://plato.stanford .edu/archives/fall2018/entries/neutral-monism/

Wolters, Gereon 1987. *Mach I, Mach II, Einstein und die Relativitätstheorie.* de Gruyter.

10

Ernst Mach and the Vienna Circle

A Re-evaluation of the Reception and Influence of His Work

FRIEDRICH STADLER

Mach in the Manifesto *The Scientific World-Conception: The Vienna Circle*

In the programmatic manifesto of the Vienna Circle at the beginning of its public phase (1929), mainly written by Rudolf Carnap, Hans Hahn, and Otto Neurath with the help of Herbert Feigl and Friedrich Waismann, we recognise a general appreciation of Ernst Mach, but also a cautious distancing from his psychological and historical foundations of the sciences. Independently, the name of the Verein Ernst Mach (Ernst Mach Society) headed by Moritz Schlick from 1928 to 1934 documents the identification with Mach's life work as a symbol for the popularisation of scientific philosophy and philosophy of science (Stadler 1992; Stadler and Uebel 2012). In this booklet, which appeared as a publication of the Ernst Mach Society dedicated to Schlick, the following paragraph describes the historical background:

> Thanks to this spirit of enlightenment, Vienna has been leading in the scientifically oriented popular education courses ... In this liberal atmosphere lived Ernst Mach (born 1838) who was in Vienna as a student and as privatdozent (1861–64). He returned to Vienna at an advanced age when a chair of the philosophy of the inductive sciences was created specially for him (1895). He was particularly intent on cleansing empirical science, and especially physics, of metaphysical notions. We recall his critique of absolute space which made him a forerunner of Einstein, his struggle against the metaphysics of the thing-in-itself and of the concept of substance, and his investigations of the construction of scientific concepts from ultimate elements, namely sense data. In some points the development of science has not vindicated his views, for instance in his opposition to atomic theory and in his expectation that physics would be advanced through the physiology of the senses. The essential points of his conception however were of positive use in the further development of science.
>
> The activity of the physicists Mach and Boltzmann in a philosophical capacity makes it conceivable that there was a lively dominant interest in the epistemological and logical problems that are linked with the

foundations of physics. These problems concerning foundations also led toward a renewal of logic.

(Stadler and Uebel 2012, p. 72)

In this general appraisal we recognise the role and function of Mach, and in part Boltzmann, as a pioneer and predecessor of the 'scientific world-conception', but we may also anticipate their limitations with regard to symbolic logic, because the Circle rejected any form of psychologism. And regarding the joint mission of the Circle around Schlick, we read:

> The Vienna Circle does not rest content with collective work as a closed group. It is also trying to make contact with the active movements of the present, so far as they are well disposed toward the scientific world-conception and turn away from metaphysics and theology. The Ernst Mach Society is today the place from which the Circle speaks to a wider public. This society, as stated in its program, wishes to 'promote and disseminate the scientific world-conception' . . . By the choice of its name, the society wishes to describe its basic orientation: science free of metaphysics. This, however, does not mean that the society declares itself in programmatic agreement with the individual doctrines of Mach.

(Stadler and Uebel 2012, pp. 80–81)

In addition, there are further references to Mach in the context of the main currents in the history of science and philosophy. Under 'Positivism and Empiricism', Mach is mentioned along with David Hume, the Enlightenment, Auguste Comte, John Stuart Mill, and Richard Avenarius, and under 'Foundations and Aims of the Empirical Sciences' he is listed in the company of Hermann von Helmholtz, Bernhard Riemann, Henri Poincaré, Federigo Enriques, Pierre Duhem, Ludwig Boltzmann, and Albert Einstein (Stadler and Uebel 2012). Later, in the recommended further reading for the overcoming of metaphysics and the analysis of reality ('*Wirklichkeitsanalyse*'), Mach's *Analysis of Sensations* and *Knowledge and Error* are mentioned, as are his *Knowledge and Error* and *Mechanics* when it comes to the foundations of physics, systems of hypotheses, and conventions.

Given this self-description, the question arises: which of the specific differences mentioned above were addressed and which of the 'essential points of [Mach's] conception' were agreed on? In order to specify these, we have to look back at the influence and reception of Mach before the foundation of the Schlick Circle (from 1929 on called the 'Vienna Circle'), beginning with the so-called First Vienna Circle.

Mach and the 'First Vienna Circle' (1907–1912)

The so-called First Vienna Circle (a term introduced by Rudolf Haller in 1985) was an informal discussion group in Vienna before World War I centred around Philipp Frank, Hans Hahn, and Otto Neurath. With the exception of the memories of Frank (1949b, pp. 1–52), which offer a short general reconstruction, we do not have further evidence of its meetings and topics (Limbeck-Lilienau 2018). Frank, who succeeded Einstein in Prague in 1912, was a lifelong adherent and admirer of Mach as a physicist and philosopher, even though he presented a critical evaluation and development of Mach's oeuvre from the perspective of modern philosophy of science:

> I used to associate with a group of students who assembled every Thursday night in one of the old Viennese coffee houses. We stayed until midnight and even later, discussing problems of science and philosophy. Our interest was spread widely over many fields, but we returned again and again to our central problem: how can we avoid the traditional ambiguity and obscurity of philosophy? How can we bring about the closest possible rapprochement between philosophy and science? By 'science' we did not mean 'natural science' only, but we included always social studies and the humanities.
>
> (Frank 1949b, p. 1)

The discussion group also included proponents of Catholic philosophy, as Neurath had also studied theology at the University of Vienna. The members focused on a synthesis of empiricism and symbolic logic covering Franz Brentano, Alexius Meinong, Edmund Husserl, Ernst Schröder, Hermann von Helmholtz, Heinrich Hertz, and Sigmund Freud. They aimed at reconciling Mach's empiricism with French conventionalism (Duhem, Poincaré, and Abel Rey) so as also to counter Immanuel Kant's apriorism and V. I. Lenin's attack on Mach's 'empirio-criticism' (1909). The wider intellectual context for the discussion group was the 'Philosophical Society at the University of Vienna', where this new approach towards an anti-metaphysical scientific philosophy was on the agenda with adherents of traditional philosophy and natural science, including quantum theory and the theory of relativity, in an interdisciplinary setting (Fisette 2014). Based on the thinking of Mach, David Hilbert, Percy Williams Bridgman, Einstein, and Bertrand Russell, they claimed to be taking an anti-idealist turn without foundationalism.

Beginning with his first article on 'The Law of Causality and Experience' (1907), Frank had tried to bring together Mach and Poincaré thus:

> To summarise these two theories in a single sentence, one might say: According to Mach the general principles of science are abbreviated economical descriptions of observed facts; according to Poincaré they are free creations of the human mind which do not tell anything about observed facts. The attempt to integrate the two concepts into one coherent system was the origin of what was later called logical empiricism.
>
> (Frank 1949b, p. 11)

And Frank continued by saying that an axiomatic system

> ... is a product of our free imagination, it is arbitrary. But if the concepts occurring in it are interpreted with some observational conceptions, our axiomatic system, if well chosen, becomes an economical description of observational facts. Now the presentation of the law of causality as an arbitrary convention ... can be freed of its paradoxical appearance.... In this way, the philosophy of Mach could be integrated into the 'new positivism' of men like Henri Poincaré, Abel Rey and Pierre Duhem.
>
> (Frank 1949b, p. 14)

This synthesis was to be facilitated by Hilbert's axiomatics of geometry as a conventional system of 'implicit definitions' in a formal mode of speech. Einstein's theory of relativity, mainly inspired by 'Mach's principle' of relativity, was integrated as a further confirming milestone in this merger, as Frank describes in the chapter on 'Mach as a philosopher of "Enlightenment"':

> Einstein derived his laws ... from very general and abstract principles, the principles of equivalence and of relativity. His principles and laws were connections between abstract symbols: the general space–time coordinates and the ten potentials of the gravitational fields. This theory seemed to be an excellent example of the way in which a scientific theory is built up according to the ideas of the new positivism. [Accordingly], the symbolic or structural system is neatly developed and sharply separated from observational facts that are to be embraced. Then the system must be interpreted, and the prediction of facts that are observable must be made and the predictions verified by observation. There were three specific observational facts that were predicted: the bending of light rays and the red shift of spectral lines in a gravitational field, and the advance of the perihelion of mercury.
>
> (Frank 1949b, p. 18)

But the theory of observation had to be part of theory proper, which entered later – with Bridgman's *The Logic of Modern Physics* (1927) – in the form of 'operational definitions': every physical theory no longer describes the 'world in itself', only its structure (or structural context) – which was also held in Schlick's 'Space and Time in Contemporary Physics' (1917) and in his *General Theory of Knowledge* (1918/1925). In this context, a relativistic, causal, and anti-vitalistic tendency in physics and biology emerged in Frank's lectures of 1907–1909, paving the way for his later booklet, *Das Kausalgesetz und seine Grenzen* (*The Law of Causality and Its Limits*, 1932/1998).

It is indeed an irony that in spite of all the impact of Mach's ideas, Frank was accused by Lenin of being a sort of reactionary 'idealistic Kantian' (Lenin 1909, p. 161) – which, by the way, saved him in Cold War Vienna from FBI charges (according to the memories of Gerald Holton (1993)). In his 1932 book, Frank countered Lenin's accusation of idealism by reminding the reader that Mach had actually eliminated the concept of the 'true world' in order to destroy any remnants of metaphysical philosophy: '[I]n my view this entire conception of Mach comes about because the supposed defenders of materialism against Mach have not thoroughly enough made their break with "school" philosophy.' And, Frank goes on, it would be a misunderstanding 'to claim that "Machism" or "positivism" "denied" the "reality" of the outer world of the matter or any auxiliary concept' (Frank 1932/1998, p. 257). In summary, Frank wrote with reference to this very point:

> Our group fully approved Mach's anti-metaphysical tendencies, and we joined gladly in his radical empiricism as a starting point; but we felt very strongly about the primary role of mathematics and logic in the structure of science ... We admitted that the gap between the description of facts and the general principle of science was not fully bridged by Mach, but we could not agree with Kant.
>
> (Frank 1949b, p. 7)

As already noted, this bridge was built with Poincaré's help, probably reinforced by the direct encounter between Mach and Frank in 1910, when Mach invited his adherent to enter into a discussion on Hermann Minkowski's four-dimensional space–time continuum applied to the special theory of relativity. He agreed with Frank's article on Einstein of the same year, and Frank successfully asked Mach to support his translation of Duhem's *L'Évolution de la Mécanique*, which appeared as *Die Wandlungen der Mechanik und die mechanistische Naturerklärung* in 1912 (Frank to Mach, 1910, cited in Blackmore and Hentschel 1985, p. 82).

In his obituary for Mach in 1917, Frank resumed his work of the pre-war years in the philosophy of science, assuming an explicitly anti-metaphysical position – further elaborated in his 'Ernst Mach – The Centenary of His Birth' (Frank 1938), and in *Das Ende der mechanistischen Physik* (*The Fall of Mechanistic Physics*, 1935) and *Interpretations and Misinterpretations of Modern Physics* (1938), leading up to the project of the *International Encyclopedia of Unified Science* from 1938 on (with his *Foundations of Physics*, 1946). He again praised Mach as a 'philosopher of Enlightenment' (Frank 1949b, p. 17) who, in his eyes, represented an anti-idealistic movement displaying similarities with Friedrich Nietzsche. He recalled the Mach–Planck controversy, and he tried to reconcile Mach and Einstein (Frank 1949a). Frank described the dispute between Mach and Boltzmann with the complementarity of phenomenalism and realism, or as a pseudo-problem, in agreement with Carnap's philosophy of science during these years.

According to Frank, the misinterpretation of Mach was caused by traditional philosophy ('school philosophy') with its antiquated conceptual system.

Therefore, he presented Mach *and* Einstein as precursors of the Vienna Circle, with Mach as the *spiritus rector* of the Unity of Science movement (Frank 1938; Tuboly 2017).

In this connection it should be added that the later Vienna Circle member Edgar Zilsel (1891–1944), who did not participate in the 'First Vienna Circle', shared Frank's views on Mach (esp. in Frank 1932/1998). He discussed an objection similar to Lenin's, with reference to Mach's neutral monism in his *Analysis of Sensations*:

> In Mach's philosophy the psychophysical problem, among others, plays a bigger part. Mach assumes the existence of 'elements' which are neither of a physical nor of a mental nature; from these neutral elements both matter as well as mental processes are then supposed to originate by way of only two different forms of combination.... Of all the 'scholarly' philosophical movements, therefore, the Bolsheviks fight most bitterly against the one which most radically denies any spiritual mythology. Materialistic Bolshevism may tolerate Samoyedic fetish worshipping, if necessary, but 'Machists' are forever faced with suspicion ... The Catholic Church, too, agitates more zealously against Protestantism than against Islam, and most zealously against modernism; and small sects, science and art cliques, etc., show the same pattern of behaviour in an even more pronounced way.... Small, hard-struggling groups guard their ideological purity more jealously than others.
>
> (Zilsel 1929, p. 178)

This interpretation was already anticipated epistemologically in Zilsel's book *Das Anwendungsproblem* (*The Problem of Application*, 1916), in which the given is linked to a probabilistic 'flow of sensations' according to Mach's epistemology:

> The world will always remain changing, partly blending things into each other, yet these changes are increasingly compensating each other, these uncertainties are distributed in so fortunate a way as to enable us humans, in spite of all vagueness, to determine very precise relationships in the world, even though we have to add to and continue these determinations ad infinitum. This fortunate distribution of uncertainties, thus, is the precondition for the knowability of the world.
>
> (Zilsel 1916, translation from Dvořák 1981, p. 28).

Given these references, it is not surprising that even much later in 'Problems of Empiricism' (1941) – his contribution to the monograph *The Development of Rationalism and Empiricism* written together with Giorgio de Santillana in the series *International Encyclopedia of Unified Science* – Zilsel still characterised Mach in his final remarks as a pioneer in overcoming mechanistic physics by epistemological and methodological means – empiricism, together with the guiding principle of economy:

All these implications were consistently developed by Mach..... Most of these problems deal rather with the deductive side of theoretical knowledge than with its empirical components. They were raised by Mach, by fictionalism and conventionalism of the late nineteenth century, and were more or less suggested by the physical revolution. Poincaré's conventionalism, however was influenced by modern mathematics as well as by the new physics. In the early twentieth century those mathematical and logical influences increased, united with the empiricist tradition, and resulted finally in logical empiricism . . .

(Zilsel and de Santillana 1941, p. 93, quoted in Zilsel 2000, p. 199)

These short references also reflect the differences between Austrian social democracy (Austro-Marxism) and Communism (Marxism–Leninism), as shown by the positive reception of Mach by the physicist–politician Friedrich Adler, the historian–politician Ludo Moritz Hartmann, the musicologist–politician David Josef Bach et al. They all preferred functionalism as opposed to causality, description as opposed to explanation, and empirical monism as opposed to a metaphysical realism or dialectic materialism. At the same time, they constitute the sociocultural background of the emergence of Logical Empiricism and the founding of the Ernst Mach Society in 'Red Vienna' after World War I. In particular, Friedrich Adler, the son of Viktor Adler, was a convinced admirer of Mach, defending him against Lenin, and translating Duhem's book *La théorie physique, son objet et sa structure* (1906) into German as *Ziel und Struktur physikalischer Theorien* (1908) with a preface by Mach about his scientific role model. This fascinating exchange is documented in the correspondence between Adler and Mach from 1903 to 1915 (Haller and Stadler 1988, pp. 258–305).

Otto Neurath (1882–1945), the second original member of the proto-circle, already admired Mach as a student. Two letters he wrote to Mach from the war front in 1915 refer to Mach's significance for the theory of relativity and his direct influence on his own work in the history of science, theory of values, and political economy (Thiele 1978, pp. 99ff.). Specifically, Neurath was inspired by Mach's writings regarding his own coherence standpoint and methodological holism in theory dynamics in the social sciences – a remarkable transfer of Machian ideas to other fields besides the natural sciences. At the same time, this is also a manifestation of the methodological and meta-theoretical potential of Mach's work for the emergence of Logical Empiricism up to the unity of science movement (beginning in 1934) – despite the already mentioned critical remarks in the manifesto. Neurath's letters focus on his research on the classification of systems of hypotheses in optics and acoustics. In the first one, we read:

It was this idea in your *Mechanics* which has fascinated me ever since I first read it and which has influenced my thoughts and ideas in the area of economics, too, in a somewhat peculiar, roundabout way. It is the tendency to derive the meaning of the particular from the whole, and not the whole from a sum of particulars. In value theory, especially, impulses like this have greatly inspired me and stimulated many associative conclusions. I have always felt a deep gratitude towards your works; your thoughts on physics have offered me ample stimulation not only in that field, but also – and even more so – in others. I would like to use this opportunity to thank you as a person *in concreto*, so to speak, sincerely from the bottom of my heart.

(Neurath to Mach around 1915 (in German) in Thiele 1978, p. 100)

In the second letter, Neurath mentioned he would be sending his article on optics ('Prinzipielles zur Geschichte der Optik', 1915) as well as 'Zur Klassifikation von Hypothesensystemen' (1916) characterising his later (holistic) philosophy of science. In the latter, he wrote programmatically:

> The theory of systems of hypothesis has been greatly advanced by men like Mach, Duhem, Poincaré. The right moment may now have come to group the systems of hypotheses of all sciences systematically and to supplement the actual hypotheses by possible ones into a more or less complete whole. It is the task of philosophical reflection to appreciate the significance of this aim; it is not the concern of the individual sciences. As we need theories to classify things, so we need theories to classify theories.

(Neurath 1916, quoted in Neurath 1983, p. 31)

Here, Mach's impact on Neurath becomes manifest in methodology and general philosophy of science, inspiring Neurath's own naturalist and non-reductive holistic version. His frame of reference is more the *Mechanics* than *The Analysis of Sensations*, although the empiricist approach dominated and determined Neurath's scientific preferences (versus the 'semantic view' in the context of realism) in later publications. Neurath shared Mach's reservations towards an exaggerated formalisation and abstraction detached from practical research, without an experiential fall-back option.

Even if Neurath continued to appreciate Mach until the end of his life, we see at the same time comments on his shortcomings, such as the underestimation of symbolic logic or the preference for neutral 'phenomenalism'. For instance, in his article 'The New Encyclopedia' in the monograph *Towards an Encyclopedia of Unified Science*, he wrote of the empiricist synopsis with logic serving as a useful scientific instrument

> preparing the ground for the ideas of Poincaré, Duhem, Abel Rey, Enriques, and especially Ernst Mach, whose analysis of the traditional fundamental concepts of physics led to considerations whose fruitfulness has now been proved by the latest research and extends beyond Mach's historical influence as a forerunner of, and stimulus to, Einstein. Although Mach himself proclaimed the unity of science and longed for a unified

language which would enable one to go from one science to another without changing one's language, he did not avail himself of the latest expedients which would have enabled him to interconnect more closely, as well as to explain the place of logic and mathematics.

(Neurath 1937, p. 134)

Nevertheless, Neurath later warned Carnap against overemphasising formalisation in the *Encyclopedia* project as a sort of metaphysical logic, but Neurath's appreciation of Mach was in some way diminished by the fact that Mach, in the preface to his posthumous *Prinzipien der physikalischen Optik. Historisch und erkenntnistheoretisch entwickelt* (*The Principles of Physical Optics*, 1921/1926), had apparently distanced himself from Einstein's relativity theory. At that time, Neurath and his colleagues could not have known that this preface was in fact written and manipulated by Mach's son Ludwig and was thus a deliberate forgery (Wolters 1987).

The third person of the proto-circle, the mathematician Hans Hahn (1879–1934), was an advocate of formal logic (an initiator for the reception of Wittgenstein's *Tractatus* in the Vienna Circle), and at the same time an ardent fighter for an anti-metaphysical world view in the tradition of Mach and Boltzmann. In this spirit, he endorsed a common-sense philosophy in order to eliminate metaphysical entities. In his brochure *Superfluous Entities. Occam's Razor* (*Überflüssige Wesenheiten. Occams Rasiermesser*, 1930), he praised and applied Mach's principle of economy in order to distinguish between two forms of philosophy – a 'world-denying philosophy' and a 'world-affirming philosophy' – according to the principle: '*Entia non sunt multiplicanda praeter necessitatem*' ('One must not assume more entities than are absolutely necessary'; Hahn 1930, p. 5), or '*Sufficiunt singularia, et ita tales res universales omnino frustra ponuntur*' ('Individuals suffice, and so it is entirely superfluous to assume universals'; Hahn 1930, p. 5). Hahn states:

> The world-affirming philosophy thus took the view that there are no entities corresponding to universals; rather, these universals are mere names, *nomina*, and this is why the adherents of this line of thought are called nominalists.
>
> (Hahn 1930, p. 5)

Subsequently, he concludes

> If we follow up this idea, which we owe to the world-affirming English philosophers and which was pursued with special emphasis by Ernst Mach, the great Viennese physicist and philosopher whose name our society bears, we learn that it is never necessary to assume behind or beneath the entities of the sensible world we can experience still other entities we cannot experience, viz. substances; and now even substances

are swept away by Occam's razor with a single stroke: their much-praised persistence does not withstand the sharpness of this razor.

(Hahn 1930, p. 17)

On the other hand, Hahn, like Frank, was convinced that Mach and Boltzmann were compatible as intellectual mentors of the Vienna Circle. In 1932, he delivered a lecture dedicated to raising funds for a monument to Boltzmann, which he gave again in the Ernst Mach Society in the same year (Hahn 1933, p. 29). While favouring Boltzmann regarding his controversy with Mach on atomism, he distinguished between a scientific and philosophical perspective towards reality:

When we establish in this way that, contrary to Mach's opinion, the use of 'atom', etc. in physics is perfectly legitimate, this does not hold in any way for certain 'philosophical' discussions about atoms.... The argument ... is used to support the thesis that the world revealed to our senses is mere appearance, and that true being, true reality belongs exclusively to atoms and their changes in position. This is a paradigm case of a metaphysical and therefore senseless proposition.

(Hahn 1933, p. 41)

By way of digression: many years later, Hahn's student, the mathematician Karl Menger (1902–1985), developed the option of a 'positivist geometry' and the mathematical concept of function applied to science with regard to Mach. In his article 'Mathematical Implications of Mach's Ideas: Positivist Geometry, the Clarification of the Functional Connection', he proposed the following:

Fifty years after Mach, some thoughts of this great Austrian philosopher–scientist have an unspent vitality. Two ideas on which I wish to concentrate lead to implication for mathematics that have not yet been fully explored: Mach's general philosophical view suggests the development of a positivistic geometry, which was hardly begun; and his maxim that causal explanations should be replaced by functional connections calls for a clarification of the method of applying the function concept and other ideas of mathematical analysis to science – a clarification which is still in its initial stage.

(Menger 1970, p. 107)

After an extensive, detailed elaboration of this claim, Menger concluded:

The clarification of the statements of classical physics, which has been briefly summarised in this paper, pointed, as far as macrocosmic laws are concerned, in this direction which certainly is in line with thoughts of Ernst Mach.

(Menger 1970, p. 124)

In addition, Menger presented a 'A Counterpart of Occam's Razor' in pure and applied mathematics (1960/1961, reprinted in 1979), interpreting it as a methodological claim that *Entities must not be multiplied beyond necessity.* 'In a more general form, often called the Law of Parsimony, the principle states that *It is in vain to do with more what can be done with less.* This law may also be construed as a maxim opposing synonyms' (Menger 1979, p. 105, emphasis in original). In this way, Menger described the development from Occam to contemporary mathematics by stating:

> Just as in medieval methodology masses of superfluous notions connected with single ideas called for Occam's Razor to shave off what is unneces-sary, so in mathematico-scientific methodology the confused conglomer-ate of ideas, all of them designated by the same term 'variable', calls for a counterpart of Occam's Razor to widen a framework that is insufficient. The paper supplies this for a two-fold use: in ontology and semantics.
>
> (Menger 1979, p. 2)

Already in his 'Introduction to the sixth American edition' of Mach's *The Science of Mechanics* (Mach 1960, p. vi), Menger had commented, 'In this century, the analysis of the mass concept and, even more, physicists' views on space and time, have advanced beyond Mach. Yet his original discussions remain classics not only of physics but also of *philosophy of science*', especially as manifestations of an operationalism (Bridgman), anti-metaphysical phenomenalism, functionalism versus causalism, and phenomenalism. Besides these achievements, said Menger,

> a third point that Mach stressed over and over again was his view that science had the purpose of saving mental effort. General laws are shorter and easier to grasp than enumerations of specific instances. Simpler theories are preferable to more complicated ones. His theory of 'economy of thought' is Mach's main point of contact with R. Avenarius, who, in his *empirio-criticism*, regarded philosophy as thinking about the world with minimum effort.
>
> (Mach 1960, p. vii, emphasis in original)

These quotations refer to the Swiss philosopher Richard Avenarius, who, in his books *Philosophie als Denken der Welt gemäß dem Princip des kleinsten Kraftmaßes. Prolegomena zu einer Kritik der reinen Erfahrung* (1876), *Kritik der reinen Erfahrung* (2 vols, 1888/1900), and *Der menschliche Weltbegriff* (1891), emerged as the philosophical twin of Mach. However, he was not received as a role model in the natural sciences because of his complex style and genuine philosophical discourse.

Given this close reading and critical reception, it is not surprising that Menger, in his posthumously published memories of the Vienna Circle and of his own Mathematical Colloquium, dedicated one section of the chapter on the philosophical atmosphere in Vienna to Mach (Menger 1994, pp. 21ff.). Here, he praised Mach's anti-Kantian radical positivism and mentioned Mach's judgement of philosophical problems as pseudo-problems. Once again, the principle of economy in science and mathematics is emphasised, whereas his anti-atomism and psychologism in arithmetic is commented on critically. Nevertheless, 'one of Mach's greatest achievements, which made him a precursor of Einstein, was his early insistence on the relativity of all observed motions, even of accelerated motions including rotations'. According to Menger, Mach, 'who had many friends and followers in the English-speaking world, exerted a great and lasting influence in Austria' (Menger 1994, p. 23).

In parallel – perhaps also in interaction – with the 'First Vienna Circle', the mathematician Richard von Mises (1883–1953), pioneer of applied mathematics, met with his friends and colleagues at coffee houses in his hometown Vienna to discuss the relation and foundation of logic, mathematics, and empiricist philosophy in modern times. From the beginning, he was a convinced admirer of Mach as a scientist in the tradition of the Enlightenment, and he provided biographies of Mach and his closest friend, the engineer, social reformer, and author Josef Popper-Lynkeus (Belke 1978). On the 100th anniversary of Mach's birth in 1938, he published *Ernst Mach und die empiristische Wissenschaftsauffassung* (*Ernst Mach and the Scientific Conception of the World*) as volume 7 of the monograph series 'Unified Science', edited by Neurath (McGuinness 1987), in which he wrote, with reference to Carnap and Frank, that

> ... the combination of Mach's empiricist attitude with certain theses in the logic of language implicit in Wittgenstein's work gave rise to the label 'logical empiricism', by which the members of the Vienna Circle liked to characterise their point of view.
>
> (von Mises 1938, p. 189)

More systematically, in his book *Kleines Lehrbuch des Positivismus. Einführung in die empiristische Wissenschaftsauffassung* (1938), published in English as *Positivism. A Study in Human Understanding* (1951), he describes the transformation of Mach's doctrine of elements in accordance with the Vienna Circle's conception of scientific language, with reference to Carnap's logic of science:

> The word 'sensations' by which Mach sometimes denotes the elements is itself not significant and can easily be replaced by 'sense impressions' or 'receptions'; what is decisive is only that here everything experienced was successfully reduced to uniform elements or (translated from the material to the formal mode) that a language was successfully created which is connectible across all boundaries of the individual sciences.
>
> (von Mises 1939, p. 84)

Consequently, von Mises continues:

> The function of Mach's doctrine of elements is a regulative one; it points out the limits within which the customary auxiliary concepts (like body, substance, mass, individual, etc.) are applicable, by disclosing their origins. All of these concepts are aggregates of elements, formed under the impulse of specific limited needs; they become detrimental to further development if they are too far beyond these limits. Notwithstanding the expediency, and even indispensability, of such auxiliary concepts in wide areas of life and science, a critical attitude toward them must, as time goes on, gain more and more ground within the ever-growing realm of human experience.
>
> (von Mises 1939, p. 90)

Departing from this conception, von Mises goes on to explicate the Vienna Circle's contested 'protocol sentences' by employing Carnap's distinction between the formal and material mode of speech:

> If we transform in this sense Mach's conception of 'elements, of which for us the world is composed', into formal language, we arrive at the concept that is considered the core of logical empiricism as represented mainly by Carnap, Otto Neurath, and Philipp Frank. Mach's 'elements' ... are replaced in the new language by simplest, not further reducible, statements, which are known as 'protocol sentences', 'element sentences', or 'atomic sentences'.
>
> (von Mises 1939, p. 91)

With his basic notion of 'connectibility' (*Verbindbarkeit*), von Mises offered a mode of transition between empirical statements with different degrees of empirical content down to protocol sentences as a regulative method by means of a set of linguistic rules. As long as statements can be connected with the accepted rules of everyday language and scientific language, the 'material mode' of expression allows for basic protocol sentences as an analysis of the world according to Mach's doctrine of elements (receptions) (von Mises 1939, p. 369). With this modernised and updated version of Mach's (logical) empiricism, von Mises praised him as 'the most effective critical philosopher of the last generation and the most typical of our time' (von Mises 1938, p. 170).

Mach in the Vienna Circle Period (1924–1936) and Beyond

This same perspective became manifest in the early work of Rudolf Carnap (1881–1917). The constitution system of his *Aufbau* programme (*Der logische Aufbau der Welt*, 1928) builds on Mach's doctrine of elements for the experiential basis, along with references to neo-Kantian philosophy, gestalt theory, Gottlob Frege, and Russell and Whitehead's *Principia Mathematica*. His ambitious project dealt with the establishment of a hierarchical constitution system of scientific concepts on an empirical basis by means of sensual perceptions in a phenomenalist language (Bonk 2003; Carus 2007; Creath 2012; Damböck 2016). As basic elements, Carnap proposed elementary experiences, and as a basic concept, he introduced the relation between basic experiences (conceived as concrete sensory data). The basic level denotes the ego-psychic (*Eigenpsychische*) in the framework of methodological solipsism, before he turned to physicalism because of the advantage of its intersubjectivity. Remarkably, much later, in 1961, in a second preface to his 1928 book, Carnap acknowledged the relevance of Mach – together with Frege, Russell, and Wittgenstein, as expressed in his 'Intellectual Autobiography' (1963) – especially with regard to the empirical basis of the *Aufbau*, a programme which he saw as in principle realisable:

> I should now consider for us as basic elements, not elementary experiences (in spite of the reasons which, in view of findings of Gestalt psychology, speak for such a choice …), but something similar to Mach's elements, e.g., concrete sense data, as for example, 'a red of a certain type at a certain visual field place at a given time'. I would then choose as basic concepts some of the relations between such elements, for example, 'x is earlier than y', the relation of spatial proximity in the visual field and in other sensory fields, and the relation of quantitative similarity, e.g., colour similarity.
>
> (Carnap 1967, p. vii)

The founder of the Vienna Circle, Moritz Schlick (1882–1936), started from a critical-realist position in his *Allgemeine Erkenntnislehre* (*General Theory of Knowledge*, 1918/1925) before, influenced by Wittgenstein, he preferred a more 'positivist' version of his philosophy (with growing criticism from Max Planck and Einstein). In this respect, the early Schlick was closer to Boltzmann than to Mach – which was also the case for Hans Reichenbach, Viktor Kraft, Herbert Feigl, Carl G. Hempel, and Karl Popper on the periphery of the Circle. In his philosophical interpretation of Einstein's relativity theory, *Raum und Zeit in der gegenwärtigen Physik* (*Space and Time in Contemporary Physics*, 1917), which Einstein himself praised as the best philosophical interpretation of his theory, Schlick already shared the anti-Kantian attitude accompanied by the principle of economy according to the characterisation in the 1929 manifesto:

> The value of relativity theory lies in describing the geometric and physical structure of the universe in the simplest and most correct way by means of general laws. The philosophical significance of relativity theory consists negatively in its critique of a-priorist philosophy (Kant) and other epistemological views, positively in its stimulation of the analysis of science with regard to its empirical and logical content.
>
> (Stadler and Uebel 2012, p. 161)

In his *Allgemeine Erkenntnislehre* (*General Theory of Knowledge*), Schlick described the nature of knowledge accessible through an investigation of the fundamental questions of science by way of philosophical clarification. Schlick maintained that 'the analysis suggested here should be clearly distinguished from the positivistic dissolution of a body into a complex of "elements" (Ernst Mach)' (Schlick 1918/1925, p. 54). In the meantime, Einstein had come to appreciate Hume's and Mach's influence on his own ideas (Einstein 1916).

With some reservations about Mach's 'principle of economy of thought' (Schlick 1918/1925, p. 98), Schlick claimed that knowledge seeks comprehensive interconnections – deductive and logical ones. Regarding the problem of reality, he described the positioning of the real (temporal determination) and the knowledge of the real (rejection of the metaphysical distinction between essence and appearance) by claiming that only quantitative knowledge, as opposed to Kant's synthetic knowledge a priori, is possible. Here, too, we can note a distancing from Mach's *Analysis of Sensations* as a version of an immanence philosophy with functionalism versus causal relations, even though these occur together with the common rejection of the Kantian thing-in-itself as the old notion of substance (Schlick 1918/1925, pp. 195, 224, 231, 246, etc.). According to Schlick, the validity of deduction is independent of the nature of the world, and the scientific method is a hypothetical-deductive one, based on a certain uniformity of the universe. Some years later, Schlick confirmed the anti-metaphysical and anti-Kantian standpoint in his distinction of '*Erleben, Erkennen, Metaphysik*' ('Experience, Cognition, and Metaphysics'). He claimed that *metaphysics* as *intuitive* knowledge of transcendent objects is impossible, as is *inductive* metaphysics:

> If the metaphysician was striving only for experience, his demand could be fulfilled, though through poetry and art and life itself.... But in that he absolutely demands to experience the transcendent, he confuses living and knowing, and, bemused of a double contradiction, chases empty shadows.... To be sure, it is not, as he supposes, an experience of the transcendent. We can see the precise sense in which there is truth in the oft-expressed opinion, that metaphysical philosophemes are conceptual poems: in the totality of culture they play, in fact, a role similar to that of poetry; they serve to enrich life, not knowledge. They are to be valued, not as truths, but as works of art. The systems of the metaphysicians sometimes contain science and sometimes poetry, but they never contain metaphysics.
>
> (Schlick 1979, p. 110f.)

With this statement, Schlick's transition to his anti-metaphysical 'positivism', inspired by Wittgenstein and Carnap, becomes apparent, and his ever-stronger commitment to Mach was confirmed when he assumed his chair in 1922. In his inaugural lecture on natural philosophy, Schlick addressed the *genius loci* with the following acknowledgement after having praised Mach's *Mechanics*:

The name Mach has ever since been associated with very strong emotions, for he was a radiant symbol of an unusual method of philosophising. This method seems to be one of the most productive ones ever known in the history of human thought. How much more intense and different my feelings would have been if I had known that I was once to teach at the same place as E. Mach did here at the university.

(Schlick 1922, translated from Stadler 2017c, p. 189)

Schlick continued to keep alive the heritage of Mach and Boltzmann in Vienna, despite their differences, as expressed in this striking statement: 'Almost *all* philosophy is natural philosophy' (Stadler 2017c, p. 191, emphasis in original). Exact thinking is not only required – and possible – in philosophy and the sciences, but also in all disciplines such as philosophy of history, epistemology, aesthetics, and ethics.

Given this confession, it is not surprising that Schlick immediately agreed to assume the presidency of the Ernst Mach Society in 1928, which he held and defended fervently, even if unsuccessfully, until it disintegrated as a consequence of the 1934 civil war in Austria (Stadler 1982b, 1992). Two years earlier, he spoke on the occasion of the unveiling of the Mach bust in the Rathauspark of the City of Vienna, as documented in newspaper articles by Einstein and the Viennese physicists Felix Ehrenhaft and Hans Thirring (*Neue Freie Presse*, 12 June 1926).

After his gradual departure from a structuralist–realist position in the *General Theory of Knowledge*, Schlick defended a new 'positivism' against his former teacher Planck and, in the context of the 'linguistic turn', privileged an explicit empiricism which he himself called '*konsequenter Empirismus*' (consistent empiricism). In his own entry to the *Philosophen-Lexikon*, which appeared posthumously only in 1950, and describing himself in the third person, Schlick stuck to a strong empiricism:

Schlick attempts to justify and construct a consistent and entirely pure empiricism, which unlike its early forms, is reached by applying modern mathematics and logic to reality . . . From there, and with the help of an analysis of the process of knowledge, the 'General Theory of Knowledge' arrives at a clear distinction between the rational and the empirical, the conceptual and the intuitive. Concepts are mere symbols that are attributed to the world in question; they appear in 'statements' ordered in a very particular way, by which these are able to 'express' certain structures of reality. Every statement is the expression of a fact and represents knowledge insofar as it describes a new fact with the help of old signs. . . . The ordering of reality . . . is determined solely by experience, for which reason there exists only empirical knowledge. The so-called rational truths, then, purely abstract statements such as logical-mathematical ones . . . are nothing more than rules of signs which determine the syntax of language (L. Wittgenstein), which we use to speak about the world.

They are of pure analytic-tautological character and therefore contain no knowledge; they say nothing about reality, but it is precisely for this reason that they can be applied to any given fact in the world. Thus, knowledge is essentially a reproduction of the order, of the structure of the world; the material content belonging to this structure cannot enter it; for the expression is, after all, not the thing itself which is being expressed. Therefore, it would be senseless to attempt to express the 'content' itself. Herein lays the condemnation of every variety of metaphysics; for it is precisely this that metaphysics has always wanted, in having as its goal the cognizing of the actual 'essence of being'.

(Schlick 1950, p. 462)

This is further striking evidence (besides Frank's and Zilsel's descriptions) of the fusion of empiricism and rationalism called 'Logical Empiricism', based on Mach's heritage, despite all of the differences in the terminology used. Schlick's early criticism of Mach, obviously under the influence of his teacher Planck, is just one more sign that within Logical Empiricism there was a group of philosophers which preferred a more realistic position, later called critical or structural realism (Neuber 2018). Kraft (1880–1975) already expressed this preference in his *Weltbegriff und Erkenntnisbegriff* (1912), and in his *Die Grundformen der wissenschaftlichen Methoden* (1925) he presented a hypothetico-deductive methodology anticipating Popper's *Logik der Forschung* (*Logic of Scientific Discovery*, 1934) with a common distancing from Mach's epistemology, which also became partly manifest in his later obituaries for Mach (Kraft 1918, 1966). Furthermore, Schlick's favourite student, Herbert Feigl (1902–1988), praised the *General Theory of Knowledge* as a cornerstone in solving the mind–body problem, which anticipated Russell's monism. In his introduction to the English edition of this book, Feigl characterised Schlick's philosophical development and his philosophical background:

Perhaps equally noteworthy, from an historical point of view, is the fact that Schlick anticipated Russell's solution . . . of the mind–body-problem.- . . . At that time Russell, influenced by William James and Mach, still held the position known as 'neutral monism'. This was an epistemological view very close to the phenomenalism of Mach's *Analysis of Sensations* and to the 'radical empiricism' of James. . . . Against these 'philosophies of immanence', Schlick offered a number of striking arguments, similar in part to those advanced by influential psychologist and critical realist Oswald Külpe and by the neglected Neo-Kantian Alois Riehl. Influenced by Carnap and Wittgenstein . . . Schlick later came to look on the issue of realism versus phenomenalism as a metaphysical pseudo-problem. Much to the chagrin of, especially, Viktor Kraft, Karl Popper, Edgar Zilsel, and Herbert Feigl, he abandoned his realism in favor of a linguistically oriented 'neutral' position.

(Feigl 1974, p. xxi)

Nevertheless, Feigl concludes that Schlick's work 'will remain a milestone in the development of a new empiricism and naturalism' (Feigl 1974, p. xxv). The story of Wittgenstein's hidden Machian roots and references in this context and of Popper's anti-Machian position cannot be told here. However, it illustrates the wide range of Mach's influence up to Paul Feyerabend (1988), despite his strong opponents – Lenin, the Frankfurt School, and neo-Kantianism. The recent promise of an empiricist and realist interpretation (Banks 2014) and the comparison with the Brentano school (Brentano 1988; Fisette et al. 2020) open up a new perspective for further investigation. As the proceedings of the big Mach centenary conference (Stadler 2019a, 2019b, 2019c) show, there are good reasons to continue research into the complex and fascinating relations between Mach, Boltzmann, and the Vienna Circle/Logical Empiricism.

Conclusion

This chapter on the relations between Mach and the Vienna Circle has shown that the former influenced in varying degrees nearly all members of the latter – before, during, and also after the classical period of the Schlick Circle (1924–1936). A lowest common denominator is described in the 1929 manifesto, where Mach is characterised as the pioneer of modern empiricism and anti-metaphysical science in the tradition of British and French Enlightenment. His basic idea of a unity of science was welcomed by the vast majority of the Vienna Circle members.

Generally, there was a strong and positive reception of Mach within the Schlick Circle, but for different reasons. Whereas Frank, Neurath, and Zilsel praised Mach for overcoming an old-fashioned aprioristic philosophy and for reformulating an anti-Kantian empiricism, Carnap and von Mises appreciated his epistemological and methodological incentives for the *Aufbau* and the unity of science conception. It is remarkable that professional physicists and mathematicians (Frank, Hahn, Menger, and von Mises) appreciated Mach – who himself did not privilege symbolic logic and formalisation – as an innovator in their fields, especially regarding his methodological principle of economy and thought experiments.

From an epistemological point of view, other members (the early Schlick, Kraft, and Feigl) favoured Boltzmann because of his more realistic position. They were reluctant in their reception of Mach, even though they acknowledged his role as a necessary step towards the formation of Logical Empiricism – independently from their judging of the differences between empiricism, phenomenalism, and positivism as pseudo-problems of an anachronistic 'school philosophy'. In any case, for all of the members of the Vienna Circle mentioned, Mach was acknowledged as a predecessor and role model for a more modernised empiricism and scientific world view. In this regard, he was praised as an anti-Kantian philosopher of science – although Mach himself did not accept the label of a philosopher or the attribute of 'positivism' for his life work, and he preferred instead to be called a *'Naturforscher'* (scholar of nature). He certainly was influential as an interdisciplinary scientist with a strong historical tendency. This historical side of philosophy and science was not a focus in the Vienna Circle, even though Neurath tried to integrate that dimension into the *International Encyclopedia of Unified Science.* In this regard, Mach can be seen as a predecessor of an integrated history and philosophy of science and a European pragmatist philosopher of science (Stadler 2017a; Gori 2018; Maddalena and Stadler 2019). *Epistemologically,* Mach combined elements of naturalism, empiricism, conventionalism, and relativism with an evolutionary approach, and *methodologically* he combined the historical-critical method with the principle of economy and thought experiments, allowing for induction, abduction, and deduction in his theory of research covering knowledge and error. This was aimed at bringing together the context of discovery with the context of justification, facilitated by a research heuristics between the poles of fantasy and economy. Recent research points to the viability of a realistic empiricism and the uniqueness of a genetic theory of knowledge and learning following Mach's heritage (e.g. Banks 2014; Siemsen 2019). This is certainly reasonable, but we cannot conclude that the Vienna Circle did not fully understand and further develop Mach's doctrines in his own sense. On the contrary, there is a scientific pluralism in Mach's reception and influence with different motives and reasons. But in the long term, we see a creative reconstruction and development of his ideas following his intentions and achievements. There is a thematic 'red thread' in the transition of Mach's empiricism and conventionalism to Logical Empiricism after the linguistic turn between the two World Wars, which should and can be analysed within the Vienna Circle, who praised mainly the empiricist side of his doctrines – although Mach was certainly placed in the context of the contested 'psychologism' (Kusch 1995). The new adjective 'logical' certainly indicated an extension and completion of empiricism (as presented in the Ernst Mach Society). This is true despite an ongoing reception of Mach in contemporary philosophy of science and historical epistemology (Stadler 2019a, 2019b). Finally, it seems that Einstein already expressed the proper significance of Mach in his obituary, when he concluded 'that the people who consider themselves opponents of Mach, scarcely know of Mach's way of thinking they have absorbed, so to say, with their mother's milk' (Einstein 1916, quoted in Blackmore 1992, p. 154).

References

Banks, Erik C. 2003. *Ernst Mach's World Elements: A Study in Natural Philosophy.* Kluwer.

    2014. *The Realistic Empiricism of Mach, James, and Russell: Neutral Monism Reconceived.* Cambridge University Press.

Belke, Ingrid 1978. *Die sozialreformerischen Ideen von Josef Popper-Lynkeus (1838–1921) im Zusammenhang mit allgemeinen Reformbestrebungen des Wiener Bürgertums um die Jahrhundertwende*. J.C.B. Mohr.

Blackmore, John T. 1972. *Ernst Mach: His Work, Life, and Influence*. University of California Press.

    1978. 'Three Autobiographical Manuscripts by Ernst Mach', *Annals of Science* 35: 401–418.

    (ed.) 1992. *Ernst Mach – A Deeper Look. Documents and New Perspectives*. Kluwer.

    (ed.) 1995. *Ludwig Boltzmann: His Later Life and Philosophy, 1900–1906*, 2 Vols. Kluwer.

Blackmore, John T. and Hentschel, Klaus (eds.) 1985. *Ernst Mach als Außenseiter. Machs Briefwechsel über Philosophie und Relativitätstheorie mit Persönlichkeiten seiner Zeit*. Braumüller.

Bonk, Thomas (ed.) 2003. *Language, Truth and Knowledge. Contributions to the Philosophy of Rudolf Carnap*. Springer.

Brentano, Franz 1988. *Über Ernst Machs 'Erkenntnis und Irrtum'*. Edited and introduced by Roderick M. Chisholm und Johann C. Marek. Rodopi.

Carnap, Rudolf 1928. *Der logische Aufbau der Welt*. Weltkreis-Verlag.

    1963. 'Intellectual Autobiography', in P. A. Schilpp (ed.), *The Philosophy of Rudolf Carnap*. Open Court, pp. 1–84.

    1967. *The Logical Structure of the World, and Pseudoproblems in Philosophy*. University of California Press.

Carus, Andre W. 2007. *Carnap and Twentieth-Century Thought. Explication as Enlightenment*. Cambridge University Press.

Cohen, Robert S. and Seeger, Raymond J. (eds.) 1970. *Ernst Mach: Physicist and Philosopher*. D. Reidel.

Creath, Richard (ed.) 2012. *Rudolf Carnap and the Legacy of Logical Empiricism*. Springer.

Damböck, Christian (ed.) 2016. *Influences on the* Aufbau. Springer.

Duhem, Pierre 1908. *Ziel und Struktur der physikalischen Theorien*. Autorisierte Übersetzung von Friedrich Adler. Mit einem Vorwort von Ernst Mach. Mit einer Einleitung Hrsg. Von Lothar Schäfer. Meiner.

Dvořák, Johann 1981. *Edgar Zilsel und die Einheit der Erkenntnis*. Löcker.

Einstein, Albert 1916. 'Ernst Mach', *Physikalische Zeitschrift* 17: 101–104. English translation in John T. Blackmore (ed.), *Ernst Mach – A Deeper Look. Documents and New Perspectives*. Kluwer, 1992, pp. 154–159.

Feigl, Herbert 1969. 'The Wiener Kreis in America', in D. Fleming and B. Bailyn (eds.), *The Intellectual Migration: Europe and America, 1930–1960*. Harvard University Press, pp. 630–673.

    1981. *Inquiries and Provocations. Selected Writings 1929–1974*. Ed. R. S. Cohen. Springer.

Feigl, Herbert and Blumberg, A. 1974. 'Introduction', in M. Schlick, *General Theory of Knowledge*. Transl. A. E. Blumberg. Springer, pp. xvii–xxvi.

Feyerabend, Paul K. 1988. 'Machs Theorie der Forschung und ihre Beziehung zu Einstein', in Rudolf Haller and F. Stadler (eds.), *Ernst Mach – Werk und Wirkung*. Hölder-Pichler-Tempsky, pp. 435–462.

Fisette, Denis 2014. 'Austrian Philosophy and Its Institutions: Remarks on the Philosophical Society of the University of Vienna (1888–1938)', in A. Reboul (ed.), *Mind, Values, and Metaphysics*. Springer, pp. 349–374.

Fisette, Denis, Fréchette, Guillaume, and Stadler, Friedrich (eds.) 2020. *Franz Brentano and Austrian Philosophy*. Springer.

Frank, Philipp 1932/1998. *Das Kausalgesetz und seine Grenzen*. Springer. English translation, *The Law of Causality and Its Limits*. Ed. R. S. Cohen. Kluwer.

  1938. 'Ernst Mach – The Centenary of his Birth', *Erkenntnis* 7: 247–256.

  1949a. 'Einstein, Mach, and Logical Positivism', in P. A. Schilpp (ed.), *Albert Einstein: Philosopher–Scientist*. Open Court, pp. 271–286.

  1949b. *Modern Science and Its Philosophy*. Harvard University Press.

Gori, Pietro 2018. *Ernst Mach. Tra scienza e filosofia*. Edizioni ETS.

Hahn, Hans 1930. *Überflüssige Wesenheiten. (Occams Rasiermesser)*. A. Wolf Verlag. English translation, 'Superfluous Entities, or Occam's Razor', in H. Hahn, *Empiricism, Logic, and Mathematics. Philosophical Papers*. Ed. B. F. McGuinness. D. Reidel, 1980, pp. 1–20.

  1933. *Logik, Mathematik, Naturerkennen*. Gerold & Co. English translation, 'Logic, Mathematics, and Knowledge', in Brian F. McGuinness (ed.), *Unified Science*. The Vienna Circle Monograph Series, originally edited by Otto Neurath, in an English edition. D. Reidel, 1987, pp. 24–45.

Haller, Rudolf 1979. *Studien zur österreichischen Philosophie. Variationen über ein Thema*. Rodopi.

  1985. 'Der erste Wiener Kreis', in R. Haller, *Fragen zu Wittgenstein und Aufsätze zur österreichischen Philosophie*. Brill, pp. 89–107.

  1988. 'Grundzüge der Machschen Philosophie', in Rudolf Haller and F. Stadler (eds.), *Ernst Mach – Werk und Wirkung*. Hölder-Pichler-Tempsky, pp. 64–86.

Haller, Rudolf and Stadler, F. (eds.) 1988. *Ernst Mach – Werk und Wirkung*. Hölder-Pichler-Tempsky.

Hentschel, Klaus 1986. 'Die Korrespondenz Einstein-Schlick: Zum Verhältnis der Physik zur Philosophie', *Annals of Science* 43: 475–488.

Holton, Gerald 1993. 'From the Vienna Circle to Harvard Square: The Americanization of a European World Conception', in F. Stadler (ed.), *Scientific Philosophy – Origins and Developments*. Kluwer, pp. 47–73.

Kraft, Viktor 1918. 'Ein österreichischer Denker: Ernst Mach', *Donauland* 2: 1209–1213.

  1966. 'Ernst Mach als Philosoph', *Almanach der Österreichischen Akademie der Wissenschaften* 116: 373–387.

  1925. 'Die Grundformen der wissenschaftlichen Methoden', *Sitzungsberichte der österreichischen Akademie der Wissenschaften, philos.-historische Klasse* 203: 1–104.

Kusch, Martin 1995. *Psychologism. A Case Study in the Sociology of Philosophical Knowledge.* Routledge.

Lenin, Vladimir I. 1909. *Materialismus und Empiriokritizismus. Kritische Bemerkungen über eine reaktionäre Philosophie.* Russian edition: Zveno. First German edition: Verlag für Literatur und Politik, 1927. English edition: *Materialism and Empirio-Criticism.* International Publishers, 1927.

Limbeck-Lilienau, Christoph 2018. 'The First Vienna Circle: Myth or Reality?', *Hungarian Philosophical Review* 62: 50–65.

Mach, Ernst 1883a. *Die Mechanik in ihrer Entwicklung historisch-kritisch dargestellt.* J. A. Barth. Translated as *The Science of Mechanics: A Critical and Historical Account of its Development.* Translated by T. J. McCormack. Open Court, 1960.

1883b. *Principles of the Theory of Heat. Historically and Critically Elucidated.* Ed. B. F. McGuinness. Kluwer.

1886. *Populär-wissenschaftliche Vorlesungen.* J. A. Barth. Translated as *Popular Scientific Lectures.* Open Court, 1986.

1911. *History and Root of the Principle of the Conservation of Energy.* Open Court.

1959. *The Analysis of Sensations and the Relation of the Physical to the Psychical.* Dover.

1960. *The Science of Mechanics: A Critical and Historical Account of its Development.* Translated by Thomas J. McCormack. New Introduction by Karl Menger. Open Court.

1976. *Knowledge and Error. Sketches on the Psychology of Enquiry.* D. Reidel.

Maddalena, Giovanni, and Stadler, Friedrich (eds.) 2019. *European Pragmatism. Historical and Contemporary Perspectives (European Journal of Pragmatism and American Philosophy).* Available from https://journals.openedition.org/ejpap/

McGuinness, Brian F. (ed.) 1987. *Unified Science.* The Vienna Circle Monograph Series, originally edited by Otto Neurath, in an English edition. D. Reidel.

Menger, Karl 1970. 'Mathematical Implications of Mach's Ideas: Geometry, the Clarification of Functional Connections', in Robert S. Cohen and Raymond J. Seeger (eds.), *Ernst Mach: Physicist and Philosopher.* D. Reidel, pp. 107–125.

1979. *Selected Papers in Logic and Foundations, Didactics, Economics.* D. Reidel.

1994. *Reminiscences of the Vienna Circle and the Mathematical Colloquium.* Eds. L. Golland, B. F. McGuinness, and A. Sklar. Kluwer.

Neuber, Matthias 2018. *Der Realismus im logischen Empirismus. Eine Studie zur Geschichte der Wissenschaftsphilosophie.* Springer.

Neurath, Otto 1937. 'Towards an Encyclopedia of Unified Science. Lectures by O. Neurath, E. Brunswik, C. L. Hull, G. Mannoury, and J. H. Woodger', in Brian F. McGuinness (ed.), *Unified Science.* The Vienna Circle Monograph Series, originally edited by Otto Neurath, in an English edition. D. Reidel, 1987, pp. 130–165.

1983. *Philosophical Papers 1913–1946*. Eds. R. S. Cohen and M. Neurath. Kluwer.

Schlick, Moritz 1917. *Raum und Zeit in der gegenwärtigen Physik. Zur Einführung in das Verständnis der Relativitäts- und Gravitationstheorie*. Springer. Translated in Moritz Schlick. *Philosophical Papers*, 2 Vols. Eds. H. Mulder and B. von der Velde-Schlick. D. Reidel, 1979, pp. 207–269.

    1918/1925/1974. *Allgemeine Erkenntnislehre*. German edition: eds. H. J. Wendel and F. O. Engler, 2009. English edition: *General Theory of Knowledge*. Translated by Albert E. Blumberg. With an Introduction by A. E. Blumberg and Herbert Feigl. Springer, 1974.

    1950. 'Schlick, Moritz', in W. Ziegenfuss and G. Jung (eds.), *Philosophen-Lexikon. Handwörterbuch der Philosophie nach Personen*. De Gruyter, pp. 462–464.

    1979. *Philosophical Papers*, 2 Vols. Eds. H. Mulder and B. von der Velde-Schlick. D. Reidel.

Siemsen, Hayo 2019. 'Transforming Thinking: Can Mach's Pedagogy be Replicated?', in Friedrich Stadler (ed.), *Ernst Mach – Life, Work, Influence*. Springer, pp. 571–600.

Stadler, Friedrich 1982a. 'History and Philosophy of Science', in M. C. Galavotti et al. (eds.), *New Directions in the Philosophy of Science*. Springer, pp. 747–768.

    1982b. *Vom Positivismus zur 'Wissenschaftlichen Weltauffassung'. Am Beispiel der Wirkungsgeschichte von Ernst Mach in Österreich von 1895–1934*. Löcker.

    1988. 'Ernst Mach – Zu Leben, Werk und Wirkung', in Rudolf Haller and F. Stadler (eds.), *Ernst Mach – Werk und Wirkung*. Hölder-Pichler-Tempsky, pp. 11–63.

    1992. 'The "Verein Ernst Mach" – What Was It Really?', in J. T. Blackmore (ed.), *Ernst Mach – A Deeper Look: Documents and New Perspectives*. Kluwer, pp. 363–377.

    2015. *The Vienna Circle. Studies in the Origins, Development, and Influence of Logical Empiricism*, 2nd edn. Springer.

    2017a. 'Ernst Mach and Pragmatism – The Case of Mach's *Popular Scientific Lectures* (1895)', in S. Pihlström, F. Stadler, and N. Weidtmann (eds.), *Logical Empiricism and Pragmatism*. Springer, pp. 3–14.

    (ed.) 2017b. *Integrated History and Philosophy of Science. Problems, Perspectives, and Case Studies*. Springer.

    2017c. 'Kommentar zu: Moritz Schlick: Vorrede zur Vorlesung "Einführung in die Naturphilosophie" (1922)', in T. Assinger, E. Grabenweger, and A. Pelz (eds.), *Die Antrittsvorlesung. Wiener Universitätsreden der Philosophischen Fakultät*. V&R unipress, Vienna University Press, pp. 189–202.

    2018a. 'George Sarton, Ernst Mach, and the Unity of Science Movement. A Case Study in History and Philosophy of Science', *Sartoniana* 31: 63–121.

    2018b. 'Kant and Neo-Kantianism in Logical Empiricism. Elements of a Research Program', in V. Waibel, M. Ruffing, and D. Wagner (eds.), *Natur und Freiheit. Akten des XII. Internationalen Kant-Kongresses.* De Gruyter, pp. 763–790.

(ed.) 2019a. *Ernst Mach – Life, Work, Influence.* Springer.

(ed.) 2019b. *Ernst Mach – Zu Leben, Werk und Wirkung.* Springer.

2019c. 'Only a Philosophical "Holiday Sportsman"? – Ernst Mach as a Scientist Transgressing the Disciplinary Boundaries', in Friedrich Stadler (ed.), *Ernst Mach – Life, Work, Influence.* Springer, pp. 3–22.

Stadler, Friedrich and Uebel, Thomas (eds.) 2012. *Wissenschaftliche Weltauffassung. Der Wiener Kreis. Verein Ernst Mach, Wien: Artur Wolf Verlag (1929).* Springer.

Tuboly, Adam Tamas (ed.) 2017. 'The Life and Work of Philipp Frank', *Studies in East European Thought*, Vol. 69, Issue 3.

Thiele, Joachim 1978. *Wissenschaftliche Kommunikation. Die Korrespondenz Ernst Machs.* Henn Verlag.

Uebel, Thomas 1992. *Overcoming Logical Positivism from Within. The Emergence of Neurath's Naturalism in the Vienna Circle's Protocol-Sentence Debate.* Rodopi.

von Mises, Richard 1938. 'Ernst Mach und die Wissenschaftliche Weltauffassung. *Zu Ernst Machs 100. Geburtstag am 18.Februar 1938*', Van Stockum. English translation in Brian F. McGuinness (ed.), *Unified Science.* The Vienna Circle Monograph Series, originally edited by Otto Neurath, in an English edition. D. Reidel, 1987, pp. 166–190.

1939. *Kleines Lehrbuch des Positivismus. Einführung in die empiristische Wissenschaftsauffassung.* 2.Aufl. Mit einer Einleitung hrsg. von F.Stadler. Frankfurt/M.: Suhrkamp 1990. English translation *Positivism. A Study in Human Understanding.* Dover Publications, 1968.

Wolters, Gereon 1987. *Mach I, Mach II, Einstein und die Relativitätstheorie.* De Gruyter.

Zilsel, Edgar 1916. *Das Anwendungsproblem. Ein philosophischer Versuch über das Gesetz der großen Zahlen und die Induktion.* J. A. Barth.

1929. 'Philosophische Bemerkungen', *Der Kampf* 22: 178–186.

2000. *The Social Origins of Modern Science.* Eds. D. Raven, W. Krohn, and R. S. Cohen. Kluwer.

Zilsel, Edgar and de Santillana, G. 1940. *The Development of Rationalism and Empiricism.* University of Chicago Press.

11

Narratives Divided

The Austrian and the German Mach

MICHAEL STÖLTZNER

The recent historiography of philosophy of science has not only demonstrated the diversity of Logical Empiricism; it has also documented the large extent to which the leaders of that movement traded in historical narratives and built strategic alliances. Many of these narratives invoked the legacy of Ernst Mach. This is most visible in the 1929 manifesto in which the Vienna Circle went public with the plea for a scientific world-conception that embraced not only modern logic and foundational physics, but also empiricism in the social sciences and modernist architecture. It was signed by Hans Hahn, Otto Neurath, and Rudolf Carnap in the name of the Verein Ernst Mach (1929) and dedicated to Moritz Schlick. The picture of Mach drawn there was unlikely to sway the majority of the German-speaking physicists and mathematicians assembled for their societies' annual conference at Prague, where Mach had taught for twenty-eight years.

By 1929, most physicists and philosophers in the German-speaking world had already settled on what they associated with Mach: anti-atomism, positivism, or phenomenalism on the negative side; breaking the path for Einstein's relativity theory or a new enlightenment on the positive side. Many of these associations, and the arguments supporting them, had remained surprisingly stable since the polemic between Max Planck and Mach two decades before. The polemic itself was a frequent point of reference in philosophical papers and academic addresses that German-speaking physicists had delivered since, mainly in response to the dramatic changes in the foundations of physics brought about by statistical physics, the relativity theories, and quantum mechanics. In these papers, I shall argue, one can discern two different understandings of Mach's philosophical legacy at work – to wit, a

I vividly remember the long discussion that Veronika Hofer and I had with Erik C. Banks while sitting on a Ghent square during the 2014 International Society for the History of Philosophy of Science (HOPOS) conference. He understood so much of the original Mach, the one deeply embedded in the biology, psychology, and physics of his time, and he was far beyond the many narratives that had evolved around Mach's thinking and that were still shaping the philosophical discourse.

German one that was largely informed by Planck's reading, focusing on the principles of theoretical physics and constructing the idea of a consistent Machian philosophical system, and an Austrian one, focusing on Mach's broader empiricist and anti-metaphysical stance. While these understandings were not necessarily tied to a positive or negative assessment, they often amounted to simplifications, transformations, or even contortions of Mach's thinking, which made it increasingly difficult to declare oneself in Mach's footsteps and simultaneously advocate scientific modernism.

This plurality of readings, to be sure, was facilitated by the fact that Mach integrated, or even explicitly endorsed, viewpoints or scientific achievements that for some philosophical interpreters pointed in opposing directions or were even contradictory. For example, later editions of his *Mechanics* (*Die Mechanik in ihrer Entwickelung historisch-kritish dargestellt*) discussed the most recent formal developments in a detailed fashion, but insisted that these theorems were 'new only in *form* and not in *matter*' (Mach 1883/1988, pp. 389/452, emphasis in original).

Mach was not a system philosopher, but one of the pioneers of a long-lasting tradition of physicist–philosophers who, in academic speeches and popular writings, prefaces, and monographs, undertook the philosophical interpretation of their own disciplines. Their goal was to address pressing issues in the foundations of science, not the defence or modification of a philosophical system, such as the attempts of Marburg neo-Kantians to adapt transcendental philosophy to a changing scientific landscape. Compared to the speeches and papers by Hermann von Helmholtz, Heinrich Hertz, or Ludwig Boltzmann, Mach stood out for his detailed and book-length historical-critical analyses of the physical disciplines. But they did not present a fixed philosophical system, even though some followers of Mach attempted to reformulate contemporary physics as a science free of hypotheses. During the same period, it is true, Logical Empiricists were debating the best methodology for a scientific philosophy, or a logic of science. But those with a scientific background especially continued to engage other physicist–philosophers following the traditional role model.

In analysing the diverging pictures of Mach and their employment in philosophical narratives, the present chapter focuses on four exemplary debates and their context. It does not attempt to depict the historical reception of Mach's thinking in Austria and Germany at large. Nor does it discuss the broader narrative of an Austrian Philosophy – from Bernard Bolzano to Ludwig Wittgenstein and the Vienna Circle – that is put forward in the manifesto and later writings of Otto Neurath, and that has played a significant role in the early historical research into the Vienna Circle.

The chapter proceeds thus: (i) Planck's 1908 Leiden speech 'The Unity of the Physical World Picture' started a polemic that prompted Mach to outline his 'Leading Thoughts'. (ii) Ironically, Planck believed that he had previously been a Machian. (iii) The polemic also had a significant influence on how the obituaries for Mach construed his legacy and significance for contemporary science and its philosophy. (iv) Planck would soon be engaged in another polemic with the Viennese Franz Serafin Exner, who advocated a distinctively Viennese synthesis between the ideas of Mach and Boltzmann, while Planck had invoked Boltzmann's atomism against Machian positivism. (v) Mach's legacy also figured prominently in the opening session of the above-mentioned 1929 Prague Congress, where Philipp Frank and Richard von Mises sparred with Arnold Sommerfeld about the value of positivism and the role of Mach's 'sloppy laws of nature' in the newest physics. (vi) In 1932, Moritz Schlick, Planck's former student and Vienna Circle member, declared that the verificationist criterion of meaning allowed a reconciliation between positivism and realism, which Planck largely accepted – not least because Schlick had accepted Planck's understanding of positivism. In a short conclusion at the end of the chapter, I summarise the main developments.

The Planck–Mach Debate and the Emergence of a German Mach

At the beginning of his Leiden speech, Planck distinguished two mutually enhancing and correcting methods in science. Careful description in the sense of Gustav Kirchhoff and Mach is confined to observations as the only legitimate basis of physics. Theoretical research, on the other hand, boldly generalises particular results and seeks a conceptual unity in the manifold of experiences. Mach replied that 'certainly no one has any objection against [unifying systems in physics], least of all representatives of the *economy of thought*' (Mach 1910, p. 230, emphasis in original).[1] But Mach could never assent to reducing one domain of experiences to another by merely theoretical means – that is, before this unity was established as a fact.

A Dispute about Boltzmann's Legacy

While Mach and the empiricist tradition defined the basic quantities of physics by reference to specific sensory experiences, such as heat and muscular effort, Planck diagnosed 'that the human-historical element in all physical definitions has significantly diminished' (Planck 1908/1944, p. 3). Today 'temperature is theoretically defined by the absolute temperature scale which is taken from the second law of thermodynamics, in the kinetic theory of gases it is defined by the living force of molecular motion, practically by the volume change of a thermometric substance' (Planck 1908/1944, p. 3). Mach, to be sure, rejected both theoretical definitions. To Planck, they expressed 'a certain emancipation from the anthropomorphic elements, in particular from the specific sense impressions' (Planck 1908/1944, p. 4).

---

[1] Unless a translation is indicated, translations from German are mine.

Having achieved this emancipation for the second law of thermodynamics was the 'life work of Boltzmann' (Planck 1908/1944, p. 14). Sadi Carnot's cyclic processes describing the workings of an ideal thermodynamic machine were still 'too much tailored for human demands' (Planck 1908/1944, p. 9); Rudolf Clausius's definition of entropy by the impossibility of a *perpetuum mobile* (i.e. the impossibility of producing work from heat) employed thought experiments based upon (de-anthropomorphised) ideal machines and processes. Boltzmann made clear that there existed a fundamental separation between reversible and irreversible processes, which 'will finally play the lead in the physical world view of the future' (Planck 1908/1944, p. 11). All reversible processes, Planck continued, are governed by the Principle of Least Action, which, according to Planck, comes closest to the ideal of scientific enquiry. It states that compared to all possible evolutions of a dynamical system consistent with the boundary conditions, the one actually realised is always that which leads to a stationary value (typically a minimum) of a certain integral quantity. Since this principle is more general than the principle of conservation of energy and yields a unique answer to all problems of reversible physics, Planck considered this side of the physical world view as completed – notwithstanding the fact that reversible processes were an idealisation. 'In the realm of irreversible processes, however ... the principle of entropy increase introduces an entirely novel element into the physical world view' (Planck 1908/1944, p. 11).

Planck had insisted on this fundamental separation between the Principle of Least Action and the second law, which he understood as an independent principle, long before he gradually adopted Boltzmann's statistical methods, in a series of attempts to interpret his formula describing black-body radiation. While this process was not completed by the time of the Leiden speech (cf. Kuhn 1987), Planck praised Boltzmann's 'general reduction of the concept of entropy to the concept of *probability*' (Planck 1908/1944, p. 14, emphasis in original). But the probabilistic character of the second law allowed local violations that were highly improbable. While Boltzmann held 'that such strange events contradicting the second law of thermodynamics could well occur in nature' (Planck 1908/1944, p. 14), Planck was glad that applying the 'hypothesis of elementary disorder' (i.e. by assuming that the microscopic events are occurring in accordance with the statistical regularities assumed by Boltzmann) restores 'the necessity of all natural events' (Planck 1908/1944, p. 14f.). But doing so introduced some spooky non-local entanglements that may 'remind us of mysterious relationships in intellectual life' (Planck 1908/ 1944, p. 17). Mach was unconvinced:

> I cannot deny my aversion to hypothetico-fictive physics.... If Boltzmann discovered that processes in accordance with the second law are very probable while those contrary to it are only very improbable, then I cannot accept that it has been proved that nature behaves according to

this theorem [*Satz*]. Also, I don't think it is right for Planck to accept the first part without wanting to accept the second part, for both halves of the conclusion are inseparable from each other.

(Mach 1910, p. 231)

Facts and Principles

Of course, Mach rejected Boltzmann's proof of the probabilistic second law from hypothetical atoms and remained committed to phenomenological thermodynamics. However, the differences reached deeper. Mach's understanding of facts and principles and his notion of causality were profoundly different from Planck's. Mach's *Mechanics* speaks about the 'principle of the lever' (not 'law' as it was commonly called), while it mentions the 'theorem of least action' – to Helmholtz and Planck the 'principle' *par excellence*.[2] To Mach, principles correspond to the most elementary facts that were accessible to direct experience, while theorems correspond to the more complex facts whose formal development aims at 'an order easy to survey … or a *system*, such that each [fact] can be found and reproduced with the *least intellectual effort*' (Mach 1883/1988, pp. 444/516, emphasis in original). The key point here was that Machian facts played on several levels, from directly intuited everyday regularities to the grand fact expressed in energy conservation (cf. Feyerabend 1984).

Planck severed this connection and put this distinction at the centre of a grand historical narrative. '[I]n all recent conflicts [between facts and theories] the great general physical principles held the field, namely, the principle of conservation of energy, the principle of conservation of momentum, the Principle of Least Action, the laws of thermodynamics' (Planck 1913/1944, p. 44). Meanwhile, well-accustomed intuitive foundations, among them the immutability of chemical elements and the continuity of all dynamical effects, had to give way. Needless to say, Mach would have been the first to admit the fallibility of our intuitions. But where Mach wanted to maintain a continuity between instinctive everyday experience, the knowledge of craftsmen, and foundational science, Planck's unity of the physical world view was exclusively construed from the top down. For Mach, establishing a fact, large or small, could not be replaced by deduction from first principles.

Mach's above-cited criticism of Planck's inconsistency as regards the second law was poignant because elementary disorder was ill-defined, and the debates would go on for a long time. Whether Planck could have it both ways depended on what was a lawful fact. If Boltzmann's statistical

---

[2] Admittedly, in *Knowledge and Error*, Mach returns to the common use for the abstract 'principles' (Mach 1905/1976). The authorised English translation by McCormack uses the word 'principle' in all cases.

mechanics were established as fact, Mach would have had no problem also considering improbable events as lawful. This was a consequence of his liberal notion of causality, outlined already in 1872, which 'is identical with the supposition that among the natural phenomena $\alpha \beta \chi \delta \ldots \omega$ there exist certain equations [of the form $f(\alpha \beta \chi \delta \ldots \omega) = 0$ or an equivalent form]' (Mach 1872/1909, pp. 35–36). The elements figuring in them could be physical, physiological, or psychological, and they were not ultimate, but defined in order to grasp the relevant facts in an economical fashion. As Banks rightly put it, 'Mach was quite capable of postulating unobservable elements when it suited him and even adding unobservables in causal relations to present sensations' (Banks 2003, p. 6).

The Creation of a Narrative

In the fourth section of his Leiden speech, Planck launched a vigorous polemic. He targeted, above all, Mach's anti-atomism, and he charged the principle of economy with being a vague and fruitless maxim for scientific research. 'By their *fruits* shall ye know them!' (Planck 1908/1944, p. 24, emphasis in original) – a biblical allusion which more than anything else was to provoke Mach. And it did:

> One sees that physicists are on the best way to becoming a church and they have already acquired the familiar means of doing so. Let me answer plain and simple: If belief in the reality of atoms is so essential to you then I renounce the physical way of thinking . . ., in short, I thank you for the community of believers. But freedom of thought is more precious to me.
>
> (Mach 1910, p. 233)

Mach's rhetoric makes the issue of atomism appear more central than it actually was. More important was his rejection of a final and definitive world view that rested exclusively upon the principles of physics and that, to Planck's mind, had already been partially achieved. 'Physics does not own the entire world; there is *biology* as well and it is part and parcel of the world view' (Mach 1910, p. 237, emphasis in original). Planck's 'concern for a physics which is valid for all peoples, including Martians, and all times, while so many physical problems of the day press upon us, seems very premature to me, indeed almost comical' (Mach 1910, p. 232). Even the Martians, to Mach's mind, would have to develop their own science step by step. As he wrote in his *Mechanics*, 'natural science does not claim to be a *completed* world view, but is aware to work on a future world view. The highest philosophy of the natural scientist consists in *bearing* an incomplete world view and to prefer it to a seemingly completed, but insufficient one' (Mach 1883/1988, p. 479, emphasis in original).

Nevertheless, it is surprising that Mach, unlike the founder of the energeticist movement Wilhelm Ostwald, never accepted Brownian motion[3] as evidence for the existence of atoms, even though (as I will argue below) Mach's conception of causality as functional dependences facilitated the acceptance of laws about fluctuations. It is thus likely that Mach's problems with atomism consisted in the fact that there were no measurable experiences for it, or at least none of a kind that the older Mach could accept.

In the end, Planck presented a narrative about Mach's place in the history of philosophy:

> Machian positivism was a philosophical expression of unavoidable disillusionment [after the fall of the mechanical world view]. In the face of a threatening scepticism, he deserves full credit for having rediscovered that the only legitimate starting point of all scientific research is sense perception. But he overshoots the mark by degrading the physical world view together with the mechanistic one. I am firmly persuaded that the Machian system, when really carried through, contains no internal contradiction, but equally it seems certain to me that its basic significance is only formalistic. It does not really touch the essence of natural science at all, the demand for a *constant* world picture which does not depend upon the changing epochs and peoples ... This constancy ... is that which we now call the real [*das Reale*].
>
> (Planck 1908/1944, p. 22, emphasis in original)

Planck's realism was more complicated than it sounds; in today's terminology, he combined a structural realism about abstract principles with a realism about the fundamental constants of nature (Stöltzner 2003). Structural realism claims that certain formal principles survive scientific revolutions largely unscathed; they are just applied to new objects. Precisely this had been the reason why, to Planck's mind, energy conservation and the Principle of Least Action stood at the top of the physical world view and would survive future revolutions. From a Machian standpoint, Planck's quest for absolute constancy simply replaced the mechanical world view with one based upon the Principle of Least Action – a topic amply treated in the *Mechanics* as a grand fact visualised as systems of pulleys.

---

[3] Brownian motion is the irregular motion of tiny particles, such as gamboge or smoke, suspended in a liquid or gas, which can be observed under a microscope. Since the first observation of the phenomenon by the botanist Robert Brown in 1827, many attempts at an explanation had failed, including those based on some form of atomism or kinetic theory. Only a proper understanding of the statistics of the collisions between the Brownian particles and water or air molecules allowed a breakthrough along these lines. The explanation, given simultaneously by Albert Einstein and Marian von Smoluchowski in 1905, not only took care of the strange phenomenon, but also provided a way to understand the observed fluctuations of the Brownian particles as a measure of the physics on the molecular scale.

In the above-quoted passage, Planck also created the idea of a coherent Machian philosophical system that would shape the understanding of many German physicists. Planck's criticism of Mach's positivism distorted the anti-substantialist understanding of elements into phenomenalism, holding 'that there are no other realities than one's own sensations and that all natural science in the last analysis is only an economic adaptation [*Anpassung*] of our thoughts to our sensations by which we are driven by the struggle for existence ... The essential and only elements of the world are sensations [*Empfindungen*]' (Planck 1908/1944, p. 20). Mach put Planck's wording right and countered with his famous slogan about the task of science: '*Adaptation of thoughts to facts and adaptation of facts to each other*' (Mach 1910, p. 226, emphasis in original). Contrary to Planck's belief, Machian *facts* were not isolated *sensations*, but they are constituted by relatively stable functional dependences between the (non-atomistic) sensational elements that were formed by us. The stability of these facts was not an ontological necessity, but a consequence of their superiority in the respective epistemological context, which could be analysed by the psychology of knowledge and evolutionary theory. Mach would typically describe this superiority as the economy of a description, meaning not only its simplicity, but also its scope and adaptability to changing contexts.

While in the Leiden speech Planck mainly attacked the scientific infertility of the principle of economy, the rejoinder to Mach's 'Leading Thoughts' took economy as an element of the practical life world. By Mach's 'generalizing it without further ado, the concept of economy ... is transformed into a metaphysical one' (Planck 1910, p. 1187). And Planck ironically reminded his readers that Mach had branded as metaphysical those 'concepts of which one has forgotten how one had arrived at them' (Planck 1910, p. 1188). Planck rejected Joseph Petzoldt's (1890) redefinition of Machian economy as stability because 'One could just as easily make variability or capacity for evolution a demand of "economy"' (Planck 1910, p. 1188). What Planck overlooked is that Mach himself had criticised Petzoldt's conception (cf. Mach 1919, p. 393f.) because it gave a restricted physical basis to what was grounded in biological and psychological evolution. Ironically, it was Planck himself who agreed that stability of our world view was a worthy goal of the scientific enterprise that could not be reduced to economy: 'Therefore the physicist, if he wants to promote science, has to be a realist, not an economist, which means that in the flow of appearances he must search above all for that which is lasting, unchanging, independent of human senses' (Planck 1910, p. 1190).

Planck had spotted an important point, but he failed to turn it into an argument against Mach by confusing Mach's Berlin ally Petzoldt with Mach himself. Mach's ontology was based on facts, large and small, that consisted of stable complexes of functional dependencies. Mach's naturalism implied that their descriptions were adaptable, within limits, to changing circumstances.

But this triggered the question of when such complexes would be sufficiently stable against variation of the circumstances to count as an established fact. Mach's answer picked up an idea advocated by Petzoldt (1895) and Ostwald (1893), and closely linked to the Principle of Least Action – to wit, that the factual state of affairs is the most determined among all possible ones. Since Mach rejected Petzoldt's claim that this was a Kantian or neo-Kantian regulative idea closely connected to stability, the problem remained unresolved.

Planck's rejoinder also contained two specific criticisms of Mach's physical works. First, in his *Wärmelehre* (*Principles of the Theory of Heat*, 1896), Mach had conflated the first and second laws of thermodynamics, a difference that had long before been clarified. Yet while for Planck this difference was the most basic trait of present and future physics, all the energeticists had played it down. Second, Planck criticised Mach's claim about the relativity of rotary motion, often dubbed 'Mach's principle'. Yet, Einstein's obituary for Mach considered precisely Mach's analyses of Newton's treatment of rotary motion as evidence of 'how close his mind was to the demands of relativity in a wider sense' (Einstein 1916/1997, pp. 103/144).

How Machian Was the Early Planck?

Planck's ardour is often linked to his late conversion to Boltzmann's statistical mechanics and to the fact that in his Kiel years (1885–1889) he had considered himself 'one of the most committed followers of Mach.... But later, I turned away from it, because I had begun to see that the glittering promise ... the elimination of all metaphysical elements from physical theory of knowledge, could in no way be carried out' (Planck 1910, p. 1187).

How Machian was Planck's 1887 *Preisschrift* (prize essay) on *The Principle of Conservation of Energy*? It took up the same theme as Mach's 1872 booklet of the same title (Mach 1872/1909), which already contained the gist of Mach's epistemology. The *Preisschrift* made repeated references to it, while Mach's *Mechanics* considered Planck's formulation of the law of causality as 'different only in form' (Mach 1883/1988, pp. 519/607). The principle of conservation of energy, Planck held, is of such a general and universal nature 'that one cannot be careful enough in purging from it all those hypothetical ideas which one is inclined to make up so easily in order to facilitate the overview over the lawful connection of the most diverse natural phenomena' (Planck 1887/1908, p. viii). The impossibility of a *perpetuum mobile* was 'a merely empirical fact, because humans more and more cared to gain work rather than lose it' (Planck 1887/1908, p. 4). Christiaan Huygens was 'instinctively convinced' (Planck 1887/1908, p. 6) of its correctness because it was so closely related to our everyday experiences. From Mach's point of view, such instinctive everyday experiences provide the most reliable justifications (cf. the description of Stevin's discovery of the principle of the inclined plane in Mach's

Mechanics). In Helmholtz's hands, Planck continued, 'the principle of conservation of energy has now become parallel to the principle of conservation of matter, a principle which we have already been familiar with for a long time and which has become, as it were, part of our instincts' (Planck 1887/1908, pp. 41–42). But when discussing the mechanical world view behind Helmholtz's analysis, Planck significantly departed from Mach. With Helmholtz 'a new epoch for the development of natural sciences began.... From now on, one was in possession of a principle which, well-tested in all known domains by careful research, provided an excellent guide also for wholly unknown and unexplored regions' (Planck 1887/1908, p. 101). This did not sound so different from the beginning of the Leiden speech.

In the systematic part of the *Preisschrift*, Planck's realism is clearly discernible. Planck provided an indirect proof of the principle of energy conservation from the empirical impossibility of a *perpetuum mobile*, which crucially depends upon 'the assumption that it is always possible *in some way* to transform a material system from a given state into any other' (Planck 1887/1908, p. 159, emphasis in original). Through this condition, Planck's proof relies heavily upon experiences *not yet had*, and thus violates the basic presuppositions of Mach's empiricism. In the first section of the *Mechanics*, Mach had strongly criticised Archimedes' geometrical proof of the law of the lever because it contained implicit assumptions about the determining conditions which, despite the argument's elegance, must have been previously intuited in nature. The 'aim of my whole book', he explained, 'is to convince the reader that we cannot make up *properties* of nature with the help of self-evident suppositions, but that these suppositions must be taken from *experience*' (Mach 1883/1988, pp. 44/27, emphasis in original). Planck was well aware of this departure: '[P]robably nobody is so inveterate an empiricist not to feel the need for another proof which, built upon a deductive foundation, lets the principle emerge in its most comprehensive meaning from some even more general truths' (Planck 1887/1908, p. 149f.). Nevertheless, Planck did not believe in the a priori validity of energy conservation. 'Natural science only knows one postulate: the principle of causality; for this is the condition of its existence' (Planck 1887/1908, p. 155).

Planck's Kantian take on causality was a far cry from Mach's functional dependencies, which emerged out of our biological interaction with the physical world. And Mach went even further – the requirement of causality dissolved once we had understood a fact:

> There is no cause nor effect in nature; nature has but an *individual* existence; nature simply is [*Die Natur ist nur einmal da*]. Recurrences of like cases in which *A* is always connected with *B*, that is, like results under like circumstances, that is again, the essence of the connection of cause and effect, exist but in the abstraction which we perform for the purpose of mentally reproducing the fact. Let a fact become familiar, and we no

longer require this putting into relief of its connecting marks ... and we
cease to speak of cause and effect.

(Mach 1883/1988, pp. 496/580, emphasis in original)

Hence, the goal of Mach's epistemology was to arrive at a direct description of
the facts in which causes, effects, and hypotheses could eventually be dispensed
with – as essential as they might have been in establishing the fact. Such a
description availed itself of mathematical equations and principles even though
it did not follow from them. But in the case of mature science, the description
of a fact hardly amounted to a simple instinctive perception. Thus, there was a
significant difference between phenomenalism (the idea that science is rooted
in sense perception (*Empfindungen*)) and phenomenology (the idea that science consists in a description of facts). It was none other than Boltzmann who,
using Mach's analysis of causality, tried to separate Mach's phenomenology
from the atomism question and drive a wedge between Mach and the energeticists. In 1897, Boltzmann contrasted atomism to mathematical phenomenology and to energeticist phenomenology, emphasising that the first two agree
to the extent that 'differential equations ... are evidently nothing but rules for
forming and combining numbers and geometrical concepts, and these are in
turn nothing but mental pictures' (Boltzmann 1905/1974, pp. 142/42).
Energeticism, on the other hand, amounted to a relapse into metaphysics,
knowing only a single substance – energy – and its modifications. As seen
above, Mach's ally Petzoldt had not shied away from such a metaphysics. He
was criticised by Mach, but Planck took him for an authorised interpreter
of Mach.

What Was Mach's Legacy? Three Obituaries

Einstein Sets the Stage

Einstein's obituary for Mach, published in the widely read *Physikalische
Zeitschrift*, emphasised his significance as a critic of Newtonian mechanics
and placed him in the broader empiricist tradition of David Hume that had
stimulated Einstein's own thinking. He quoted passages from the *Mechanics* in
the form of a dialogue between Mach and Newton. The significance of minds
such as Mach, Einstein asserted, lies in the criticism of purported necessities of
thought: 'Concepts that have proven useful in ordering things can easily attain
an authority over us such that we forget their worldly origin and take them as
immutably given' (Einstein 1916/1997, pp. 102/142). Einstein praised Mach's
'desire to find a point of view from which the various branches of science, to
which he dedicated his lifelong labour, can be seen as an integrated endeavour.
He comprehended all science as a striving for order among individual elementary experiences which he called "sensations" [*Empfindungen*]. This choice
of word is probably the reason why those who are less familiar with his works

often mistook the sober and careful thinker for a philosophical idealist and solipsist' (Einstein 1916/1997, pp. 104/145). While Einstein's defence tacitly corrected several aspects of the picture drawn by Planck, including the fact that Mach was not a philosopher but 'an ardent researcher of nature' (Einstein 1916/1997, pp. 104/144), Einstein himself slipped into a phenomenalist choice of words: 'Science is, according to Mach, nothing but the comparison and orderly arrangement of factually given contents of our consciousness' (Einstein 1916/1997, pp. 102/142).

Duelling Obituaries

By 1916, *Die Naturwissenschaften* had become the main venue for debates among physicist–philosophers. The obituary 'Ernst Mach's Lifework' penned by the Jena physicist Felix Auerbach largely adopted Planck's reading (most importantly, the idea of a Machian system and phenomenalism), but gave it a positive twist. It was not choice of words, as Einstein held, but rather Mach's 'fear of metaphysics' (Auerbach 1916, p. 181) which prevented an appropriate reception of his thinking. His philosophical 'system' rested upon two pillars: the principle of economy and the idea that 'the given' consisted of elementary sensations. As motivation for the principle of economy, Auerbach identified the Principle of Least Action – which he called the 'principle of the least quantity of force'. This showed, to his mind, that great care was required when applying the concept of economy, and even more when extending it to intellectual processes. But 'the principle comes like a shot, almost as a teleological maxim; similar to Kant ... Mach comes from the a priori given to experience. And this is surely metaphysics, even though critical metaphysics' in a Kantian sense (Auerbach 1916, p. 179). This largely matched Planck's criticism and his complete neglect of the principle's naturalist basis in Mach's psychology of enquiry and Lamarckian evolutionary biology. Although Auerbach, in contrast, properly appraised Mach's neutral monism, using a Kantian perspective as comparison rendered Mach, at bottom, a phenomenalist, according to whom sensations were the 'final elements' (Auerbach 1916, p. 181). This interpretation again missed the flexibility present in Mach's elements and the subtle interrelation between facts, or stable complexes, and the elements they consist of. Hence, in effect, Auerbach came rather close to Planck's reading of Mach's epistemology.

But when it came to the classical polemics, Auerbach took an intermediate position. He rejected Planck's claim that our world view was already partially complete. Was Planck not aware 'that his electrodynamical–thermodynamical theory including its quantum consequence ... is, on the one hand, too special and, on the other hand, too complicated to count as an ultimately satisfactory system? ... Doesn't he, who has grown up under the wings of Helmholtz and Kirchhoff, and who has acted alongside Heinrich Hertz, know well that all these are just pictures which we make of the world ... in order to arrive at coherent knowledge?' (Auerbach 1916, p. 181). And Auerbach wondered how Planck could have condemned wholesale Mach's *Principles of the Theory of Heat*, 'a book from which so many mature scholars and thousands of maturing younger scholars have learnt infinitely much' (Auerbach 1916, p. 181). On the field of historical-critical-epistemological investigations, Mach was 'the qualified if not the only existing guide' (Auerbach 1916, p. 182). In understanding Mach's approach not just as history, but – as we would say today – historical epistemology, Auerbach not only criticised Planck's diagnosis of the present state of physics as being already partially complete. He also criticised Planck's presenting of undeniable progress as an approximation to a final universal theory that would perhaps never be reached, but would at least contain the general principles of contemporary physics. This was indeed the core of the Leiden speech.

The paper published by Frank in the following year in *Die Naturwissenschaften* deliberately countered the idea of a Machian system. 'There is something fascinating about his simple, straightforward teachings.. ... There are indeed but few thinkers who can provoke such sharp differences of opinion, who are so inspiring to some and so utterly repugnant to others' (Frank 1917/1961, p. 69). Frank's specific targets were Planck and the mathematician Eduard Study, who had criticised Mach is his book *Die realistische Weltansicht und die Lehre von Raume* (*The Realistic World-Conception and the Theory of Space*) (Study 1914).

Frank had criticised Planck before. In a 1910 review of the Leiden lecture, he spotted various misunderstandings. The real conflict, Frank thought, was that Planck – in contrast to Mach – assumed that our present physical world view possesses some lasting traits, which are counted as real. Frank rejected Planck's assumption of realism as a guiding principle, it being 'even less admissible to repeat now what had happened with God, freedom, and immortality in favour of atoms and electrons' (Frank 1910, p. 47). In a review of the third edition of Planck's *Preisschrift*, Frank rejected the idea that energy conservation was both the most important guiding principle of scientific research and had empirical content. 'There is still a breach through which the skilfully expelled "conventionalism" can intrude into this [general] form of the energy law and this lies in the concept of "the same state"' (Frank 1916, p. 18). Frank's line of reasoning shows that also he was no stranger to embedding the Mach–Planck debate into a historical narrative, for it was precisely French conventionalism, outlined in the works of Pierre Duhem and Henri Poincaré, that early on allowed the Vienna Circle to circumvent Mach's Lamarckian biology (and Boltzmann's mechanical Darwinism), which they considered as problematic relics of the nineteenth century that impeded the development of a scientific world-conception based on twentieth-century logic.

Frank's 1917 article provided ample historical evidence about scholars, including Michael Faraday and James Clerk Maxwell, successfully taking a phenomenalist approach to physics, which Planck had deemed fruitless. But were their successes – their ability to imagine new physics – not just owed to the fact that they behaved as normal physicists and in contradiction with their phenomenalist methodology? Frank considered this the wrong approach to understand Mach's project. Taking Study's comparison of Mach with Leopold Kronecker's project of reducing all mathematical theorems to statements about integers, Frank emphasised that the working mathematician would never formulate his theorems and proofs in such a way. In the same vein, 'it is not a question of actually expressing all physical statements as statements about relations among sense perceptions [*Sinnesempfindungen*]. It is important, how-ever, to establish the principle that only those statements have a real meaning that *could* in principle be expressed as statements about the relations among our sense perceptions' (Frank 1917/1961, pp. 67/74, emphasis in original). It is quite ironic that Frank here adopts the very terminology that Einstein had seen as responsible for Mach being misunderstood as a follower of Berkeley. Recall that Mach had reprimanded Planck precisely for speaking of the adaptation of thoughts to sensations (*Empfindungen*) rather than to facts. Moreover, Frank largely identified Mach's positivism with phenomenalism, counterbalancing it with the principle of economy. 'The phenomenalist conception becomes a danger only in those cases in which the requirement of economy is not realised with equal intensity' (Frank 1917/1961, pp. 69/74). The extreme phenomenalist Goethe succumbed to this danger. Frank's notion of phenomenalism seems rather vague, and it seems to follow the rather broad and syncretistic under-standing of Hans Kleinpeter (1913), who is credited in Frank's paper for having drawn analogies between Mach and the positivist Nietzsche.

This shows that Frank's intention was not a subtle analysis and defence of Mach's epistemology. Rather, he saw Mach's main achievement as his having adapted the great project of Enlightenment to the present epoch. Its main tenet – as Mach amply described in the *Mechanics* – was 'the protest against the misuse of merely auxiliary concepts' (Frank 1917/1961, pp. 70/80) as absolute foundations because this bore the danger of conceiving any change in the foundations of physical theory as a bankruptcy of science as a whole. Of course, each epoch creates its own auxiliary concepts, which may in turn transcend their own domain of definition. 'The work of Mach is therefore not essentially destructive . . . but on the contrary it is an attempt to create an unassailable position for physics' (Frank 1917/1961, pp. 68/75), despite con-stant change of theories. While Planck believed that parts of theories or general principles would eventually become lasting truths, Mach held that it was the functional dependencies that would remain, when the theories were long gone. 'The known connections among phenomena form a network; the theory seeks to pass a continuous surface through the knots and threads of the net. Naturally, the smaller the meshes, the more closely is the surface fixed by the net. Hence, as our experience progresses the surface is permitted less and less play, without ever being unequivocally determined by the net' (Frank 1917/1961, pp. 66/72). Frank did not posit any analogue of Mach's principle of uniqueness in order to guarantee the integrity of the facts constituted by this network. Instead, he affirmed that all of our theories were empirically under-determined and contained an irreducible conventional element. Moreover, he and the Vienna Circle would increasingly concern themselves with the logical structure of this network and openly admit the existence of pragmatic elements in science.

Frank also took a stand on atomism. Emphasising that Mach, above all, strove after concepts that were applicable in all sciences, he concluded 'that Mach allowed himself to be misled by this argument into attacking the use of atomism in physics more sharply than can be justified.... His followers, as is generally the case, often saw in this weakness of the master his greatest strength.... I believe that one can completely free the nucleus of Mach's teachings from this historically and individually conditioned aversion to atomism' (Frank 1917/1961, pp. 68–69/77–78).

Exner's Synthesis between Mach and Boltzmann

On 15 October 1908, Exner delivered his inaugural address as rector of the University of Vienna. His basic question was: why do natural laws exist at all? For one thing, 'these laws do not exist in nature, only man formulates them and avails himself of them as linguistic and calculatory means' (Exner 1909, p. 7). Through this Machian tack, the second law of thermodynamics came out on top. Looking around us, we find that all natural processes are directed. Boltzmann 'was the first to give a definite and clear interpretation of this direction ... showing that the world ceaselessly develops from less probable into more probable, and hence more stable, states' (Exner 1909, p. 9). In the molecular dynamics of a gas, we 'observe regularities produced exclusively by chance' (Exner 1909, p. 13). But highly probable states (viz. stable laws) are only possible for an extremely large number of individual events. However, in virtue of the law of large numbers, this limit is only reachable in practice if external conditions do not change until sufficiently many individual events have occurred. For this reason, disciplines such as biology and geology, let alone the humanities, never attain exact laws.

This shows that Exner had adopted the relative frequency interpretation of probability, or Gustav Theodor Fechner's *Kollektivmasslehre*, instead of restricting probability to the range of possibilities (*Spielraum*) that was left open by the laws of nature – as Johannes von Kries had done. The physical basis for this interpretational move was that Exner's 'thought collective' (as Ludwik Fleck (1935/1979) would have called it) had made substantial achievements in the study of fluctuations that existed between the microscopic and the macroscopic domains when the thermodynamic limit was not yet reached. Among them were the discovery of radioactive fluctuations by Austrian physicist Egon von Schweidler, which Exner's group quickly took as a proof of indeterminism, and the Polish physicist Marian von Smoluchowski's explanation of Brownian motion, arrived at independently from Einstein (Stöltzner 2012). One can condense Exner's philosophical outlook, which I have called Vienna Indeterminism (Stöltzner 1999), into three basis convictions: (1) the highly improbable events admitted by Boltzmann's statistical derivation of the second law of thermodynamics exist; (2) the burden of proof rests with the determinist who has to provide a sufficiently specific theory of microphenomena; and (3) the relative frequency interpretation of probability allows an objective foundation for physical phenomena.

A precondition for (1) was to accept Mach's definition of causality as functional dependencies. It provided the freedom to find an ontological basis for statistical laws and the non-strict regularities found outside of physics as long as one could specify the elements for functional dependencies. Tenet (2) required consistent empiricism. In his *Lectures*, Exner endorses Mach's analysis, but he gives it a new twist insofar as it 'expresses nothing else but the fact that natural processes, to the extent we can observe them macroscopically, that is on average, are lawful' (Exner 1922, p. 674). Yet studying the microscopic realm required Exner to go beyond Mach's narrow conception of theory. 'The kind of natural study which had as its final aim only a *description* of nature in terms of systems of equations is unsatisfactory. And even if this was in place for a while, today research is directed toward a molecular-mechanical understanding of natural processes' (Exner 1922, p. 721, emphasis in original). This research might arrive at an indeterminist micro-theory, which was acceptable given Mach's understanding of natural laws.

While Planck's Leiden speech had already made clear that he did not accept Tenet (1), he attacked Exner's position in his 1914 rectoral address 'On Dynamical and Statistical Regularities':

> This dualism which has inevitably been carried into all physical regularities by introducing statistical considerations, may appear unsatisfactory to some, and there have been attempts to remove it ... by denying absolute certainty and impossibility at all and admitting only higher or lower degrees of probability. Accordingly, there would no longer be any dynamical laws in nature, but only statistical ones; the concept of absolute necessity would be abrogated in physics. But such a view should very soon turn out to be a fatal and short-sighted mistake.
>
> (Planck 1914/1944, p. 63)

Exner responded to Planck point by point. While Planck assumed absolute causality as a necessary precondition for understanding nature, 'nature does not ask whether man understands her or not, nor are we to construe a nature adequate to our understanding, but only to reconcile ourselves as much as possible with the given one' (Exner 1922, p. 709). Exner also criticised Planck's unjustified trust in our habits of thought, which makes it likely 'to fall into a sort of physical mythology' (Exner 1922, p. 709) about an inaccessible micro-theory. His empiricist approach prompted Exner to reject Planck's distinction between reversible and irreversible processes because in nature we only encounter 'irreversible processes which can come, however, arbitrarily close to reversibility' (Exner 1922, p. 710). Between the extremes, there are many intermediate cases. 'Whether a process is reversible or irreversible in fact only depends upon whether the recurrence of a certain state is practically observable' (Exner 1922, p. 711).

Exner and Planck also disagreed about probability theory: 'It is claimed that in its applications probability calculus cannot dispense with the assumption of absolutely dynamical laws for the elementary processes' – here Exner almost quotes Planck (1914/1944, p. 64) – '[but] the assumption suffices that the elementary processes be equally characterised by average laws' (Exner 1922, p. 712). While Planck called for a dynamical explanation of statistical laws, Exner asserted the position he had taken in the rectoral address: 'Nothing prevents us from regarding the so-called dynamical laws as the ideal limit cases to which the real statistical laws converge for the highest degrees of probability' (Exner 1922, p. 713).

Exner's synthesis of Mach and Boltzmann allowed many scholars who had grown up in the context of Vienna's Physics Department to accept the indeterminism inherent in quantum mechanics early on.[4] But in doing so, Exner also opposed Planck's attempts, starting from the Leiden speech, to paint the revolutions of modern physics simplistically as a vindication of Boltzmann's realism and a rejection of Mach's positivism. In an interview with Thomas Kuhn, Boltzmann's former student Frank was quite explicit about the fact that most Viennese physicists did not see such irreconcilable differences: '[S]trange as it was, in Vienna the physicists were all followers of Mach *and* Boltzmann. It wasn't the case that people would hold any antipathy against Boltzmann's theory because of Mach.... I was always interested in the problem, but it never occurred to me that because of the theories of Mach one shouldn't pursue the theories of Boltzmann' (quoted from Blackmore et al. 2001, p. 63, emphasis in original). In a letter to Arthur Eddington written in 1940, Erwin Schrödinger, who had been Exner's assistant until 1920, gave a similar testimony, succinctly explaining the intricacies of Boltzmann's picture realism:

---

[4] I have pursued this line of reasoning in several places, among them Stöltzner (2011).

[W]e did not consider them irreconcilable. Boltzmann's ideal consisted in forming absolutely clear, almost naively clear and detailed 'pictures' – mainly in order to be quite sure of avoiding contradictory assumptions. Mach's ideal was the cautious synthesis of observational facts that can, if desired, be traced back to the plain, crude sensual perception.... However, we decided for ourselves that these were just different methods of attack, and that one was quite permitted to follow one or the other provided one did not lose sight of the important principles ... of the other one.

(Quoted from Moore 1989, p. 41)

Prague: From Enlightenment to the Scientific World-Conception

The shadows of the Planck–Mach debate extended into the days of quantum mechanics. The opening session of the above-mentioned 1929 Prague congress of German physicists and mathematicians was composed of three philosophical talks featuring the congress president Frank, his friend the mathematician Richard von Mises, and the Munich physicist Arnold Sommerfeld.

Reclaiming a Local Hero for a New Movement

Frank's paper 'What do the present physical theories imply for general theory of knowledge?' presented the new scientific world-conception, as outlined in the manifesto, as a consistent application of scientific methods to philosophical analysis, contrasting it with outdated, paradox-laden school philosophy (*Schulphilosophie*). The latter pretended the existence of a separate domain of philosophical truths investigated by genuinely philosophical methods. But insisting on a 'purely physical point of view' was the best guarantee of rehearsing tacitly such 'a philosophy that contains a fossilization of the earlier physical theories' (Frank 1929/1961, pp. 991/119). Frank's specific target was the idea that there existed truths independently of any possible experience of them:

One who considers it obvious that an electron must have at every instant a definite position and velocity – though the measurement of them may be impossible – ... is forced to interpret the quantum-mechanical calculations, which he uses nevertheless, in such a way that these definite positions and velocities of the electron do not determine the future. Since, on the other hand, the doctrines of the school philosophy in the field of mechanical phenomena require strict determinism, one is forced to assume for the motion of the electron some mystical vital causes, similar to organic life.

(Frank 1929/1961, pp. 973/102)

The only solution is to abandon the idea of a correspondence between our thoughts and the real world altogether. As a standard-bearer for such a consistent empiricism, Frank enlisted the local Prague hero Ernst Mach. 'His fundamental point of view was that all principles [*Sätze*] of physics are principles concerning the relations between sense perceptions [*Sinnesempfindungen*], hence principles that state something about concrete experiences [*Erlebnisse*]. All concepts such as atom, energy, force, and matter are . . . only auxiliary concepts' (Frank 1929/1961, pp. 973/102). After a short description of Mach's economy of thought and the metaphysical dangers of auxiliary concepts, Frank concluded with the critical distance familiar from his 1917 paper:

> Neither Mach himself nor his immediate students have systematically carried further his point of view. . . . On the contrary, Mach's teaching, through many presentations, has been washed out into something indefinite rather than built up to a consistent scientific conception of the world. It has even been interpreted again in line with the school philosophy, sometimes more realistically, sometimes more idealistically.
>
> (Frank 1929/1961, pp. 975/105)

While the above terminology indicates that Frank again classified Mach as a phenomenalist, he considered this distinction as an instance of meaningless metaphysics that had no place in the scientific world-conception. In the same way as in the manifesto, Frank showed that its ancestry was broader: the conventionalism of Pierre Duhem and Henri Poincaré, the pragmatism of William James, the logical works of Gottlob Frege, David Hilbert, Bertrand Russell, and Ludwig Wittgenstein. This represented a major change in Frank's assessment compared to 1917: there was a systematic implementation of Mach's approach – to wit, the kind of philosophy of science that Logical Empiricists wanted to develop.

Among the positive results of this programme was Schlick's dissolution of the correspondence theory of truth into the uniqueness of coordination (*Zuordnung*). 'Every verification of a physical theory consists in the test of whether the symbols coordinated[5] to the theory are unique' (Frank 1929/1961, pp. 987/111). However, uniqueness did not imply realism. The agreement of various determinations of h did not warrant the inference to its real existence, as Planck and school philosophy held. Hence, 'the concept of a really existing quantum of action is only an abbreviation for the group of experiences which yield one and the same numerical value for h' (Frank 1929/1961, pp. 989/114).

---

[5] Frank's own English translation reads 'assigned', which is not the standard translation of '*Zuordnung*' in the text of other Vienna Circle members.

Interestingly, Frank never related this uniqueness back to Mach's discussion of the issue.[6]

At least in principle, classical physics allowed one to maintain the idea that exact knowledge of the initial state of a system was attainable by increasing measurement precision at will. But if we stay on the level of possible experiences, Frank contended, each measurement of length ultimately reaches into atomic dimensions and requires radiation of arbitrarily small wavelengths, hence arbitrarily high energy. But this also disturbs the measured object through Compton scattering. This was Werner Heisenberg's famous disturbance argument. As a Machian, Frank rejected the finality claim that is often associated with the Copenhagen interpretation and remained open to future deterministic modifications of quantum mechanics. But simply setting up causal mechanical equations did not amount to actual experiences.

When it comes to ontology, Frank considered statistical collectives as entities that could figure in natural laws. They were ideal objects, but within Frank's conception all theoretical concepts represented abstract entities that were uniquely coordinated to experiences by certain definitions or correspondence rules. Thus, collectives can be coordinated to single experiences (*Erlebnisse*) and, accordingly, represent a possible ontology for physical laws that map probabilities onto probabilities. This shows that Frank, notwithstanding the phenomenalist terminology, allowed rather abstract experiences. These collectives were the basic objects of the relative frequency interpretation, whose mathematical foundations had been developed by Richard von Mises. He spoke next.

Von Mises explained that the principle of causality is '*changeable*, and will *subordinate itself to the demands of physics*' (von Mises 1930, p. 146, emphasis in original). Already 'Newtonian mechanics only provides a useful means of causal explanation of nature as long as *relatively simple force laws entail more complex motions.* . . . Explanation just means reduction to something simpler' (von Mises 1930, p. 146, emphasis in original). Otherwise, Mach's principle of economy would be violated. For the same reason, von Mises had long advocated a consistently statistical approach, even in cases where a deterministic micro-theory was at least in principle available. Hydrodynamics, Brownian motion, and Boltzmann's various attempts to provide a mechanical foundation for the kinetic theory all show that '[t]he transition between the physics of the single elementary body, atom, proton, electron, etc., to the macroscopic phenomena is simply *obtained only by statistics*' (von Mises 1930, p. 148, emphasis in original). By advocating a purely statistical approach, von Mises used the principle of economy to go beyond Tenet (2) of Vienna

---

[6] The career of the idea of uniqueness is not often discussed. For my own view, in the context of the Principle of Least Action, see Stöltzner (2003); for a broader perspective, see Howard (1992).

Indeterminism. Moreover, the declared Machian von Mises (cf. his 1939) rejected the singular nature of quantum mechanics in contrast to all previous statistical theories and, above all, the need to impose a positivist meaning criterion, emphasised by Heisenberg, which was very much at odds with Machian positivism. For Heisenberg, a measurement of a quantum mechanical observable could be analysed not as a physical process, but as a subject witnessing the collapse of the wave packet and registering a measurement value. Excluding the subject from any law-like description was a far cry from Mach's functional dependences which embrace both psychological and physical elements. To a Machian like von Mises, the so-called Heisenberg cut simply dressed traditional metaphysical dualism up as positivism.

Against Mach's Sloppy Laws of Nature

The third speaker, Sommerfeld, emphasised that he differed 'widely in the evaluation of the philosophical background' (Sommerfeld 1929, p. 866). He conceded that

> ... the great Mach inspired the creation of the new phase of quantum theory. It was fully in line with Mach's philosophy when Heisenberg, in his first quantum mechanical paper, declared as his *leitmotifs* the renunciation of the model, the restriction to observable quantities, the creation of an abstract quantum formalism. And in this case, as we all know, Mach's philosophy, as an exception, had fertile effects.... In actual fact [though,] I believe that its effect on physics is typically the opposite.
>
> (Sommerfeld 1929, p. 866)

Sommerfeld also rehearsed the list of failures well-known from Planck's writings, among them Mach's anti-atomism, his support for energeticism, and the short lifetime of Mach's principle within Einstein's relativity theory. Sommerfeld's praise shows how much the Copenhagen interpretation had cemented the understanding of positivism as phenomenalism.

While Frank had tried to separate Mach, the advocate of a renewed enlightenment and consistent empiricism, from the Machians and their anti-atomism and scepticism about relativity theory, Sommerfeld threw these all right back at him. Together with positivism, he rejected the Jamesian pragmatism that had figured prominently in Frank's talk. 'The physicist does not *invent* laws of nature, but has to be thankful to have the privilege to *discover* a fraction of the magnificent unity and harmony of natural laws. The conviction about a mathematical order of nature that is independent of the researching subject and rises above any conventionalism, presents itself in the development of the new physics' (Sommerfeld 1929, p. 866, emphasis in original).

This was vintage Planck (cf. the debate with Exner). The only difference was that Sommerfeld was sceptical about assigning ontological commitments to constants of nature – or at least as far as Eddington's attempts to deduce the fine structure constant were concerned.

Not that Sommerfeld considered the Copenhagen interpretation satisfactory. Heisenberg's matrix elements, it is true, allowed a systematic derivation of the transition probabilities, but they contained the initial and final state, such that we

> 'must take the final state into account as a co-determining factor ... a certain anticipation of admitted possibilities.... In order to exclude the idea of purposiveness [*Zweckmäßigkeit*] that is often linked to the word "finality", we prefer to speak of a *qualified* or *extended* form of causality.... I also believe that in this way we meet the requirements of biology which apparently cannot make do with pure mechanism and must give more scope to probability than classical physics did'
>
> (Sommerfeld 1929, p. 868, emphasis in original).

To approach this problem, Sommerfeld contemplated a broadened version of Niels Bohr's much-debated notion of complementarity – that is, the impossibility of simultaneously measuring certain (non-commuting) observables, among them position and momentum. 'Can this dualism be overcome? It does not appear as if this will be possible in the physical arena. Perhaps rather through some kind of philosophical synthesis' (Sommerfeld 1929, p. 870), by extending complementarity to the relationship between physiology and psychology. And this brought him ultimately back to the basic problem of traditional philosophy: 'Neither materialist nor spiritualist monism has hitherto solved satisfactorily the hybrid nature of organic existence in body and soul' (Sommerfeld 1929, p. 870). Indeed, Bohr's remarks about the extension of quantum mechanical complementarity to the relationship between the animate and inanimate world caused some of the metaphysical misunderstandings that Frank would continue to combat in the following years in searching for a philosophy of biology suitable for the scientific world-conception. Absent such a naturalist basis, positivism was merely a criterion about the semantics of scientific theories.

Schlick and Planck: Rapprochement under the Banner of Verificationism

In a letter to Schlick of October 1932, Sommerfeld wrote: 'Positivism leads to infertility.... Take Philipp Frank who despite all his acumen never tackled a physical problem. I am not a dogmatist in a religious sense, but I am a dogmatist on the issue of natural laws. I cannot tolerate Mach's "principle of sloppy laws of nature" [*Prinzip der schlampigen Naturgesetze*], despite the uncertainty relation.' In his response, Schlick assented to the rejection of Mach's definition of natural law in terms of functional dependencies, but defended Frank and the value of philosophical enquiry.

Sommerfeld's letter was most likely a reaction to Schlick's 1932 attempt to reconcile logical positivism and realism under the banner of verificationism: '[A]nyone who acknowledges our principle must actually be an empirical realist' (Schlick 1932/1979, pp. 30/283). The verificationist principle asserted that 'the meaning of every proposition is exhaustively determined in the given' (Schlick 1932/1979, pp. 29/283), or, more precisely, by the state of affairs in which it is true. Verificationism was, to Schlick, the legitimate core of the positivist schools because the idea that only the given is real was a meaningless metaphysical claim. 'I should charge with folly . . . every philosophical system that involved the claim . . . that the chair against the wall ceases to exist every time I turn my back on it' (Schlick 1932/1979, pp. 14/274). Verificationism, on the other hand, also permitted Schlick to bar Kant's transcendental idealism because natural science always remains in the realm of what Kant called empirical realism. 'Hence the external world of the physical realist is not that of traditional metaphysics. He employs the technical term of the philosopher, but what he designates by means of it has seemed to us the external world of everyday life, whose existence is not doubted by nobody, not even the "positivist"' (Schlick 1932/1979, pp. 25–26/274–275). Already Kant was aware that if we 'claim of an object that it is real, then all this asserts . . . is that it belongs to a law-governed connection of perceptions' (Schlick 1932/1979, pp. 18/274).

Schlick's paper triggered a short correspondence with his former teacher Planck in which they largely reached agreement about this point. Schlick and Planck also agreed that so-called statistical laws are not genuine laws, but must be separated into strict laws and pure randomness. While Planck would remain sceptical about the Copenhagen interpretation, Schlick believed that a consistent application of the verificationist meaning criterion would do the job in the sense that quantum mechanics represented a law-like limit to the knowability of nature.

Conclusion

The far-reaching agreement that Frank and Schlick displayed in their assessment of modern physics during the 1930s shows that the German interpretation of Mach's positivism had finally won the day, not least because the Vienna Circle could not provide a satisfactory naturalistic basis for their epistemology. When the Circle, in the 1930s, went international, they favoured the name 'Logical Empiricism' over 'Logical Positivism'. While many of the Austrians in the Circle, among them Frank, did not consider this change significant (Uebel 2013), the choice of wording eventually paid tribute to that fact that the association of positivism with phenomenalism had ultimately carried the day. Empiricism was a broader framework and allowed them, in their historical narratives, to pay tribute to Mach in a rather general way.

Let me summarise the developments discussed in this chapter in six themes.

(1) Mach's conception of causality and natural laws was liberal enough to accept statistical laws as genuine laws, which became one of the cornerstones of the Vienna Indeterminism. Yet it appeared unsatisfactory and sloppy from the Kantian perspective that informed most German physicist–philosophers.

(2) Following Planck's lead, Mach's rather intricate conception of world-elements (cf. Banks 2003) became viewed as phenomenalism. While many Austrians rejected this association, they agreed that it had become unattractive for a new physics based upon abstract principles rather than a sense-physiological foundation of physical quantities.

(3) The focus on physical theory stripped Mach's epistemological principles of their biological and physiological basis, most importantly the principle of economy and the variation of the determining circumstances that constituted the basis of Mach's experiment-orientated approach. Conventionalism and logical analysis could not make up this deficit.

(4) Mach's attitude towards the mathematics required for twentieth-century physics was ambiguous. While his *Mechanics* provided a detailed discussion of all principles of analytical mechanics, Mach insisted that axiomatisation was unable to squeeze out any physical facts, and that mathematics was nothing but a sophisticated 'economy of counting'. This was unacceptable to both sides.

(5) Over two decades, positivism changed its meaning from a rather general anti-metaphysical stance – a *philosophie positive* – into a meaning criterion that demanded that the empirical sciences restrict themselves to observable quantities. This meaning criterion figured prominently in the interpretations of relativity theory and quantum mechanics, and it was even seen as a way to eliminate the opposition of positivism and realism. Since the Austrians had emphasised Mach's role in a rather general fashion, they did not protest.

(6) In the debates about Mach's legacy, the analysis of historical developments was often intertwined with philosophical interpretations, including the construction of the movement's own narratives (cf. Hofer and Stöltzner 2013). But none of those historical analyses reached the comprehensiveness and detail that had characterised Mach's books. While the Viennese in the Circle certainly felt this lacuna, it was not at the top of their agenda.

References

Auerbach, Felix 1916. 'Ernst Machs Lebenswerk', *Die Naturwissenschaften* 4: 177–183.

Banks, Erik C. 2003. *Ernst Mach's World Elements: A Study in Natural Philosophy.* Kluwer.

Blackmore, John T., Itagaki, Ryoichi, and Tanaka, Setsuko (eds.) 2001. *Ernst Mach's Vienna 1895–1930, or Phenomenalism as Philosophy of Science*. Kluwer.

Boltzmann, Ludwig 1905/1974. *Populäre Schriften*. J. A. Barth; partly translated in *Theoretical Physics and Philosophical Problems*, ed. B. F. McGuinness. D. Reidel, 1974.

Einstein, Albert 1916/1997. 'Ernst Mach', *Physikalische Zeitschift* 17: 101–104; English translation in *The Collected Papers of Albert Einstein, Volume 6*. Princeton University Press, 1997, pp. 141–145.

Exner, Franz S 1909. *Über Gesetze in Naturwissenschaft und Humanistik*. Alfred Hölder.

Exner, Franz S. 1922 *Vorlesungen über die physikalischen Grundlagen der Naturwissenschaften*, 2nd edn. Franz Deuticke.

Feyerabend, Paul K. 1984. 'Mach's Theory of Research and its Relation to Einstein', *Studies in History and Philosophy of Science* 15: 1–22.

Fleck, Ludwik 1935/1979. *Genesis and Development of a Scientific Fact*. Trans. F. Bradley and T. J. Trenn. University of Chicago Press.

Frank, Philipp 1910. 'Review of Planck "Die Einheit des physikalischen Weltbildes"', *Monatshefte für Mathematik und Physik* 21: 46–47.

1916. 'Review of the third edition of Planck's 'Das Prinzip der Erhaltung der Energie'', *Monatshefte für Mathematik und Physik* 27: 18.

1917/1961. 'Die Bedeutung der physikalischen Erkenntnistheorie Machs für das Geistesleben der Gegenwart', *Die Naturwissenschaften* 5: 65-72; English translation in Philipp Frank, *Modern Science and its Philosophy*. Collier, 1961, pp. 69–85.

1929/1961. 'Was bedeuten die gegenwärtigen physikalischen Theorien für die allgemeine Erkenntnislehre', *Die Naturwissenschaften* 17: 971–977 and 987–994; English translation 'Physical Theories of the Twentieth Century and School Philosophy', in Philipp Frank, *Modern Science and its Philosophy*. Collier, 1961, pp. 96–125.

Hofer, Veronika and Stöltzner, Michael 2013. 'Vienna Circle Historiographies', in M. C. Galavotti, E. Nemeth, and F. Stadler (eds.), *European Philosophy of Science – Philosophy of Science in Europe and the Viennese Heritage*. Springer, pp. 295–318.

Howard, Don 1992. 'Einstein and *Eindeutigkeit*. A Neglected Theme in the Philosophical Background to General Relativity', in J. Eisenstaedt and A. J. Kox (eds.), *Studies in the History of General Relativity*. Birkhäuser, pp. 155–243.

Kleinpeter, Hans 1913. *Der Phänomenalismus. Eine naturwissenschaftliche Weltanschauung*. J. A. Barth.

Kuhn, Thomas S. 1987. *Black-Body Theory and the Quantum Discontinuity*. Chicago University Press.

Mach, Ernst 1872/1909. *Die Geschichte und die Wurzel des Satzes von der Erhaltung der Arbeit*. J. A. Barth.

1883/1988. *Die Mechanik in ihrer Entwicklung. Historisch-kritisch dargestellt*, eds. R. Wahsner and H.-H. von Borzeszkowski. Akademie-Verlag, 1989 (first published 1883); English translation, authorised by Mach, *The Science of Mechanics: A Critical and Historical Account of its Development.* Open Court, 1988.

1905/1976. *Erkenntnis und Irrtum. Skizzen zur Psychologie der Forschung.* J. A. Barth; English translation, *Knowledge and Error.* D. Reidel, 1976.

1910. 'Die Leitgedanken meiner naturwissenschaftlichen Erkenntnislehre und ihre Aufnahme durch die Zeitgenossen', *Scientia* 7: 225–240.

1919. *Die Prinzipien der Wärmelehre. Historisch-kritisch entwickelt.* J. A. Barth (first published 1896); English translation, *Principles of the Theory of Heat. Historically and Critically Elucidated.* D. Reidel, 1986.

1921. *Die Prinzipien der physikalischen Optik.* J. A. Barth.

Moore, Walter 1989. *Schrödinger – Life and Thought.* Cambridge University Press.

Ostwald, Wilhelm 1893. 'Ueber das Princip des ausgezeichneten Falles', *Berichte über die Verhandlungen der Königlich Sächsischen Gesellschaft der Wissenschaften zu Leipzig (Mathematisch-Physische Klasse)* 45: 599–603.

Petzoldt, Joseph 1890. 'Maxima, Minima und Oekonomie', *Vierteljahrsschrift für wissenschaftliche Philosophie* 14: 206–239, 354–366, 417–442.

1895. 'Das Gesetz der Eindeutigkeit', *Vierteljahrsschrift für wissenschaftliche Philosophie* 19: 148–203.

Planck, Max 1887/1908. *Das Prinzip der Erhaltung der Energie.* Teubner (first edition 1887).

1908/1944. 'Die Einheit des physikalischen Weltbildes', as reprinted in Max Planck, *Wege zur physikalischen Erkenntnis.* S. Hirzel, 1944, pp. 1–24.

1910. 'Zur Machschen Theorie der physikalischen Erkenntnis. Eine Erwiderung', *Physikalische Zeitschrift* 11: 1180–1190.

1913/1944. 'Neue Bahnen der physikalischen Erkenntnis' ('New Pathways of Physical Knowledge'), as reprinted in Max Planck, *Wege zur physikalischen Erkenntnis.* S. Hirzel, 1944, pp. 42–53.

1914/1944. 'Dynamische und statistische Gesetzmässigkeit' ('On Dynamical and Statistical Regularities'), as reprinted in Max Planck, *Wege zur physikalischen Erkenntnis.* S. Hirzel, 1944, pp. 54–67.

Schlick, Moritz 1932/1979. 'Positivismus und Realismus', *Erkenntnis* 3: 1–31; English translation in *Philosophical Papers, Vol. II*, eds. Henk L. Mulder and Barbara F. B. van de Velde-Schlick. D. Reidel, 1979, pp. 259–284.

Sommerfeld, Arnold 1929. 'Einige grundsätzliche Bemerkungen zur Wellenmechanik', *Physikalische Zeitschrift* 30: 866–870.

Stöltzner, Michael 1999. 'Vienna Indeterminism: Mach, Boltzmann, Exner', *Synthese* 119: 85–111.

2003. 'The Principle of Least Action as the Logical Empiricist's Shibboleth', *Studies in History and Philosophy of Modern Physics* 34: 285–318.

2011. 'The Causality Debates and their Preconditions. Revisiting the Forman Thesis from a Broader Perspective', in C. Carson, A. Kojevnikov, and

H. Trischler (eds.), *Quantum Mechanics and Weimar Culture. Selected Papers by Paul Forman and Contemporary Perspectives on the Forman Thesis*. World Scientific, pp. 505–522.

2012. 'Zur Genese der Schweidlerschen Schwankungen und der Brownschen Molekularbewegung', in S. Fengler and C. Sachse (eds.), *Kernforschung in Österreich. Wandlungen eines interdisziplinären Forschungsfeldes 1900–1978*. Böhlau, pp. 309–340.

Study, Eduard 1914. *Die realistische Weltansicht und die Lehre vom Raume*. Vieweg.

Uebel, Thomas 2013. '"Logical Positivism"–"Logical Empiricism": What's in a Name?', *Perspectives on Science* 21: 58–99.

Verein Ernst Mach 1929. *Wissenschaftliche Weltauffassung. Der Wiener Kreis*. Artur Wolf.

von Mises, Richard 1930. 'Über kausale und statistische Gesetzmäßigkeit in der Physik', *Die Naturwissenschaften* 18: 145–153.

1939. *Kleines Lehrbuch des Positivismus. Einführung in die empiristische Wissenschaftsauffassung*. van Stockum.

12

Phenomenalism, or Neutral Monism, in Mach's *Analysis of Sensations*?

JOHN PRESTON

Introduction: Two Ways of Reading Mach

The more epistemological works of Ernst Mach, notably his great books *The Analysis of Sensations* (1886) and *Knowledge and Error* (1905), have often been interpreted as presenting some version of, or view related to, idealism, such as phenomenalism. Philosophers from diverse traditions – some close to Mach, others very negative about his work – have attributed to him views of this kind. Such readings of Mach unite allies of his such as Richard Avenarius, Hans Kleinpeter, the Logical Positivists (e.g. Rudolf Carnap and Philipp Frank) and their descendants (e.g. Rudolf Haller), with the harshest of his critics, such as V. I. Lenin, Edmund Husserl, Karl Popper, Gerald Holton, and John Blackmore.[1]

There has long been another way of understanding Mach's epistemological works, though. Bertrand Russell, for example, counted Mach as a 'neutral monist', by which term he meant the view that 'the things commonly regarded as mental and the things commonly regarded as physical do not differ in respect of any intrinsic property possessed by the one set and not by the other, but differ only in respect of arrangement and context' (Russell 1914/1956, p. 139).

The latest and best exponent of the neutral monist reading was Erik C. Banks, whose work did more than anyone's to establish that Mach saw himself as, and should be thought of as, a neutral monist and *not* a phenomenalist (Banks 2003, 2014). I think Banks was right that something *like* neutral monism is Mach's best-considered position (or the best formulation of his position). However, that still leaves those who read Mach this way with a problem: how to explain why so

---

I am grateful to Erik C. Banks, a conversation with whom in Vienna during the Ernst Mach Centenary Conference in June 2016 supplied much of the impetus for this chapter. For helpful comments on an earlier version, I am grateful to Pietro Gori and Luca Guzzardi.
[1] A very small sample of the many others who identify Mach as a phenomenalist, either in passing or at length, includes Robert Cohen (1968/1970), Joseph Agassi (1978/1988), and Klaus Hentschel (1985). Blackmore and his co-editors even subtitled one of their books on Mach 'Phenomenalism as Philosophy of Science' (Blackmore et al. 2001).

many intelligent and thoughtful readers, some of them sympathetic to Mach, have thought of him instead as a phenomenalist. I propose to set out the factors which tempt people into reading Mach thus and then to assess the strengths and weaknesses of these two readings.

In this chapter, I consider *The Analysis of Sensations* only, treating that as a relatively self-contained text.[2] Recent scholars making the case for a neutral monist reading tend to do so by referring, entirely legitimately, to unpublished sources as well as to Mach's entire published oeuvre. What I hope to do here is to present an 'internal' view, showing that the text of *The Analysis of Sensations* itself by no means mandates the phenomenalist reading, and that a case for something more like the neutral monist reading can be made from within that book, indeed largely from within its famous but much misunderstood first chapter.

Mach's Argument against Traditional Philosophical Views and the Appearance/Reality Distinction

Mach situates his own view as preferable to traditional philosophical views, and his argument against those turns on the idea that certain familiar tendencies of thought, when pursued, lead to intolerable conclusions, in the form of pseudo-problems. Mach is clear that designating relatively permanent complexes of sensations using singular terms is a perfectly acceptable instance of 'the partly instinctive, partly voluntary and conscious economy of mental presentation and designation' (p. 3), and that it is useful (p. 6). But he thinks its usefulness extends only so far. When we have moved beyond what he calls a 'first survey' of our substance-concepts (p. 5), this initial habit comes into conflict with a *second* movement of thought, 'a more exact examination of the changes which take place in these relatively permanent existences' (p. 5), a movement that involves 'the tendency to isolate the component parts' (p. 6). When it comes to bodies, for example, a natural line of philosophical reasoning then tempts us into a certain misconception:

> The vague image which we have of a given permanent complex, being an image which does not perceptibly change when one or another of the component parts is taken away, seems to be something which exists in

---

[2] Unless otherwise stated, all quotations are from (Mach 1886/1914). I have not hesitated to quote from the English translation, the whole of which was read by Mach (p. xxxiv), whose English was very good (Blackmore 1972, p. 9). But, wherever the wording is crucial, I have also given the original German. Throughout, page references to the German text (in the form 'S. n') are to the relevant volumes of the recent *Ernst Mach Studienausgabe*. No doubt I ought also to apologise for concentrating, as so many have done, on the first chapter of *The Analysis of Sensations*. My excuse is simply that it is one of my favourite philosophical texts.

itself. Inasmuch as it is possible to take away singly every constituent part without destroying the capacity of the image to stand for the totality and to be recognised again, it is imagined that it is possible to subtract *all* the parts and to have something still remaining. Thus naturally arises the philosophical notion ... of a 'thing-in-itself', different from its 'appearance', and unknowable.

(p. 6, emphasis in original)

This idea is, for Mach, the end point of a *reductio ad absurdum* since, like so many thinkers in the wake of Immanuel Kant, he finds the idea of the thing-in-itself to be intellectually intolerable. Here he calls that idea 'monstrous' (p. 6).

Similarly, when it comes to the ego, a dilemma is posited for the supposition that the ego is a *real* unity, the first horn of which is that we would have to 'set over against the ego a world of unknowable entities (which would be quite idle and purposeless)' (p. 28). And much later in the book, Mach distances himself from any conception of 'a transcendental, unknowable ego, which many philosophers perhaps still think it impossible to eliminate as a last remnant of the thing-in-itself' (p. 359, note).

These are among the 'series of troublesome pseudo-problems' (p. xl) that Mach takes his book to be intended to eradicate. He says explicitly: 'That protean suppositious philosophical problem of the single thing with its many attributes, arises wholly from a misinterpretation of the fact, that summary comprehension and precise analysis, cannot be carried on simultaneously' (p. 7),[3] and that 'the ego itself ... gives rise to similar pseudo-problems' (p. 8). He later explains that although an encounter with Kant's *Prolegomena* made an enormous impression upon him around the age of fifteen, two or three years later he came to see the superfluity of the 'thing-in-itself' (p. 30, note). Avoiding the suggestion of any such propertyless and unknowable entities is a *leitmotiv* of his philosophy.

Mach's opposition to the appearance/reality distinction is also based on this same *horror noumena*. Discussing Plato's parable of the cave, which deploys that distinction to such great effect, Mach urges that it has not been thought out to its final consequences, 'with the result that it has had an unfortunate influence on our ideas about the universe. The universe, of which nevertheless we are a part, became completely separated from us and removed an infinite distance away' (pp. 11–12). I take this to be related to the idea that postulating an unknowable thing-in-itself will be inevitable.

---

[3] I have amended the published English translation here, since it doesn't bring out the fact that Mach's words ('*Das vielgestaltige vermeintliche philosophische Problem ...*' (S. 15)) convey the idea that it is the *problem* that is suppositious (not its being philosophical).

Why Mach's View Was Not Phenomenalism

'Phenomenalism' has been the name for different views in the history of philosophy. When Mach's follower Hans Kleinpeter advocated it, for example, he had in mind a 'natural-scientific world view' incorporating various claims, some of which overlap with what I mean here by 'phenomenalism' (e.g. that sensations are the ground of our knowledge), others of which do not (e.g. the rejection of the mechanical world view and the idea that concepts are mere labels; see Gori 2012).[4] I shall be concerned with what I take to be the root of the term's principal modern meaning, according to which everything concrete, including physical objects, can be 'reduced to' 'sensations', where these latter are conceived of as purely psychological phenomena. (Thus, I do not mean to confine it to a thesis about *language*, but rather to capture both linguistic and non-linguistic phenomenalism – this is what most of those who read Mach as a phenomenalist (or a 'sensationalist') have had in mind.)

There is one absolutely central feature of Mach's published views that doesn't fit phenomenalism, and this is the nature of his 'elements'. In *The Analysis of Sensations*, this is stressed several times, first in the following passage, which warns readers against a misunderstanding of the term 'sensation':

> In what follows, wherever the reader finds the terms 'sensation', 'sensation-complex', used alongside of or instead of the expressions 'element', 'complex of elements', it must be borne in mind that it is *only* in the connection and relation in question, *only* in their functional dependence, that the elements are sensations. In another functional relation they are at the same time physical objects. We only use the additional term 'sensations' to describe the elements, because most people are much more familiar with the elements in question *as* sensations (colours, sounds, pressures, spaces, times, etc.), while according to the popular conception it is particles of mass that are considered as physical elements, to which the elements, in the sense here used, are attached as 'properties' or 'effects'.
>
> (p. 16, emphasis in original)

This is reiterated in a similarly insistent passage on p. 44, and then again in a passage within the 1902 'Preface to the Fourth Edition' (p. xl).[5] These passages should leave us in no doubt that Mach did not mean to be the usual kind of 'phenomenalist' – that is, the kind who, after identifying 'elements', would

---

[4] Luca Guzzardi pointed out to me that Kleinpeter sometimes took 'phenomenalism' to be the view that the laws of nature are descriptions of phenomena (rather than attempts to specify real causes).

[5] In Mach's later article 'Some Questions of Psycho-Physics', we are told that 'the same A B C ... are both physical and psychical elements' (Mach 1891, p. 398).

characterise them as psychological and *not* physical. (Obviously, they also rule out, in the same way, his being a materialist.)

There are still two ways of putting the resulting view, though. The first way, which to me more explicitly courts Russell's designation '*neutral* monism', is to say that the nature of the basic reality is *neither* mental nor physical. Mach does sometimes speak this way, saying of forms and colours that they are 'in themselves neither psychical nor physical' [*an sich weder psychisch noch physisch*] (S. 67). The second way of putting matters is to say that the basic reality is always *both* mental *and* physical. Mach seems to have preferred this latter way of putting things, as in the long quotation above. (The original reads '*Sie sind in anderer funktionale Beziehung zugleich physikalische Objekte*' (S. 23).) Another passage giving the same impression, and where '*A*' refers to the green of a leaf, runs:

> Now in its dependence upon *B C D* [sensations of space, touch, and sight] ... *A* is a physical element, in its dependence on *X Y Z* [the elements of a retinal process] ... it is a sensation, and can also be considered as a psychical element. The green (*A*), however, is not altered at all in itself, whether we direct our attention to the one or to the other form of dependence. I see, therefore, no opposition of physical and psychical, but simple identity as regards these elements. In the sensory sphere of my consciousness everything is at once physical and psychical [*ist jedes Objekt zugleich physisch und psychisch*].
>
> (p. 44, S. 48)

This, I think, might most naturally be called a 'dual-aspect' view. However, and when distancing himself from other views, Mach takes care to point out that his is not a conception like that of Gustav Theodor Fechner, which is what Mach *himself* understood by the phrase 'dual-aspect view'.[6] Mach says:

> The view here advocated is different from Fechner's conception of the physical and the psychical as two different aspects of one and the same reality.... We refuse to distinguish two aspects of an unknown *tertium quid*; the elements given in experience, whose connection we are investigating, are always the same, and are of only one nature, though they appear, according to the nature of the connexion, at one moment as physical and at another as psychical elements.
>
> (p. 61)

However, whether we state Mach's view in terms of the basic reality being neither mental nor physical, or alternatively in terms of its being *both* mental and physical, the 'neutrality' of that reality is preserved: it is *equally* physical *and* mental.

---

[6] See Heidelberger (2010), where it is explained that this was Fechner's pre-1855 view only.

A second kind of evidence for Mach's not being a phenomenalist is supplied by his several attempts to situate his own view and his explicit attempts to distance himself from idealism. We now distinguish between idealism and phenomenalism, as Mach did not, but the reason he gives for not being an idealist would make an equally good case for his not being a phenomenalist, in the sense specified above.

In the main text of *The Analysis of Sensations*, although Mach uses the term 'element' throughout the book's first chapter, in the place where he most explicitly justifies using this term in his proprietary sense, he says, 'Usually, these elements are called sensations. But as vestiges of a one-sided theory inhere in that term, we prefer to speak simply of elements' (p. 28). The 'one-sided theory' in question here is idealism. Mach admits that his mature view developed *from* an earlier idealism of his own. When defending himself from the accusation that his view is a Berkeleyan idealism, he says:

> This misconception is no doubt partly due to the fact that my view was developed from an earlier idealistic phase, which has left on my language traces which are probably not even yet entirely obliterated. For, of all the approaches to my standpoint, the one by way of idealism seems to me the easiest and most natural.

(pp. 361–362)

He subsequently explains how a Kantian idealism had been the 'starting-point' of his critical thought, but that he soon gravitated 'towards the views of Berkeley', which he regarded as lying latent within Kant's writings (pp. 367–368). He then mentions that his views went on to develop towards those of Hume (who can also be thought of as an idealist).

None of this idealist intellectual ancestry, however, should make us characterise the view Mach eventually arrived at as idealism (whether phenomenalist or not). In the 1901 'Preface to the Third Edition' of the book, he took the trouble to point out that 'Some passages of the second edition have been cast in clearer form, since they were often understood in a one-sided idealistic sense, an interpretation which I in no wise intended' (p. xxxix).[7] There should be no doubt that by speaking of idealism as 'one-sided' Mach had in mind the way idealism portrays the basic nature of reality as mental (rather than physical). Materialism, for him, was equally 'one-sided' (see Mach 1895/ 1896, p. 385). But this accusation of one-sidedness would apply equally as well to phenomenalism as to any other version of idealism, since phenomenalists, too, in trying to 'reduce' physical things to sensations, conceived of as purely mental, make the basic nature of reality mental. When Mach commented on the phenomenalism which would have been most familiar to

---

[7] See also Mach (1905/1976, p. 114).

him – John Stuart Mill's conception of objects as 'permanent possibilities of sensation' – he expressly distanced himself from it (p. 363).[8]

Finally, if Banks is right, Mach's admittedly occasional (and largely unpublished) gestures towards the existence of 'world-elements' – that is, 'elements that are *not* interpretable as anyone's sensations' (Banks, this volume, p. 263) – would also point decisively away from phenomenalism. No one who thinks the world is constructed out of 'elements' which have to be psychological could acknowledge elements which exist beyond (human, or any) mentality.

Why Mach's View Has Been Thought to Be Phenomenalism

None of these declarations definitively refute the suggestion that Mach was a phenomenalist of some kind. They are all compatible with his view's being a version – a *non*-idealist version – of that view. So the more careful way to put my thesis is that if, but only if, phenomenalism interprets 'elements' as mental but not physical, then Mach would have rejected it.[9] But I hope to show now that the main factors that tempt people into seeing Mach as a phenomenalist (or an idealist of any kind) are in no way decisive in that direction – that they can all be explained by a correct understanding of what he meant in each case. Those main factors in *The Analysis of Sensations*, as I see it, are as follows:

(1) The way in which he understands our existing 'substance-concepts' of bodies and egos as 'complexes of elements'
(2) The way in which he ties 'sensations' (*Empfindungen*) to the subject
(3) His characterisation of sensations and elements as 'given'
(4) His actual examples of 'sensations' and 'elements'
(5) The way in which his 'sensations' and 'elements' are supposed to depend on the physiology of the senses
(6) His characterisation of spaces and times as 'sensations'

Let us look at each of these in turn.

Factor 1: Substance-Concepts as Referring to 'Complexes of Elements'

Mach speaks often of both bodies and egos being complexes of elements. For the former, we have his declarations that what get called bodies *are* 'complexes of colours, sounds, pressures, and so forth' (p. 2), that 'thing, body, matter, are

---

[8] Evidence against the phenomenalist reading from elsewhere includes Mach (1891, pp. 397–398, 1895/1896, pp. 208–209, 1910/1970, pp. 38–40).

[9] As long as the phenomenalist *doesn't* say that, but thinks of all 'phenomena' as equally mental and physical, commentators such as Blackmore and Agassi might be right to characterise Mach as a phenomenalist *as well as* a neutral monist (thinking of neutral monism as a kind of phenomenalism) (Blackmore 1972, p. 64; Agassi 1978/1988, p. 23).

nothing apart from the combination of the elements' (p. 6), that 'Bodies do not produce sensations, but complexes of elements (complexes of sensations) make up bodies' (p. 29), and that 'all bodies are but thought-symbols for complexes of elements (complexes of sensations)' (p. 29).

When it comes to the latter, the provocative ideas are that each ego is a 'complex of memories, moods, and feelings, joined to a particular body' (p. 3), or that 'It is out of sensations that the subject is built up' (p. 26).

Further, both bodies and egos, Mach is keen to insist, are only of *relative* permanency (pp. 2–4), and in 'a more exact examination of the changes that take place in these relatively permanent existences' these complexes are seen to be 'made up of common elements' and can be 'disintegrated into elements' (p. 5).

There is no doubt that these pronouncements are high on the list of things that tempt commentators to compare Mach with Berkeley, for example. However, they should lead us to think of Mach as a phenomenalist or idealist of some kind only if he had *not* told us that 'elements' are always both mental *and* physical. As long as one bears this insistence in mind, there is nothing inherently idealist here.

Factor 2: 'Sensations' and the Subject

Mach talks sometimes of elements and sometimes of sensations (*Empfindungen*). However, some of the things he says *about* sensations (but not about elements) seem to tie them to the subject, and this tie can easily be given an ontological reading. Here, I am not referring to his merely using the *word* 'sensation'. We already saw that he warned his readers about the way he was using that word, and that he tried to accustom us to call them 'elements' instead (p. 23). If the phenomenalist reading relies on ignoring that warning, so much the worse for it. Sensations, for Mach, are not *merely* psychological, but always also physical.

Nevertheless, Mach quite often ties 'sensations' *to the subject* (or to plural subjects). So, for example, in a crucial place where he is presenting his own view, Mach says, 'The assertion, then, is correct that the world consists only of our sensations. In which case we have knowledge *only* of sensations' (p. 12, emphasis in original) [*Es ist dann richtig, dass die Welt nur aus unseren Empfindungen besteht. Wir wissen aber dann eben nur von der Empfindungen*] (S. 20).[10] The word 'our' is what should worry us here. Despite identifying his 'elements' with sensations, Mach doesn't use the phrase 'my elements' and only uses the phrase 'our elements' (e.g. Mach 1905/1976,

---

[10] Elsewhere, the phrase 'my sensations' occurs in 'Some Questions of Psychophysics' (p. 394), and 'our sensations' in Mach (1872/1911, p. 91, SM, p. 559, 1895/1896, pp. 200, 209; 'the world is our sensation').

p. 12, note 7) to mean 'the elements *we* have supposed'. Why, then, can sensations be characterised as mine, yours, or ours? (Unless this merely means 'the *sensations* we have supposed', which seems very unlikely in the context.) And why, if there exist 'world-elements', doesn't the world instead consist of *our* sensations *plus* these others?

Although this is a good question, and this phrase might create a presumption against Banks' reading, it is not good evidence for any idealist reading. By calling them 'our' sensations, Mach may just have been characterising their *accessibility* to us, not their *ownership* by us. The world consists in *our* sensations because those are the sensations we *all* have access to. What's important is that Mach never characterises *elements* as mine, yours, or ours. He can allow that 'sensations' can be owned, but that's because sensations are elements in their *psychological* connections. The laws of *psychology* (but not the laws of physics, of course) relate them to the *ego* (*non*-substantially conceived). I think that whenever Mach refers to 'our sensations' he can be understood in this way.

The import of his claim that 'the world consists of our sensations' is to deny the knowability (and thereby the meaningfulness of any assertion of the existence of) of things-in-themselves, *not* to deny the existence of physical phenomena. As we shall later see, Mach clearly states that 'sensations' *are* partly (or in certain respects) physical.

In the second place where this world-constitution thesis appears, there is no mention of *our* sensations. But something else problematic takes its place:

> [P]erceptions, presentations, volitions, and emotions, in short the whole inner and outer world, are put together, in combinations of varying evanescence and permanence, out of a small number of homogeneous elements. Usually, these elements are called sensations. But as vestiges of a one-sided theory inhere in that term, we prefer to speak simply of elements.
>
> (p. 22)

Here, one problematic idea is that the *outer* world might comprise items from the list: perceptions, presentations, volitions, and emotions. Given that all of those items would normally be thought of as psychological (and *not* physical), Mach's declaration that the 'whole inner and outer world' [*die ganze innere und aüßere Welt*] (S. 28) is composed of them sounds idealist. (Unless by 'the outer world' he means merely our representation thereof, which is unlikely in the context.) But the important thing here is that these items are in turn put together out of his 'elements', which he explicitly tells us are *not* exclusively psychological. Thus, there is no reason here to suspect him of phenomenalism.

However, there is still a problem. How can perceptions, presentations, volitions, and emotions exhaust the contents of the (i.e. be the *whole*) *outer* world? It's difficult to see any way of taking this that isn't idealist, or which would not preclude the existence of genuine 'world-elements' (understood as elements that are not interpretable as 'sensations'). Neutral monist readers might seem well-advised to regard this as one of the more careless things Mach says, an example of what Erwin Hiebert called Mach's 'philosophically unbuttoned' comments.

I suspect that the best construction that can be put upon this remark of Mach's is as follows. 'Elements' are related to one another in two ways. 'Sensations' are merely elements as seen in their *psychological* connection. The whole world consists of 'sensations', then, only in the sense that each element stands in *some* psychological relation to other elements. *Elements* exhaust the content of the world, and each element can be viewed not only in its physical connection to other elements, but also in its psychological connection to them.

This supports the neutral monist reading. Unfortunately, though, I think it must impact negatively either on Banks' characterisation of 'world-elements' as 'elements that are *not* interpretable as anyone's sensations' or on the supposition that Mach postulated such elements, for it puts 'world-elements' *beyond* mentality. It not only makes their physical connections more basic than their psychological connections; it also means that not all elements are psychologically connected. This would wreck the neutrality of Mach's neutral monism, and perhaps it explains why almost all of his remarks on world-elements remained confined to his notebooks.[11]

Factor 3: Characterising Sensations and
Elements as 'Given'

A third factor, linked with this second one, is surely Mach's quite frequent characterisation of sensations and/or elements as 'given'. There should be little doubt that Mach's empiricism involved *some* kind of foundationalism about 'sensations'.[12] So, for example, he says that '*I* have the sensation green, signifies that the element green occurs in a given complex of other elements (sensations, memories)' (p. 23, emphasis in original) [*in einem gewissen Komplex von anderen Elementen*] (S. 30), that 'elements form the real, immediate, and ultimate foundation [*die eigentliche, nächste und letzte Grundlage*], which it is the task of physiologico-physical research to investigate' (p. 29 (S. 34)), that 'colours, sounds, spaces, times ... are provisionally the ultimate elements, whose given connexion it is our business to investigate' (pp. 29–30) [*vorläufig*

---

[11] In fact, when Mach himself used the term '*Weltelemente*' in print (S. 37), he explicitly means the elements he has *already* introduced in the book.

[12] In his chapter for this volume, Banks denies this, and appeals to Feyerabend (1970/1999). But what's said there suggests (rightly or wrongly) only that Mach was not a 'radical' or classical foundationalist.

die letzten Elemente, deren gegebenen Zusammenhang wir zu erforschen haben]
(S. 35), and most explicitly, referring to those elements which compose what
we normally think of as physical objects other than our own body, he calls
them 'immediately and indubitably given' (p. 45) [*unmittelbar und unzwei-
felhaft gegeben*] (S. 49). His empiricist preference for what is 'given' is also
reflected negatively in certain dismissals of what is *not* 'given', such as in the
idea that 'reference to unknown fundamental variables which are not given
(things-in-themselves) is purely fictitious and superfluous' (p. 35) [*die
Beziehung auf unbekannte, nicht gegebene Urvariable (Dinge an sich) eine rein
fiktive und müßige ist*] (S. 39).

However, Mach's willingness to characterise sensations/elements as 'given'
should not be taken to mean that, for him, they must be purely mental or
psychological, which is what would be necessary to sustain any idealist reading. That they are 'given' to us is (again) only an epistemological relation, and
it is not supposed to include or support any ontological classification. That
they are the sorts of things we experience does not mean, for Mach, that their
nature is solely psychological. Later on, in his debate with Max Planck, Mach
was absolutely explicit that 'There can be no question of interpreting
sensations in a purely subjective way, as Planck seems to assume' (Mach
1910/1970, p. 40).

Factor 4: Mach's Examples of 'Sensations' and 'Elements'

As we saw, Mach does warn his readers that the terms 'sensation' and
'element' should not be used entirely interchangeably (p. 23). But the actual
examples of 'sensations' and 'elements' which he gives often have a curious
character. In the first chapter of *The Analysis of Sensations*, all of the following
things are said to be 'elements':

> A stain on a coat or a tear in its fabric (p. 3)
>
> Colours, tastes, and hedonic qualities (usually thought of as properties)
> that are common to different complexes (complexes usually thought of as
> bodies) (p. 5)
>
> Colours, sounds, and other features of bodies which are usually thought
> of as their attributes (pp. 6–7)
>
> The constituents of 'those complexes of colours, sounds, and so forth,
> commonly called bodies' (p. 8)
>
> The constituents of the complex commonly known as our own body
> (p. 9)
>
> The constituents of the complex composed of volitions, memory-
> images, and the rest' (p. 9)
>
> A pencil, its being seen by us as straight, water, the pencil's being seen
> as crooked when half-immersed in the water, a bright surface, and how
> that surface appears (p. 10)
>
> A prick from a pin, and the resulting pain (p. 13)

In the 'white ball' example (discussed below), sounds, colours, pressures, spaces, times, and maybe even the ball itself (p. 16)
    Colours, sounds, spaces, times, motor sensations, etc. (p. 21)
    Green (pp. 23, 43)
    Sensations, memories (p. 23)
    Colours, sounds, spaces, times (pp. 29–30)
    The green of a leaf (p. 44)

When it comes to the term '*Empfindung*', Mach uses that to cover individual
physical bodies such as a table and a tree (p. 13, note), sounds, colours,
pressures, spaces, times (p. 16), as well as the colour green, and the green of
a specific leaf (pp. 23, 43). Having explained his concepts, Mach often uses the
compound expression 'elements (sensations)' (pp. 23, 25, 26) in order to draw
attention to the overlap.

As well as this overlap, various things should be noted about these
examples. Firstly, Mach's examples of elements and *Empfindungen* are entirely
homely phenomena, and not obscure, unfamiliar entities from the theoretical
reaches of any science. No such entity figures as an example. So although
Mach tells us that colours, spaces, and times are only '*provisionally* the
ultimate elements' (p. 29, emphasis added), no future discovery of any more
ultimate basis should be taken as withdrawing these examples.

Secondly, they are phenomena drawn just as much from what philosophers
would think of as the category of the physical as from that of the mental.

Thirdly, they pointedly slew *across* the appearance/reality distinction which,
as we saw, Mach scouted (p. 10). They include what we would normally think
of as phenomena a person merely *thinks* they perceive, as well as phenomena
they really do perceive. In this respect, his elements and sensations might be
thought of simply as *data*.[13]

What tempts people towards an idealist reading here is simply our tendency
to think of many of these things as psychological but *not* physical. For
philosophers, at least, volitions, memory-images, and motor sensations might
seem to be paradigm examples of purely mental events, and an effort would
often be made to squeeze colours and sounds into that same category.

The thing to remember here is that Mach also counts physical things, and
spaces, and times, as complexes of elements. But this is not the end of the
argument. The underlying worry is that when it comes to telling us what
bodies, spaces, and times *are*, Mach can refer only to other 'elements', the ones
that look more eminently and exclusively psychological. This may be so, but

---

[13] Cf. Richard von Mises' intriguing suggestion that 'All the elements, which Mach calls
"sensations", could perhaps be called by a more neutral term such as is used in photog-
raphy, "takes"' (von Mises 1938/1970, p. 261).

for Mach, none of these elements are exclusively psychological. Even our paradigm examples of mental phenomena – no matter how 'inner' or 'phenomenal' they may be – Mach thought of as having both a psychological *and* a physical nature. That was the legacy of his deep immersion in psychophysics: always finding a physiological aspect to *any* psychological phenomenon. Because his interest was in *functions* and *connections*, any way in which a psychological phenomenon could be affected by altering something physical counts as a respect in which that phenomenon is physical. Seen thus, Mach might look to us more like a liberal and non-reductionist kind of 'physicalist', and he might be ranked against more recent thinkers such as Thomas Nagel and David Chalmers, who postulate, or suppose they can discern, mental phenomena which have no physical aspect. His psychophysical project was the investigation of physical (physiological) and psychological aspects of the *same* events or qualities. The events or qualities in question are related to one another in the physical way by certain laws, and in an entirely different, psychological way by a different set of laws. In one of the two available 'directions of investigation' he mentions (p. 18), they are investigated by physiology, and in the other by psychology (whose method he conceived of as introspective).

Factor 5: Dependence on the Physiology of the Senses

This brings us to another factor also falling under the heading of 'things Mach says *about* (what he calls) "sensations" or "elements"': the way he supposes these to vary and to depend upon 'the physiology of the senses'. Consider his discussion of perceiving a cube:

> A cube when seen close at hand, looks large; when seen at a distance, small; its appearance to the right eye differs from its appearance to the left; sometimes it appears double; with closed eyes it is invisible. The properties of one and the same body, therefore, appear modified by our own body; they appear conditioned by it.
>
> (p. 9)

To conclude that properties of physical objects '*appear* modified' by the conditions of the perceiver's body is suitably modest and unobjectionable. Contrast this, though, with a passage from only a few pages later:

> A white ball falls upon a bell; a sound is heard. The ball turns yellow before a sodium lamp, red before a lithium lamp. Here the elements (*A B C* ...) appear to be connected only with one another and to be independent of our body (*K L M* ...). But if we take santonin [a drug which turns yellow on exposure to light], the ball again turns yellow. If we press one eye to the side, we see two balls.
>
> (p. 16)

The general conclusion Mach is driving at here he later expresses thus:

> [T]he elements which belong to the sensible world . . . stand in a relation
> of quite peculiar dependence to certain of the elements *K L M* – the nerves
> of our body, namely – by which the facts of sense-physiology
> are expressed.

(pp. 35–36)

As long as by 'the ball' we mean the physical object in question, a white ball
doesn't literally 'turn yellow' or red depending on whether it is illuminated by
a coloured light or whether one has just taken a drug. Only if by 'the ball' one
is speaking loosely, and meaning something like the way the ball *looks* or
appears, is it correct to say that 'the ball' changes colour under such conditions. This consideration surely plays directly into the suspicion that Mach
must be a phenomenalist since, for this argument for the dependence of
'elements' on environmental or physiological conditions to work, he must be
identifying 'the ball' with its appearances.

Of course, Mach might protest against such an invocation of the appear-
ance/reality distinction (which, we saw, he criticises (pp. 10–12)). But he has to
make his view intelligible to us in terms of the ways of thinking and speaking
we all (by his own account) start from and, given the incorrectness of saying
that the ball really does turn yellow, we are at a loss to find something that
really turns yellow *unless* we make the familiar philosophical move of conjur-
ing up something mental (e.g. a collection of appearances, sense-data, or
whatnot). This, I conjecture, is one of the more powerful reasons people have
had for taking him to be a kind of idealist.

Is there any way of defending Mach against the accusation of incipient
phenomenalism here? Merely deploying his excuse that an earlier idealism left
traces on his way of expressing himself (p. 362) does not look promising. It's
not merely that his language implies the existence of mental phenomena such
as appearances (e.g. in the way sense-datum theorists do). His argument won't
work unless he *identifies* the ball with such phenomena. And that puts him
firmly in the phenomenalist camp.

If there *is* a way to resist categorising Mach thus, it involves insisting again
that to say that something depends upon facts about sensory physiology isn't
yet to identify it as *exclusively* mental. For Mach, that the 'elements' seen alter
in the transition between what we would ordinarily describe as these two
situations:

- one in which we see a white ball
- one in which a sodium lamp is shone on that same white ball

is undeniable, for these elements are all (or almost all) that we experience.
But even to identify them as what is experienced is not yet to make them
exclusively mental. For Mach, what we experience is as much physical as it

is mental. When he tells us that certain matters depend on the physiology of the senses, this should not make us think of those matters as exclusively mental.

Factor 6: Spaces and Times as 'Sensations'

A final factor which makes people think of Mach as a phenomenalist or idealist is, I believe, of even greater importance. Early on in *The Analysis of Sensations* comes the striking declaration that 'The physiology of the senses . . . demonstrates that spaces and times may just as appropriately be called sensations [*Empfindungen*] as colours and sounds' (p. 8),[14] and at this point Mach says that he will explain this idea later in the book. What subsequently happens in the chapters devoted to space (notably chapters VI, VII, and IX) and to time (chapter XII), though, supplies little that goes towards this end. Mach does argue that certain memory-phenomena can be more easily understood if we conceive of time as a sensation (p. 246), and that our idea of an invisible, immovable space must be based on and secondary to the space of motor-sensations (p. 139), but these considerations are surely not weighty enough to warrant the very substantial conclusion that spaces and times *are* sensations.

Mach's conception of sensation (or rather of what counts as an *Empfindung*) is, it has to be said, strange. He remarks:

> Ordinarily pleasure and pain are regarded as different from sensations. Yet not only tactual sensations, but all other kinds of sensations, may pass gradually into pleasure and pain. Pleasure and pain also may be justly termed sensations.

> (p. 21)

From the point of view of our everyday conceptual scheme and the English term 'sensation', this has matters backwards. Pangs of pleasure and pain are paradigms of sensations in the ordinary sense. Tactual sensations and temperature sensations are in the same position. But this does not mean that the other things Mach usually calls 'sensations' (colours and sounds) deserve that title, let alone that it can be applied to spaces and times. The English term 'sensation' is conceptually connected to that of *feeling*, sensations being what is felt in the relevant organ by the person concerned. Pleasure and pain are felt phenomena, there being no unfelt (or 'unowned') pleasures or pains. Touching things with the relevant sensitive parts of one's body, too, does typically issue in feelings, tactile sensations. Sounds *can* produce feelings, and thus sensations, if they are very loud or piercing, but normally they are heard, not felt,

---

[14] See also Mach (1883/1960, p. 611).

and thus involve no sensations. Colours, though, simply cannot be felt in vision, and (*pace* a host of empiricist philosophers) produce no *sensations* in the eye.

Did Mach, despite aiming to free himself (and science) from past philosophies, simply buy into the mainstream empiricist tradition in this respect? Perhaps not. I suspect that he used the term '*Empfindung*' in a much broader way, to mean something like 'anything we can be said to sense' (or even 'anything we can be said to even *think* we sense', since he explicitly includes what we would think of as misleading ways in which things seem to be).

His claim that spaces and times are *Empfindungen* might also be illuminated by going outside *The Analysis of Sensations*, back to his book *History and Root of the Principle of the Conservation of Energy*, where he tells us that 'Space and time are not here conceived as independent entities, but as forms of the dependence of the phenomena on one another' (Mach 1872/1911, p. 95, note). His ambition to move away from the idea that space and time are independent entities (substances) is creditable, and puts some distance between him and any metaphysical view such as idealism. Mach consistently thought of space and time in relational rather than substantival terms. Since ways in which phenomena depend on one another (relations) are neither inherently mental nor inherently physical, no form of idealism is in the offing. Mach's negative suggestion therefore has merit, although there may be no direct connection between relationalism and his view that spaces and times are sensations. Further, if I am right about the way in which he uses the term *Empfindung*, his positive claim amounts merely to the idea that spatial and temporal relations are things we sense. This sounds right, but of course, *pace* Mach, no sensory physiology is required to demonstrate it.

The Physical and the Mental

None of these six factors mean that the sensations/elements Mach has in mind are exclusively or even dominantly mental, though (by his lights). Unfortunately, they do look as if they must be mental when considered by the lights of certain philosophers – those who would classify perceptions, volitions, emotions, etc., as purely mental. Mach is committed to rejecting this, and to rejecting certain familiar associations which attach to our terms. He speaks of a 'great gulf between physical and psychological research [which] persists only when we acquiesce in our habitual stereotyped conceptions' (p. 17).

So far, the single move of denying that 'elements' are solely psychological has served to rebut all of the suggestions that Mach must be a phenomenalist. For Mach, all 'elements' are physical in certain connections and mental in others. However, this throws all of the weight onto his account of these terms, of what we mean when talking of phenomena being either physical or mental.

In virtue of what are 'elements' mental and physical, then? The answer lies in Mach's notion of 'connection' or, more strictly, functional dependence. Unfortunately, he puts this answer in different ways, not all of them satisfactory. Some imply mysteriously that dependence upon something *physical* (viz. the retina) could make an element psychological, and at the same time make it sound as though the *same* set of elements are physical (but *not* psychological) when 'considered as' connected with one another in one way, and psychological (but not physical) when 'considered as' connected with one another in another way (e.g. p. 17). Since 'considering something as so-and-so' is a psychological notion, the idea of aspects of reality either being or even *appearing* mental or physical because of how they are 'considered' (by humans) slants the ground in favour of idealism. The lesson is that Mach's story about what makes 'elements' mental and physical had better not rely on anything which is *drawn from* either of those categories. If it violates this condition by appealing to something psychological, it will inevitably give ammunition to those who think of him as some kind of idealist.

Other formulations, such as when we're told that 'not the subject-matter, but the direction of our investigation, is different in the two domains' (pp. 17–18), court a similar objection. What makes phenomena physical or mental shouldn't depend on matters as contingent and variable as the way in which we are currently investigating.

Happily, though, Mach has a better way of putting matters. What I take to be his official view is best put in one of the passages we saw already, where he says that 'it is *only* in the connection and relation in question, *only* in their functional dependence, that the elements are sensations. In another functional relation they are at the same time physical objects' (p. 16, emphasis in original). This is the way of putting things that gets reiterated in the following passage:

> The fundamental constituents of [(outer) bodies and human mental phenomena] are *the same* (colours, sounds, spaces, times, motor sensations, etc.); only their character of their connection is different.
>
> (p. 21, emphasis in original)

This point is also reiterated in the important Preface to the Fourth Edition, where 'sensations' are said to be 'the common elements of all possible physical and psychical experiences, which merely consist in the different kinds of ways in which these elements are combined, or in their dependence on one another' (p. xl).

The relations adverted to here, of 'connection', and functional dependence are *objective* relations, but not physical or mental relations, so the objections above will not apply to them. Nevertheless, there is something suspiciously solipsistic about a certain 'picture' which Mach thinks results from the basic neutral monist idea:

> In this way ... we do not find the gap between bodies and sensations
> above described, between what is without and what is within, between the
> material world and the spiritual world. All elements $A\ B\ C\ldots, K\ L\ M\ldots,$
> constitute a *single* coherent mass only, in which, when any one element is
> disturbed, *all* is put in motion; except that a disturbance in $K\ L\ M\ldots,$ has
> a more extensive and profound action than one in $A\ B\ C\ldots$ A magnet in
> our neighbourhood disturbs the particles of iron near it; a falling boulder
> shakes the earth; but the severing of a nerve sets in motion the *whole*
> system of elements. Quite involuntarily does this relation of things suggest
> the picture of a viscous mass, at certain places (as in the ego) more firmly
> coherent than in others.
>
> (p. 17, emphasis in original)

The problem here is that it is *only* true that 'the severing of a nerve sets in
motion the *whole* system of elements' if this 'system' is that of a single given
person, since *my* having a nerve severed won't typically affect *any* elements
accessible to you, unless we're nearby and I simply can't keep it to myself.
Officially, Mach had little time for solipsism, but this proposed 'picture' does
seem to court it, as does a closely related picture which came to him as the
result of a sort of conversion experience in his later youth: 'On a bright
summer day in the open air, the world with my ego suddenly appeared to
me as one coherent mass of sensations, only more strongly coherent in the ego'
(p. 30, note 1).

Mach as a *Radically* Revisionist Metaphysician?

Mach, notoriously, did not consider himself a philosopher (p. 30, note, plus
p. 368), or as putting forward any philosophical system (p. xl).[15] He
did publish in philosophy journals, though, and he eventually accepted the
designation 'philosophizing scientist' [*philosophierende Naturwissenschaftler*]
(Hentschel 1985, p. 391).

He would *certainly* have disavowed the title of metaphysician. His scorn for
what he thought of as metaphysics appears absolutely undiminished through
his career (pp. xxxviii, xl, 27, note, 30, note, 35, 369).[16] And yet, when we look
at the things that tempt people into thinking he was a neutral monist, it is as a
metaphysician that Mach now most naturally appears. In fact, in terms of
Peter Strawson's famous dichotomy, he might seem to be the most radically

---

[15] Elsewhere, in the original Preface to *Knowledge and Error*, Mach again protests that he is
not a philosopher (Mach 1905/1976, p. xxxi), and he explains that he aimed not at
introducing a new philosophy into science, but at removing an old one from it (p. xxxii),
although he does express the hope that philosophers might one day recognise his
endeavour as a philosophical clarification of scientific methodology (p. xxxiii).

[16] See also Mach (1872/1911, pp. 9, 17, 1896/1986, pp. 1, 3).

revisionary of metaphysicians, intent on sweeping away the conceptual scheme common to our ordinary activities *and* science and replacing it entirely with a new science-inspired scheme, his scheme of elements. Of course, Mach might appear thus (and did not appear so to himself) because our idea of metaphysics has changed. For us, the notion has changed by being broadened through the idea of *naturalistic* metaphysics (and the associated naturalistic rejection of any 'first philosophy'). Could we bring him back today, Mach might well have come to see that he *was* a metaphysician, a metaphysician of this new, naturalist kind.

What, after all, did Mach understand by 'metaphysics' and the 'metaphysical'? When he used these terms, he usually did so in what he took to be Kant's sense, meaning that which transcends experience (pp. xl, 359, note), and he associated this with the activities of scholastics, rationalist philosophers, and Kantians, all of whom postulated ontologies that were thus 'metaphysical'. This does not rule out thinking of Mach as also presenting an ontology; it means only that his ontology was not a metaphysical one.

We might therefore think of Mach as (1) associating metaphysics with *philosophy* (but not with science), (2) rejecting the notion that science (as an empirical discipline) could really answer metaphysical questions (since those questions are themselves suspect), (3) rejecting *any* philosophical (i.e. a priori) attempt to say what the fundamental nature of reality is or what the fundamental components of reality are, and finally (4) insisting that his own scheme of 'elements' was an attempt *from within scientific research* (p. 31) to supply science (but not necessarily anything outside of science) with a temporary platform across which the various sciences he was concerned with could best communicate. An ontology, therefore, but not a metaphysical one – and an ontology postulated in the spirit that Mach saw science as having; that is, tentatively, and as subject to further analysis.

It is true that when Mach says 'The assertion, then, is correct that the world consists only of our sensations' (p. 12) this looks like the conclusion of an a priori argument from the untenability of the usual way of thinking, *not* a consideration deriving from the need for a platform across which sciences can communicate. However, the latter is the way in which Mach presents matters when he is at his best, as I see it. *The Analysis of Sensations* begins with him bemoaning the current state of science because of the 'unwonted prominence' of ways of thinking derived from physics and an associated and inappropriate loss of way on the part of sensory physiology (p. 1). Mach's anti-reductionism (and the way he understood his own idea of the 'unity of science') then becomes evident in his reminder that physics 'constitutes but a portion of a *larger* collective body of knowledge' (p. 1, emphasis in original), that it *cannot* exhaust the subject matter of science, and in his suggestion that the physiology of the senses could 'afford physical science itself powerful assistance' (p. 2). Later in the same first chapter, this is most explicit in a wonderful passage in which Mach advertises his scheme of elements:

> If we regard sensations, in the sense above defined, as the elements of the world, the problems referred to appear to be disposed of in all essentials, and the first and most important adaptation to be consequently effected. This fundamental view (*without any pretension to being a philosophy for all eternity*) can *at present* be adhered to in all fields of experience; it is consequently the one that accommodates itself with the least expenditure of energy, that is, more economically than any other, to the present temporary collective state of knowledge. Furthermore, in the consciousness of its purely economical function, this fundamental view is eminently tolerant. It does not obtrude itself into fields in which the current conceptions are still adequate. *It is also ever ready, upon subsequent extensions of the field of experience, to give way before a better conception.*
>
> (p. 32, emphasis added)

That this is how Mach understood his proposed scheme is also evident from a similar passage in his Preface to the Fourth Edition of this same book. This passage makes it clear that he was not in the business of giving metaphysical questions naturalistic answers, but of disposing of such questions, since he regarded them as, or as posing, 'troublesome pseudo-problems' (p. xl). 'The aim of this book', he assures us

> is not to put forward any system of philosophy, or any comprehensive theory of the universe . . . [Rather, a]n attempt is made, not to solve all problems, but to reach an epistemological position which shall prepare the way for the co-operation of special departments of research, that are widely removed from one another, in the solution of important problems of detail.
>
> (pp. xl–xli)

Mach's scheme of elements is both a (naturalistic) ontology – an attempt to say what exists – and a pragmatic programme or proposal for linking sciences with one another. Whether that scheme really has all of the virtues he advertised it as having I cannot say. But it does not fit with our *contemporary* philosophical naturalism, which usually endorses with so much glee the imperialism of physics (especially mechanics) that Mach was most concerned to resist.

Conclusion, and a Doubt about Whether Mach Was a Neutral Monist

On balance, I think the evidence from *The Analysis of Sensations*, at least, favours Mach's being more like a 'neutral monist' than a phenomenalist.

The things that make him sound like a phenomenalist are incompatible with his basic viewpoint, as well as with his objecting to idealism as 'one-sided'.

Towards the end of the first chapter of *The Analysis of Sensations*, though, Mach sounds rather a different note, insisting that 'nothing would be changed in the actual facts or in the functional relations, whether we regard all the data as contents of consciousness, or as partially so, or as completely physical' (p. 36). This should come as a shock both to phenomenalist and to neutral monist readers. Phenomenalists can make no sense of it unless they drop the idea that elements (sensations) are purely psychological. But neutral monists must have trouble with it, too, for it implies that the categories of mental and physical don't really *matter*, and that the mental/physical 'distinction' is of no permanent applicability or integrity. The same note is sounded in a chapter from the second edition, when Mach endorses Rudolf Wlassak's account of Avenarius's proposal to abolish the terms 'physical' and 'psychical', Wlassak then crediting both thinkers with the idea of 'the untenability of the old conception of the psychical' (p. 51). What was crucial for Mach is that his 'elements' and their connections can be investigated both by physical sciences and by psychology. In *The Analysis of Sensations*, his concern was to draw science away from philosophies that generate pseudo-problems and to ensure that the physical and the psychological sciences can communicate, interact, and benefit from one another. His real monism, the 'monistic point of view' (p. 14) [*monistische Standpunkt*] (SS. 21–22) that he mentions, turns out to be his thesis of the unity of science.

References

Agassi, Joseph 1978/1988. 'Mach on the Logic of Enquiry: Taste-Maker Philosopher of Science', as reprinted in his *The Gentle Art of Philosophical Polemics: Selected Reviews and Comments*. Open Court, pp. 21–32.

Banks, Erik C. 2003. *Ernst Mach's World Elements: A Study in Natural Philosophy*. Kluwer.

    2014. *The Realistic Empiricism of Mach, James, and Russell: Neutral Monism Reconceived*. Cambridge University Press.

Blackmore, John T. 1972 *Ernst Mach: His Work, Life, and Influence*. University of California Press.

Blackmore, John T., Itagaki, Ryoichi, and Tanaka, Setsuko (eds.) 2001. *Ernst Mach's Vienna 1895–1930 or, Phenomenalism as Philosophy of Science*. Kluwer.

Cohen, Robert S. 1968/1970. 'Ernst Mach: Physics, Perception and the Philosophy of Science', *Synthese* 18: 132–170, as reprinted in Robert S. Cohen and

Raymond J. Seeger (eds.), *Ernst Mach: Physicist and Philosopher*. D. Reidel, 1970, pp. 126–164.

Cohen, Robert S. and Seeger, Raymond J. (eds.) 1970. *Ernst Mach: Physicist and Philosopher*. D. Reidel.

Feyerabend, Paul K. 1970/1999. 'Philosophy of Science: A Subject with a Great Past', as reprinted in his *Knowledge, Science and Relativism: Philosophical Papers, Volume 3*, ed. J. M. Preston. Cambridge University Press, 1999, pp. 127–137.

      1984. 'Mach's Theory of Research and Its Relation to Einstein', *Studies in History and Philosophy of Science* 15: 1–22.

Gori, Pietro 2012. 'Nietzsche as Phenomenalist?', in H. Heit, G. Abel, and M. Brusotti (eds.), *Nietzsches Wissenschaftsphilosophie: Hintergründe, Wirkungen und Aktualität*. de Gruyter, pp. 345–355.

Heidelberger, Michael 2010 'Functional Relations and Causality in Fechner and Mach', *Philosophical Psychology* 23: 163–172.

Hentschel, Klaus 1985. 'On Feyerabend's Version of "Mach's Theory of Research and Its Relation to Einstein"', *Studies in History and Philosophy of Science* 16: 387–394.

Mach, Ernst 1872/1911. *History and Root of the Principle of the Conservation of Energy*. Open Court.

      1883/1960 *Die Mechanik in ihrer Entwickelung historisch-kritisch dargestellt*. F.A. Brockhaus. 9th German edition translated by T. J. McCormack as *The Science of Mechanics: A Critical and Historical Account of its Development*, 6th American edition. Open Court.

      1886/1914. *Beiträge zur Analyse der Empfindungen*. Gustav Fischer. Translation of the 1st German edition by C. M. Williams, revised and supplemented from the 5th German edition by C. M. Williams and S. Waterlow as *The Analysis of Sensations and the Relation of the Physical to the Psychical*. Open Court, 1914, reprinted Dover Publications, 1959.

      1891. 'Some Questions of Psycho-Physics: Sensations and the Elements of Reality', *The Monist* 1: 393–400.

      1895/1896. *Populär-wissenschaftliche Vorlesungen*. J. A. Barth. Translated by T. J. McCormack as *Popular Scientific Lectures*. Open Court.

      1896/1986. *Die Prinzipien der Wärmelehre, historisch-kritisch entwickelt*. J. A. Barth. Translated by T. J. McCormack, P. E. B. Jourdain, and A. E. Heath as *Principles of the Theory of Heat, Historically and Critically Elucidated*, ed. B. F. McGuinness. D. Reidel.

      1905/1976. *Erkenntnis und Irrtum. Skizzen zur Psychologie der Forschung*. J. A. Barth. Translated by T. J. McCormack and P. Foulkes as *Knowledge and Error: Sketches on the Psychology of Enquiry*. D. Reidel.

      1910/1970. 'The Guiding Principles of My Scientific Theory of Knowledge and Its Reception by My Contemporaries', as translated in S. E. Toulmin (ed.), *Physical Reality: Philosophical Essays on Twentieth-Century Physics*. Harper Torchbooks, 1970, pp. 28–43.

Russell, Bertrand 1914/1956. 'On the Nature of Acquaintance, II: Neutral Monism', *The Monist* 24: 161–187, as reprinted in his *Logic and Knowledge: Essays 1901–1950*, ed. R. C. Marsh. George Allen & Unwin, 1956, pp. 139–159.

von Mises, Richard 1938/1970. *Ernst Mach und die empiristische Wissenschaftsauffassung.* Van Stockum, translated as 'Ernst Mach and the Empiricist Conception of Science', in Robert S. Cohen and Raymond J. Seeger (eds.), *Ernst Mach: Physicist and Philosopher.* D. Reidel, 1970, pp. 245–270.

13

The Case for Mach's Neutral Monism

ERIK C. BANKS

Introduction

For many years, the received view of the Viennese physicist and philosopher Ernst Mach (1838–1916) has been that he was a phenomenalist who only believed in given human sensations and who held that everything else was unverifiable metaphysical nonsense. This view comported well with Mach's historical influence on members of the Vienna Circle and seemed to explain both his negative scepticism about atoms and his positive scepticism about Newton's absolute space. In recent years, however, this phenomenalistic view of Mach's work has given way to a more realistic and nuanced 'neutral monist' view, far more in line with his contemporary reception by Paul Carus, Hans Kleinpeter, William James, Bertrand Russell, and American Realists such as Ralph Barton Perry, leading to Herbert Feigl and Wilfrid Sellars and beyond to the contemporary neutral monist movement. There are now two main traditions in the literature, one tying Mach to positivism and the other to neutral monism. In this chapter, I will defend the neutral monist tradition and show that it is actually a form of scientific realism, not positivism.

I start with a characterisation of what I believe to be some tenets of neutral monism in general, many of which were shared by James and Russell, both deeply influenced by Mach. I will then go point by point and find evidence for these views in Mach's texts (including his notebooks and other documents). Seeing Mach as a kind of realist also casts much light on his scientific views and corrects a number of historical misconceptions regarding both atomism and Mach's philosophy of space and time. Finally, I will discuss Mach's place in the neutral monist movement of James, Russell, and the American Realists, and the revival of these views in recent philosophy of mind.

Much of the background material for this chapter can be found in my two books on Mach (Banks 2003, 2014). On Mach and the Vienna Circle, see Banks (2013). On Mach versus James on pragmatism and other differences, see Banks (2019).

What Is Neutral Monism?

(1) *The elements.* Neutral monism holds that objects and the human mind are both concatenations of neutral elements (e) neither exclusively mental nor physical in nature. These elements are transient event particulars that never recur. They are bound up in functions with one another such that the same element (e/s) may partake of both mental variations (memory, association) and physical variations (connections to other objects, physical laws). Objects and egos are functional complexes of elements, not substrata or substances prior to events.

(2) *Unobserved elements.* In addition to elements which obey both mental and physical variations (e/s), there are also elements that make up the rest of the physical world, other minds, animal sensations, etc. These are found by causal continuity with our experienced sensations, by extending the functions we find in perception to complete unobserved objects 'in thought' and by adding extra (*hinzugedachte*) elements. We know these elements only through their causal functional relations, but we do not experience their quality directly, so we can say little concrete about them, barring future experiments, or what Mach called 'an extension of biology'.

(3) *Functions.* Elements are not atomistic or logically independent simples, as in the case of Hume's impressions or the simples of Russell's logical atomist period. They are always bound up in functions that supervene on their causal powers to affect each other. Polyadic functional relationships from complex to complex are also possible. The division between mental and physical functional relationships is provisional only, due to our incomplete knowledge, and will eventually be overcome as mental variations are traced to the physical states of the brain. There will then be one set of elements and one set of natural variations in a future state of the science. Functions are naturalistic, not merely logical or combinatorial possibilities without a physical grounding in the qualities of natural events.

(4) *Forceful qualities.* Elements are events that express qualities, or powers with causal force to affect other events. They are not inert sense-data or logically independent simples. Nor are their qualities mere 'epiphenomenal qualia' with no causal efficacy. Even physical events possess qualities of their own, whether or not these look anything like the manifested qualities of our sensations.[1] The qualities of elements are occurrent and particular, not universals.

---

[1] In panpsychism, the unobserved qualities are analogised to our sensations, so instead of blue, we have 'protoblue' inside atoms (or perhaps neurons) that combine to make blue when enough of them are combined in a certain way. See Chalmers (2002) for the introduction of proto-phenomenal qualities.

(5) *The ego.* Consciousness is a set of complex functional relations, not a single unitary phenomenon. The unity of consciousness is an illusion. Sensations and elements always occur in functional complexes supervening on their causal relations to each other, and one of these complexes is the conscious mind, but the elements do not *acquire* their quality, or existence, from being beheld by the mind. It is rather the reverse: the mind acquires its features through the complex functional relation of its contents in perception and also in judgements.[2]

(6) *Matter.* Physical objects are also functional constructions out of events and are not absolutely permanent. Underlying the notion of the physical object, however, is the much more solid permanence of functions and conserved systems of elements. Conservation first establishes the possibility of identity of systems through time and over spatial transformations in place, rotations, and velocity boosts. It is also the basis for identifying properties preserved through transformations such as energy, momentum, and angular momentum. These are the only true universals. Other semi-permanent things (species, organisms, tables, chairs) are quasi-'objects' only as long as their semi-permanence and ultimate composition out of elements is recognised. So-called properties established by similarity relations,[3] or by correlation with logical subject terms, predicates, or sentences, are merely linguistic or psychological in nature.

(7) *Naturalised epistemology-direct realism.* Elements are not a 'given' basis for a foundational construction of knowledge such as in Rudolf Carnap's *Aufbau* (Carnap 1967) or Russell's early theory of knowledge by acquaintance. Rather, elements are provisionally the smallest divisions of experience we are capable of making at the present time. Analysis of the elements may, and should, continue indefinitely. Mental events have no intrinsic intentional power to represent other objects. Where no causal relations can be established to objects, mental events do not refer to anything beyond themselves. Where causal relations can be established,

---

[2] This is a second difference with panpsychism, which holds that qualities always require conscious egos to behold them. Chalmers and Strawson are panpsychists of this sort. It isn't ruled out that a higher-order judgement couldn't affect a quality that is part of the complex we call the judgement. But judgements are nothing besides the complexes to which they belong. It would be wrong to think of judgement as some special mental act determining its content. It would also be wrong to think of judgement as somehow responsible for determining the quality of the content merely because it can affect that quality (which is surely the case). For example, Daniel Dennett's attempt to muddle the issue with his example of the Maxwell House tasters would certainly have to be rejected by the neutral monist.

[3] As will be discussed below, similarity is not an equivalence relation. It therefore is not suitable for physical properties determining a partition of individuals that is, in turn, invariant through transformation groups.

mental events may lead to knowledge or error along the same causal links and even using the same methods of enquiry. There are no established canons of inductive or even hypothetical deductive or any other scientific methods guaranteed to lead to knowledge. Where the object can be connected up with mental events, our knowledge of the object is *directly realistic* because the whole object will then include events that are part of it, both physically and also as part of the knower. The only methods for attaining knowledge are fallible scientific methods of tracing causal chains in the whole set of naturalistic functions, *not* knowledge via propositions and logical relations of truth and falsity to linguistically delineated facts.

Not all neutral monists would subscribe to all of these points (e.g. see Stubenberg 2016). In fact, James and Russell dissented on points (3) and (7), respectively (see Banks 2014, 2019). So far as I know, Mach is the only philosopher who would match up point by point to this ideal list. I will now seek to prove this reading.

Mach's Neutral Sensation-Elements

In *The Analysis of Sensations*, Mach expresses the neutrality of the elements thus:

> A colour is a physical object as soon as we consider its dependence, for instance upon its luminous source, upon other colours, upon temperatures, upon spaces and so forth. When we consider however its dependence upon the retina, it is a psychological object, a sensation. Not the subject matter but the direction of our investigation is different in the two domains.
>
> (Mach 1886/1959, p. 18; cf. Mach 1905/1976, pp. 15–16)[4]

He also says that the term 'sensation' is only relative to the set of mental variations. Otherwise the same colours, sounds, etc., are to be considered physical elements:

> In what follows, wherever the reader finds the terms 'sensation' 'sensation complex' used alongside of or instead of the expressions 'element' 'complex of elements' it must be borne in mind that it is only in the connexion

---

[4] Where exactly neutral monism originates is hard to tell. Similar passages about the two orders and the neutral stuff occurred in writings by C. S. Peirce, W. K. Clifford, and Richard Avenarius. Mach's view still has priority because it seems to have originated in writings of the 1860s and may have appeared in the first draft of *The Analysis of Sensations* in 1864. G. T. Fechner's refusal to let Mach dedicate the book to him in 1864 may have delayed publication for twenty years (*The Analysis of Sensations* was first published in 1886, long after Mach had formulated his ideas) (see Heidelberger 2004 for this interesting story).

and relation in question, only in their functional dependence that the elements are sensations. In another functional relation they are at the same time physical objects. We only use the additional term 'sensation' to describe the elements, because most people are much more familiar with the elements in question as sensations (colours, sounds, pressures, spaces, times), while according to the popular conception it is particles of mass that are considered as physical elements, to which the elements, in the sense used here are attached a 'properties' or 'effects'.

(Mach 1886/1959, p. 16)

It is clear from the last quote that elements are not objects, but events. It is also clear from the early pages of the *Analysis* that elements are not types or universals, but event particulars that do not recur (see Mach 1886/1959, pp. 2–7, 29, 331).

As many authors have observed, a colour patch does not seem like a physical object, like a particle or a field; it seems more like a sensation only – a 'secondary quality' in the mind with no connection to the real properties of objects such as the reflected wavelengths of light. In boldly calling sensory qualities physical events, some like V. I. Lenin sensed a word game:

Mach and Avenarius secretly smuggle in materialism by means of the word 'element', which supposedly frees their theory of the 'one-sidedness' of subjective idealism, supposedly permits the assumption that the mental is dependent on the retina, nerves and so forth, and the assumption that the physical is independent of the human organism. In fact, of course, the trick with the word 'element' is a wretched sophistry, for a materialist who reads Mach and Avenarius will immediately ask: what are the 'elements'? It would, indeed, be childish to think that one can dispose of the fundamental philosophical trends by inventing a new word. Either the 'element' is a sensation, as all empirio-criticists, Mach, Avenarius, Petzoldt, etc., maintain – in which case your philosophy, gentlemen, is idealism vainly seeking to hide the nakedness of its solipsism under the cloak of a more 'objective' terminology; or the 'element' is not a sensation – in which case absolutely no thought whatever is attached to the 'new' term; it is merely an empty bauble.

(Lenin 1909, pp. 48–49)

Quine (1966, p. 667) and Thomas Nagel (at least in 2000) also attacked neutral monism as not 'really neutral', but a disguised attempt to reduce the physical to the mental (see also Stubenberg 2016, section 7.2, for more examples of this criticism).

Myself (Banks 2014, chapter 5), I think it is clear that the sensation-elements such as blue are assumed to be complex electrochemical brain events, and that blue is simply how the physical powers manifested in these sorts of configured events appear *to us*. From outside the brain, the same event of our seeing the blue will appear to an external observer as a complex

firing pattern of 10,000 complexly configured neurons. If the external observer siphons off the neural energy into 10,000 electrodes, the event of seeing
blue will vanish for the internal observer, which is what you would expect if
they were identical in some way. If the external observer then reconfigures
the 10,000 events along wires in exactly the same way with exactly the same
causal powers, I would also expect the blue sensation to be recreated within
the complex of powers and manifestations in the wires. (See Banks 2014,
chapter 5, for a more complete neutral monist view of sensation qualities in
terms of identical powers and non-identical manifestations.) If any such
explanation is workable, or even just logically coherent, I see absolutely no
reason why it can be claimed a priori that sensory qualities such as blue *could
not possibly be* physical events, as is often done. It seems to me that a priori
arguments or intuitions have nothing to do with the question of whether
sensations are physical.

Passages about the two orders and the neutrality of the sensation-elements
occur in both James's 1904 'Does "Consciousness" Exist?', the first of his
radical empiricist essays, and of course in Russell's 1919 essay 'On
Propositions' and his 1921 book *The Analysis of Mind*. Russell saw Mach's
basic insight about the physicality of sensation as a breakthrough and a new
truth for breaking the stranglehold of the mind–body problem: 'Mach argued
that our sensations are part of the physical world and thus inaugurated the
movement toward neutral monism' (Russell 1914/1984, p. 16). Even before his
own conversion to the view in 1918, Russell accepted as 'a service to philosophy' Mach's and James's idea that 'what is experienced may itself be a part of
the physical world and often is so' (Russell 1914/1984, p. 31), or that 'constitu-
ents of the physical world can be immediately present to me' (Russell 1914/
1984, p. 22).

This brings us to feature (2) and a fundamental question for neutral
monism. Are all physical elements *also* sensations (e/s), or are there world
elements (e) that are *not* interpretable as anyone's sensations? You could say –
and Mach did say sometimes – that the question is moot if minds are not
fundamental anyway and if sensations are *already* physical in any desired
sense of 'mind-external physical objects'. Since there is no distinction for him,
he may have considered the question to be nonsense. A more modest reading
would perhaps be that even if this is true, Mach never actually goes beyond
assuming those mind-independent physical elements that *also* bear an interpretation as sensations. Indeed, this is by far the most common view of Mach's
neutral monism.

So a clearer position is still necessary on whether there *are* further elements
in causal relations to the (e/s) that fill out perceived objects and objects in
causal relations to us that are simply (e) and never themselves (e/s), or a denial
that there are *no* such elements. Fortunately, there is evidence that Mach took
the further step to the pure (e) elements as well.

The Existence of the World Elements[5]

In *The Analysis of Sensations*, Mach talks about the need to add elements in thought (*hinzudenken*) to those observed in order to complete our experience of objects, or to complete 'half-observed facts' (1883/1960, p. 587, 1886/1959, p. 333). This will involve adding elements not directly experienced by us, such as the elements of other human beings (by analogy), animals, and elements of physical objects such as the backs of chairs, distant stars, and so forth. Mach is clear that we do not experience the qualities of these added elements, but he is also clear on their causal powers and linkage to experienced elements. They are not *mere* thought additions without some kind of causal functional efficacy backing them up. If they are linked to elements with causal powers, then they too must have causal powers and cannot be inert thought-things, even though we must add them in thought to our experiences. The merely imagined back of a chair, for example, is not what establishes the solidity and permanence of the object of perception. Mach also says that the analogy to the sensations of another person is a 'causal' not an imaginary analogy:

> When I speak of the sensations of another person, those sensations are not, of course, exhibited in my optical or physical space; they are mentally added and I conceive them causally, not spatially, attached to the brain observed or rather functionally presented.
>
> (Mach 1886/1959, p. 361)

He also says that in a causal presentation one need not even distinguish the sensations of one ego from another anyway – all sensation-elements can be represented by their causal connections to each other in a kind of map, and then one completes the causal relations on the map by adding the necessary elements required to complete the causal relations, or 'partially observed facts', even if the further assumed elements are not observable:

> From the standpoint which I here take up for purposes of general orientation, I no more draw an essential distinction between my sensation and the sensations of another person than I regard red or green as belonging to an individual body. The same elements are connected at different points of attachment, namely the egos. But these points of attachment are not anything constant. They arise, they perish, and are incessantly being modified. But where there is no connexion at a given point there is no perceptible reciprocal influence. Whether it may or may not prove possible to transfer someone else's sensation to me by means of nervous connexions, my view is not affected one way or the other.
>
> (Mach 1886/1959, p. 27)

---

[5] The term 'world elements' is not Mach's, but Friedrich Adler's (1908). I also used it in the title of my 2003 book *Ernst Mach's World Elements*.

There may be many ways to complete the map, of course, all consistent with observation, but there are no in principle theoretical difficulties, since all of the components are of the same homogeneous kind and we are connecting like with like. We never, for example, assume elements in no possible causal relation to experience. This would be a dreaded extra-causal *Ding an sich*, permanently isolated from our experience, which he thinks science cannot tolerate.

As I showed in my 2003 book, Mach's commitment to mind-independent elements had a very long background in his early career when he was studying Johann Friedrich Herbart and Gustav Fechner and working his way towards his own position. For a time, Mach says, he was an idealist, a monadologist (1886/1959, p. 30, note), and even a panpsychist (1886/1959, p. 362), all views he later abandoned. These early views and Mach's development intrigued me, so I investigated many of these early writings, including the *Vorträge über Psychophysik* of 1863, where, influenced by Fechner's panpsychism, he seems to have held that even atoms had some kind of qualitative inner nature:

> We cannot attribute to atoms an outer side. If we must think anything, we must attribute to them an inner side, an inwardness analogous in some respects to our own soul. In fact, where could the soul come from in a combination of atoms in the organism if the kernel did not already lie in the individual atom?

> (Mach 1863, p. 364)

In an 1866 paper 'Über die Entwicklung der Raumvorstellungen', Mach speculated about inner states, or qualities, of pressure making up the phenomenon of spatially extended matter, similar to the Leibnizian constructions of matter and space in Herbart's metaphysics from forceful qualities pressing upon each other (see Banks 2003, chapter 3). Mach concluded that space and time extension were unnecessary to physics and that physics could instead confine its attention to functionally and combinatorially related 'inner states' such as forces or pressures, from which spatio-temporal phenomena would arise by construction, as in Herbart's works.

Finally, in lectures entitled 'Über einige Hauptfragen der Physik' (Ernst Mach Nachlass, Deutsches Museum, 1872 NL 174/1/003), Mach put together these ideas and asked whether sensation could be considered 'a general property of matter'. If so, then nothing in principle stood in the way of unifying the mental with the physical, since both would consist of the same kind of qualities in different sorts of causal relations:

> Sensation is a general property of matter, more general than motion. Let us seek to set down this proposition clearly. An organism is a system of molecules. Electrical currents run into the interior and come back again into the muscles. Everything is physically explainable. But not that the person should have sensations. What we can investigate physically is

always merely physical. We find no sensation. And yet the human being senses. The material flows forward and through and through him. The old departs. The new comes in. We have therefore the problem of finding something fundamentally new in the whole that is not in the parts. We escape this difficulty when we consider sensation a general property of matter . . .

What scientific value this assumption of a general sensation of matter has, this can only be decided by how much better we can deduce and understand physical phenomena through them. Rules for deduction of our sensations with the help of other sensations added in thought and in causal relation to them. Thus, as the most immediate goal of science the construction of the world out of sensations. The appearance of matter, in so far as it is sensation, built up similar to the way physics has built up material added in thought (the atom).

(Haller and Stadler 1988, p. 173)

What seems clear is that world elements played a very significant role in the development of Mach's neutral monism. They were not a later add-on to a phenomenalist view, as it might appear from a cursory read of *The Analysis of Sensations*. Recognising the role qualitative elements could play in the construction of matter, as well the construction of mind, appeared to convince Mach that the elements were ideally suited to play a role that is *neutral* between the two and that the two categories could henceforth be abandoned in favour of the elements in one natural array capable of constructing both mind and matter. That view made its first appearance at the end of Mach's *History and Root of the Principle of the Conservation of Energy* in 1872.

Mach continued to write this way even after the development of neutral monism, referring always to a 'future state of science' in which the sameness in kind between sensations and physical elements would be recognised in the way he hoped by 'building a tunnel' through from the one to the other. There are letters to Friedrich Adler and Gabriele Rabel (see Banks 2003, pp. 6–7) in which he says that he refused to limit himself to the observed element/ sensations only, but reserved speculation on world elements for the future. Adler asked him point-blank: '*Do you consider necessary the assumption of elements that belong to an object but not to a subject?* Obviously such elements are not directly given, but do you assume them hypothetically as it appears to me you do from passages cited in *The Analysis of Sensations*?' (Banks 2003, p. 6, emphasis added). Mach wrote to Adler that indeed he

'assumed analogous elements in animals, plants and inorganic bodies. This hypothesis serves only to round off the world view provisionally and in hope of the future construction of biology . . . Healthy biological research must teach if this hypothesis has any worth and if so, what worth. Speculation cannot manage this. Provisionally it appears to me that we completely overlook a side of our experience when we overlook this hypothesis . . . I have not further cultivated all of these matters, for

I feared the nearness of the metaphysical abyss where there is no experiential foundation'

(Banks 2003, pp. 6–7).

There are also published passages in the 1883 *Mechanics*, in Mach's 1882 lecture 'On the Economical Nature of Physical Inquiry', and in the 1905 *Knowledge and Error* to this effect:

> Careful physical research will lead to an analysis of our sensations. We shall then discover that hunger is not so essentially different from the tendency of sulfuric acid for zinc, and our will not so different from the pressure of a stone as it now appears. We shall again feel ourselves nearer nature without its being necessary that we should resolve ourselves into a nebulous mass of molecules or make nature a haunt of hobgoblins.
>
> (Mach 1883, p. 559)

> If the ego is not a monad isolated from the world but a part of it, in the midst of a cosmic stream from which it has emerged, and into which it is ready to dissolve back again, then we shall no longer be inclined to view the world as an unknowable something and we are then close enough to ourselves and in sufficient affinity to other parts of the world to hope for real knowledge.
>
> (Mach 1905, p. 361)

Clearly, 'real knowledge' of nature would involve some sort of actual acquaintance with mind-independent natural qualities, which Mach thought might be possible through the further development of biology, perhaps by adapting the nervous system so that previously unexperienced natural qualities actually fell within our experience. In other words, the existence of the world elements is an experimental question, or so Mach hoped. It is not a question to be blocked off by philosophical positions such as phenomenalism. This is one of the errors one makes in interpreting Mach as a traditional philosopher building a closed and incorrigible system. No such thing is going on here. His terms are simply starting points for future enquiries, nothing more.

Natural Functions and Forceful Qualities

Mach was aware that elements and functions would seem to many people too flimsy a foundation to build up egos and solid objects. He wrote:

> My world of elements, or sensations, strikes not only men of science but also professional philosophers as too unsubstantial. When I treat matter as a mental symbol standing for a relatively stable complex of sensational elements, this is described as a conception which does not make enough of the material world. The external world, it is felt, is not adequately expressed as a sum of sensations; in addition to the actual sensations, we ought at least to bring in Mill's possibilities of sensations. In reply to this,

I must observe that for me also the world is not a mere sum of sensations. Indeed, I speak expressly of functional relations of the elements. But this conception not only makes Mill's 'possibilities' superfluous, but replaces them with something much more solid, namely the mathematical concept of function.

(Mach 1886/1959, pp. 362–363)

It should be pointed out that Mach had held this view at least since his *History and Root of the Principle of the Conservation of Energy* in 1872, where he described the idea that objects should be complexes that exhibit a sturdy functional dependence of the elements: $f(\alpha, \beta, \gamma, \ldots) = 0$. Mach also pointed out that elements are always bound up in complexes and never occur in isolation. They obey a general principle called 'the reciprocal functional dependence of elements on each other', which Mach substitutes for the principle of causality, and also for ideas of conservation, which otherwise would require a substratum of conserved substances, stuffs, or fluids, all of which he rejected as 'metaphysical additions to physics'. It is also clear from these and other writings that Mach thought of elements as having causal force or power to abut upon each other and generate those reciprocal variations (an idea which Mach may have retained from Herbartian psychology, where qualities behaved just like forces; see Banks 2003, chapter 3). I have also ventured the idea that the elements are like energy-potential differences equalised by forces, a view that fits in with Mach's view of the natural processes as tending to the maximum (or minimum) possible equalisations of potential differences by the law of least action (Banks 2003, 2014).[6] If elements are really direct qualitative manifestations of energy differences, manifested in natural events, then calling them forceful would make much sense, since this is what forces are, according to physics.

The functional dependence of the elements is guaranteed because the functional behaviour is grounded in the qualities and causal forces expressed by the elements themselves. This is indeed a solid foundation, more so than any impermanent object which merely seems solid to us: bodies become 'embodied laws of conservation' such as $f(\alpha, \beta, \gamma, \ldots) = 0$.

James and Russell further developed Mach's causal–functional connections into spatio-temporal perspectival systems. The elements, which James called 'pure experiences' and Russell called 'event-particulars', represented an object by breaking it up into all of the perspectival views – or causal interactions – it might have with observers and other objects. This has the extra benefit of giving the causal relations some further perspectival structure and implicitly defining the objects as the invariants of perspectival transformations from one perspective to another.

---

[6] I mean that the Principle of Least Action is an expression that the differences between the time averages of the kinetic energy *minus* the potential energy is a minimum.

It is clear that the functions used by Mach, James, and Russell to give the world structure are not mere mathematical functions or combinatorial relations or other logical devices such as the spurious quasi-analysis relations of Carnap. These are naturalistic functions grounded in the behaviour of the elements and leading to natural ideas of conservation and invariance. I do not find this to be true of Carnap's later constructions of similarity classes in the *Aufbau*. Similarity, if a relation does not supervene on causation, is not a natural relation, but a purely psychological or perceptual similarity in the eye of the beholder. It is not even clear that there exists in nature a real relation of similarity that is reflexive, symmetric, and non-transitive. It is certainly not an equivalence relation of the kind that would indicate the presence of a real physical property over systems or events. In fact, there is a close relationship between equivalence relations and groups: for every equivalence relation producing a partition on a set there is a group of transformations that preserves the partition. For example, if we have a system of bodies classified by their momenta with respect to a reference body at rest, we can construct a partition of bodies with the same momenta. We can then introduce a group of velocity transformations on the partition and preserve the identified property of momentum through that group. Groups in turn follow naturally from the adoption of a system of perspectives (see Frank 2001, chapter 5 and pp. 84–85). This, I believe, is how the functional relations should be parsed out in a Machian way.

'Das Ich ist unrettbar': *The Self Cannot Be Saved*

One of Mach's most famous and characteristic ideas is the rejection of the self or consciousness. He still retained the idea that consciousness was a kind of complex functional relation of associated sensations, images, memories, spaces, times, and the like. In fact, *The Analysis of Sensations* goes a long way towards characterising these mechanisms as a series of machine-like reflexes conditioned largely by the evolutionary history of the organism. In that sense, Mach was already a completely modern psychologist in an age when bizarre ideas still held sway over that field of study. Mach did not believe that the contents of the mind depended upon consciousness as a kind of circumambient medium. A sensation of blue retains its existence and quality even if it is not being attended to or embedded in a conscious act of observing the blue:

> Consciousness is not a special mental quality or class of qualities different from physical ones; nor is it a special quality that would have to be added to physical ones in order to make the unconscious conscious. Introspection as well as observation of other living things to which we have to ascribe consciousness similar to our own shows that consciousness has its roots in reproduction and association: their wealth, ease,

speed, vivacity and order determine its level. Consciousness consists not
in a special quality but a special connection between qualities ... A single
sensation is neither conscious nor unconscious: it becomes conscious by
being ranged among the experiences of the present.

(Mach 1905/1976, pp. 31–32)

Of course, Mach's famous and rather chilling 'headless body' illustration
(Mach 1886/1959, p. 19) makes the absence of the ego abundantly clear. He
was known for this idea even outside philosophy, in cultural circles. This
picture may be one inspiration for Robert Musil's *The Man without
Qualities* (Musil wrote his doctoral dissertation on Mach). Of course, James
soon followed suit in discarding consciousness in 'Does "Consciousness"
Exist?' in 1904, and Russell, in 1919, was led by Mach and James's example
to abandon his theory of acquaintance, which was fundamental to his theory
of knowledge up to that time (Russell 1959, pp. 134–135).

Space and Time

Mach's scepticism of the basic concepts of physics is well established in the
literature. He is usually considered a relationist in the philosophy of space and
time because of his famous critique of Newton's bucket experiment in *The
Science of Mechanics*. According to Mach, accelerated and rotating reference
frames should be exchangeable for a rest frame if the rest of the bodies in the
universe accelerate or rotate in the opposite direction. The fictitious 'inertial
forces' that appear to work on a body in these resting reference frames (K) are
to be replaced by gravitational forces at a distance or through a medium due to
the other bodies in the universe moving around it instead (A, B, C). Einstein
famously enshrined this idea as 'Mach's principle' and originally thought
general relativity upheld it. However, as became clear over time (see Janssen
2014), the gravitational field has its own independent existence in general
relativity and does not depend upon generating source masses, as proved by
the cogency of the de Sitter solution and the rejection of Einstein's 'masses at
infinity'. The gravitational field is, however, the source for both gravitational
and inertial effects, which can be transformed into each other, and so the field
can be described as a combined inertio-gravitational field. What is hardly ever
mentioned is that Mach *also* admitted the possibility of an inertio-
gravitational field independent of, or collateral to, source masses, in defiance
of Mach's principle,[7] *provided* it did not need to be embedded in an absolute

---

[7] Einstein described 'Mach's principle' in a letter to Mach in 1913. Shortly thereafter is dated
the preface to the *Optik* in which Mach disavows being a forerunner to the theory of
relativity. According to Gereon Wolters (1987), the preface is a forgery by Mach's son
Ludwig, and Mach had accepted relativity and atomism by this time. Wolters has made a

space and time. That is, inertial effects could be due to accelerated motion and rotation with respect to the field, but would also presumably occur if the field itself moved or rotated relative to the body as well. This passage actually goes back to the 1883 first edition of the *Mechanics*:

> It might be indeed that the isolated bodies A, B, C . . . play a collateral role in the determination of the body K and that this motion is determined by a medium in which K exists. In such a case we should have to substitute this medium for Newton's absolute space.
>
> (Mach 1883/1960, pp. 282–283)

So far as I know, this passage has played *no role whatever* in the vast discussion of Mach's principle – a huge oversight – but it should be more widely known, since it shows that Mach was not at all wedded to relationism, even if he expresses a strong preference for the determination of motion with other bodies as reference points (Mach 1883/1960).

I think Mach's well-known relationism (like Leibniz's) hid a much deeper eliminationist view about space and time which takes us back to the germination of his theory of elements and functions, specifically in two papers: the aforementioned 1866 'Raumvorstellungen' paper and the 1871 paper 'Über die physikalische Bedeutung der Gesetze der Symmetrie'. There, Mach indicated his desire to eliminate spatio-temporal representation from physics altogether, not just to reduce space and time to spatio-temporal relations between bodies. This, too, is an oversight deserving correction, especially from scientists. In the 'Raumvorstellungen' paper (reprinted in Mach 1872/1910, pp. 88–90), Mach suggested replacing forces as functions of distance by distance as a function of the intensity of forces. Space and time were to be entirely replaced by causal functional dependencies among the elements, expressed in abstract form and not dependent on any prior extended representation. As he put it in the *Principle of the Conservation of Energy*:

> I think I must add, and have already added in a previous publication [the 'Raumvorstellungen' paper], that the express drawing of space and time into the law of causality is at least superfluous. Since we only recognise what we call space and time by certain phenomena, spatial and temporal determinations are only determinations by means of other phenomena.
>
> (Mach 1872/1910, p. 60)

case for his view with many supporting documents. For me, however, the timing is too close between the two documents to be a coincidence, and since I do not think Mach would have accepted Mach's principle as the only reconstruction of his thoughts on Newton, Mach may have been trying to head off Einstein's all too neat summary of his ideas. The forbidding tensor calculus field equations may have been another reason Mach did not want to be associated with theories he did not understand mathematically.

In the 'Symmetrie' paper (Mach 1871, p. 147), he wrote that it made no sense to force all physical phenomena into the three-dimensional space of our imagination. What if natural phenomena were discovered that had more degrees of freedom to consider? He suggested in the *Principle of the Conservation of Energy* that the spectral lines of different chemical elements might present such a case. He added in a footnote: 'It follows from this that the dependence of natural phenomena be expressed through relations of number, not spatially or temporally' (Mach 1871, p. 147, note).

This programme in his early papers was amplified in later works. Mach came to believe that spatio-temporal representations in physics were based in anthropomorphic human visualisation and psychology, and that a future physics would be of his more abstract and sparse element-and-function ontology. His deeper theory of elements explains why he did not think relativity was the ultimate answer to the space–time problem, since it still assumed an – albeit relativised – spatio-temporal format and did nothing to reduce it to non-spatio-temporal foundations. Likewise, Mach could not accept little spatio-temporally extended atoms, but rather thought that his theory of elements could go deeper and explain the energy changes *within* atoms as evidenced by the spectral lines. As Paul Feyerabend pointed out in a seminal paper: '[E]lements as envisaged by Mach are more fundamental than atoms' (Feyerabend 1984, p. 11). And as I have shown, Mach continued to hope for a constructive *elimination* of space and time from physics to the end of his career (Banks 2003, chapter 14). Like so much else about Mach, this programme has received almost no attention at all – a very puzzling state of affairs. It is assumed by philosophers of science that Mach's relationism in physics was simply a result of his phenomenalism in philosophy, thus compounding one error with another.[8]

Knowledge and Error

Mach may have been one of the first naturalised epistemologists, *avant la lettre*, since he believed that the only methods for reaching knowledge were scientific methods.[9] He still held on to introspection in his psychology, of course, but supplanted it with careful measurements. Mach certainly was not a

[8] I have often wondered why so many misinterpretations proliferate about Mach's works. I have been told that Mach is hard to read in the original and that most readers consult anthologies or 'CliffsNotes' versions by other authors or in erroneous histories of philosophy which seek to pigeonhole authors in neat historical movements. I urge authors considering writing about Mach to read his original works and also to become familiar with the scientific issues he discusses.

[9] Although Mach was not like Peirce at all and did not canonise scientific methods as an epistemology. The difference is subtle; see Banks (2017).

foundationalist and did not start with a set of indubitable givens from which to construct the world (like Carnap and Schlick – see Feyerabend 1970/1999, pp. 133–135; Uebel 2007; Banks 2013). He insisted that his elements were 'provisional only . . . the smallest divisions science is capable of making at any given time' (Mach 1905/1976, p. 12, note), hence his view that the elements of spectral lines were a level of analysis *beneath* atoms. The functions grouping the elements into systems were the naturalistic functions they obeyed by virtue of their causal powers, not second-order tortured logical or linguistic constructions with little relation to reality. The Ramsey sentence or Craig's theorem, for example, are the farthest things possible from a Machian approach to science, even though many would take these developments to be the apotheosis of the Machian view.

Mach's theory of knowledge as represented in *Knowledge and Error* was biological and evolutionary in nature. He proposed the idea that thoughts accommodated themselves to experience and to each other in an evolutionary or adaptive process. He also saw conceptual and linguistic knowledge as resulting naturally and continuously from pre-conceptual habit and reflex, enhanced by social communication. In some ways, perhaps, he does not do justice to the abstract structure of logic and mathematics or the generality of abstract concepts. (I now believe this is a serious gap in his philosophy, though not one that cannot be filled.) Mach did, however, recognise that hypotheses could not be arrived at by Baconian enumerative induction, but involved a kind of abductive leap which could then be compared with experience. He also believed that nature exhibited patterns or 'great facts' which could become accessible to human intellect, such as various partial differential equations that showed up across various domains in physics. He even hoped for a 'phenomenological physics' linking all of the great facts under abstract, master principles such as the conservation of energy, the second law of thermodynamics, and the law of least action, all of which rested upon experience for their evidence. It is very wrong to say that Mach was sceptical of laws or abstract principles and believed only in economical lists of particulars. His view of economy is much more subtle than that (see Banks 2004).

In *Knowledge and Error*, Mach suggested that enquiry took many paths, calling his book a series of 'sketches' of different techniques, some of which might lead to knowledge or else error depending on the circumstances of the problem, most of which are out of our control:

> Knowledge and error flow from the same mental sources, only success can tell the one from the other. A clearly recognised error, by way of correction, can benefit knowledge just as a positive piece of knowledge can.
>
> (Mach 1905/1976, p. 84)

As he says pointedly, this catalogue of approaches is all we have: there are *no codified scientific methods* guaranteed to lead to truth. Notice that knowledge and error flow along the causal links between our thoughts and ideas and circumstances in which an expectation is met or thwarted or even found to be illusory. Mach holds a causal theory of knowledge and not a propositional theory, in which a proposition has an intrinsic power to picture a state of affairs beyond itself and to correspond truly or falsely to it. James, who also rejected any intrinsic intentionality for mental images and judgements, held a very similar view, first presented in his 1894 address 'The Knowing of Things Together', later printed as 'The Tigers in India' in the *Meaning of Truth*:

> The pointing of our thought to the tigers is known simply and solely as a procession of mental associates and motor consequences that follow on the thought and would lead harmoniously, if followed out, into some kind of ideal context, or even into the immediate presence of the tigers. It is known as our rejection of a jaguar if that beast were shown to us as a tiger; as assent to a genuine tiger if so shown ... In all of this there is no self-transcendency in our mental images taken by themselves. They are one physical fact; the tigers are another; and their pointing to the tigers is a perfectly commonplace physical relation ... I hope you may agree with me now that in representative knowledge there is no special inner mystery, but only an outer chain of physical or mental intermediaries connecting thought and thing.
>
> (James 1977, pp. 155–156)

I have presented my own reconstruction of how I think James analyses away phenomenal intentionality in chapter 3 of Banks (2014). Incidentally, Bertrand Russell, who also converted to neutral monism, never fully accepted this epistemology or naturalistic theory of meaning and reference. He still believed that one needed images and image propositions to 'mean' non-present states of affairs, and to correspond to them truly or falsely, of which the images could not be considered direct copies or memories. There is thus a residue of propositions and propositional judgement in Russell that was never eliminated (for which see Banks 2014, chapter 4).

I have also argued elsewhere that it is a mistake to attribute the modern observable–unobservable distinction to Mach (Banks 2013). Mach emphasised a different distinction based upon causal continuity with experience. Unobservable elements are at least causally continuous with observation and allow for analogies, and even the assumption of the unobservable, if that can be shown to play a role in extending organising experience. Mach said as much in his famous discussion with Einstein in 1907 (related in Frank 1947). What Mach rejected as 'metaphysical' were:

(1) An isolated *Ding an sich* with no possible causal connection to other experiences. He originally thought atoms were like this, forever unobservable things of thought. When he began to think of them as continuous with observation, his opinion seemed to change, although he never

regarded them as ultimate. Mach probably would have been much happier with Heisenberg's abstract matrix atom made up of energy transitions, very much like his elements, in fact.

(2) Intuitive pictures or visualisations based on human psychology being substituted in thought where human intuition could not be expected to reach. Again, remember his remark in 'Symmetrie' about a natural system with more degrees of freedom than will fit into three-dimensional space. Mach wanted to purge physics of psychological visualisations, substances, extended space, and time, which interfered with its content, and he called this 'metaphysics'.

Conclusion: The Two Traditions of Mach Interpretation and Their Relation to Contemporary Philosophy of Mind

So, to return to the beginning, there are two main strands of Mach interpretation. The first 'phenomenalist' strand runs from Lenin, Schlick, Carnap, and Popper, to John Blackmore and many scientific writers and historians of ideas. In fact, to put things in the strongest terms, most historians of philosophy, and the vast majority of philosophers (and scientists, unfortunately), still associate Mach's views closely with the Vienna Circle as a primitive early form of what later became twentieth-century Logical Positivism. I wish to add also that our understanding of the Vienna Circle has advanced considerably in recent years, however, and the old 'received view' of Logical Positivism (e.g. in Suppe 1977) is currently under attack. Friedrich Stadler (2001) has questioned the Mach–Vienna Circle association in some detail, and Thomas Uebel (2007) has set apart the *Protokollsatz* debate in the Vienna Circle from anything having to do with Mach's elements. Other active investigations seek to link Mach to European pragmatism, or even Nietzsche's evolutionary naturalism (see Gori 2009), rather than with positivism. I hesitate to declare this positivist strand *dead*, since so many people still believe it and articles still often appear declaring Mach a phenomenalist. It is like a myth that has become so widely believed it is not possible to change anyone's mind who has not looked into the matter. But these are usually superficial accounts using Mach for some subsidiary purpose rather than interpreting him directly, and usually citing the same recycled quotes from *The Analysis of Sensations* or the *Mechanics* that are printed and reprinted in historical anthologies, but rarely going into details or citing the secondary literature at all.

The second 'neutral monist' strand of interpretation I have been defending here runs from Mach, James, and Russell to the American Realists (see Banks 2014, 2019). Paul Carus and Hans Kleinpeter also made a special effort to insist that Machian elements were *really* physical and that the realistic side of his work had to be fully acknowledged for any understanding of his views, as Russell understood so well. All of these early writers at least seem to have grasped the realistic tenor of Mach's neutral elements. For example, Ralph Barton Perry, one of the founding American Realists, claimed that 'Mach's book *The Analysis of Sensations* deserves to be numbered among the classics of realism', *not* positivism (Perry 1925, p. 79).

After Russell, neutral monist ideas lived on in the work of Wilfrid Sellars and Herbert Feigl, but the recent revival of neutral monist ideas in the philosophy of mind takes its cue from Russell's *Analysis of Matter*. Recently, however, this tradition has been traced back further to Mach and James (e.g. by Stubenberg 2016). Michael Lockwood and Grover Maxwell are most directly responsible for reviving Russellian ideas in philosophy of mind. Russellian monism, as it is called, was first offered as an alternative to Saul Kripke's modal arguments against physicalism in *Naming and Necessity* and later against David Chalmers' 'zombie' argument in *The Conscious Mind* (1996). The basic idea behind both arguments is that if pain and C-fibres are both rigid designators, identifying the same state across all possible worlds, or counterfactual situations, then the identity should be necessary, but there should *also* be no conceivable way to imagine their non-identity. Another way to put it *à la* Chalmers is that if two rigid designators such as pain and C-fibres agree in their secondary intensions, or references, they should also agree in their primary intensions, or meanings. This is not true for, say, the meaning of non-rigid 'watery stuff': clear, colourless liquid in lakes and ponds which might pick out H_2O on Earth and XYZ on Twin Earth in a counterfactual situation. By contrast, really grasping what it means to be in pain is the same thing as *being* in pain, while grasping the meaning of C-fibres is just knowing what C-fibres *are*, and thus there should be a conceptual identity between the meanings, too, however complicated, which someone could potentially work out through the concepts alone – on paper, as it were – even if the derivation was extremely complicated. Since this is clearly not the case for the identity between pain and C-fibres, they are not identical and physicalism is false.[10] So runs that argument.

Roughly, Russellian monists respond to this valid – but I think unsound – argument by defending a kind of enhanced physicalism in which the outer structural aspect of physics is actually grounded, or realised, in some kind of qualitative interior or categorical properties of physical objects. Hence, when you physically designate C-fibres, you might fail to designate their interior aspect of pain, as Maxwell argued (1978). Russellian monists then argue that as the atoms and molecules combine physically, their qualitative interiors, or categorical properties, also combine internally into minds. Sometimes these views veer off into panpsychism, attributing proto-phenomenal qualities to all bits of matter (as we saw in the early Mach before he embraced neutral monism).

---

[10] The focus on conceivability is a bit of a red herring, since the real issue is the existence of an a priori conceptual connection between the secondary intensions.

Chalmers and Galen Strawson have both offered compositional panpsychist versions of Russellian monism. Both authors also think qualities of experience imply awareness or an ego to behold them, thus I do not hold this view to be a neutral monist view. In my 2014 book defending neutral monism, I rejected *both* the panpsychism and also the a priori composition of minds out of proto-phenomenal qualities of matter, which I believe to be the more basic error. Yet, despite these differences, the realist strand of neutral monism lives on today in a variety of forms. Who knows, just as Mach once evolved from a panpsychist to a neutral monist and discarded the ego on the way, the present movement may end up retracing his steps, and soon we might even see a revival of Mach's work parallel to Russell's.

References

Adler, Friedrich 1908. 'Die Entdeckung der Weltelemente: Zu Ernst Machs 70. Geburtstag', *Der Kampf* 5: 231–240.

Banks, Erik C. 2003. *Ernst Mach's World Elements: A Study in Natural Philosophy.* Kluwer.

2004. 'The Philosophical Roots of Ernst Mach's Economy of Thought', *Synthese* 139: 23–53.

2013. 'Metaphysics for Positivists: Mach versus the Vienna Circle', *Discipline filosofiche* 23: 57–77.

2014. *The Realistic Empiricism of Mach, James, and Russell: Neutral Monism Reconceived.* Cambridge University Press.

2019. 'Empiricism or Pragmatism? Ernst Mach's Ideas in America 1890–1910', in F. Stadler (ed.), *Ernst Mach – Life, Work, Influence* (Vienna Circle Institute Yearbook). Springer, pp. 485–500.

Bhattacharya, Manjulekha 1972. 'Ernst Mach: Neutral Monism', *Studi Internazionali di Filosofia* 4: 145–182.

Blackmore, John T. 1972. *Ernst Mach: His Work, Life, and Influence.* University of California Press.

Blackmore, John T. (ed.) 1992. *Ernst Mach – A Deeper Look: Documents and New Perspectives.* Kluwer.

Carnap, Rudolf 1967. *The Logical Structure of the World.* University of California Press.

Carus, Paul 1893. 'Professor Mach's Term "Sensation"', *The Monist* 3: 298–299.

1906. 'Professor Mach's Philosophy', *The Monist* 16: 331–356.

Chalmers, David 1996. *The Conscious Mind.* Oxford University Press.

2002. 'Consciousness and its Place in Nature', in D. Chalmers (ed.), *Philosophy of Mind: Classical and Contemporary Readings.* Oxford University Press, pp. 247–272.

Dennett, Daniel C. 1988. 'Quining Qualia', in A. J. Marcel and E. Bisiach (eds.), *Consciousness in Contemporary Science.* Oxford University Press, pp. 42–77.

Feyerabend, Paul K. 1970/1999. 'Philosophy of Science: A Subject with a Great Past', as reprinted in his *Knowledge, Science and Relativism: Philosophical Papers, Volume 3*, ed. J. M. Preston. Cambridge University Press, 1999, pp. 127–137.

    1984. 'Mach's Theory of Research and Its Relation to Einstein', *Studies in History and Philosophy of Science* 15: 1–22.

Frank, Philipp 1947. *Einstein: His Life and Times*. Knopf.

Frank, Stephanie F. 2001. *Symmetry in Mechanics*. Birkhäuser.

Gerhards, Karl 1914. *Machs Erkenntnistheorie und der Realismus. Münchner Studien zur Psychologie und Philosophie*. Verlag W. Spemann.

Gori, Pietro 2009. 'The Usefulness of Substances: Knowledge, Science and Metaphysics in Nietzsche and Mach', *Nietzsche-Studien* 38: 111–155.

Haller, Rudolf and Stadler, Friedrich (eds.) 1988. *Ernst Mach: Werk und Wirkung*. Hölder-Pichler-Tempsky.

Heidelberger, Michael 2004. *Nature from Within: Gustav Theodor Fechner and His Psychophysical Worldview*. Pittsburgh University Press.

Hentschel, Klaus and Blackmore, John T. (eds.) 1985. *Ernst Mach als Aussenseiter*. Braumüller.

Holt, Edwin B., Marvin, Walter T., Montague, W. P., et al. 1910. 'Program and First Platform of Six American Realists', *Journal of Philosophy, Psychology and Scientific Methods* 7: 393–401.

James, William 1977. *The Writings of William James*. Ed. J. J. McDermott. University of Chicago Press.

Janssen, Michel 2014. '"No Success Like Failure . . .": Einstein's Quest for General Relativity, 1907–1920', in M. Janssen and C. Lehner (eds.), *The Cambridge Companion to Einstein*. Cambridge University Press, pp. 167–227.

Kleinpeter, Hans 1906. 'On the Monism of Professor Mach', *The Monist* 16: 161–168.

Lenin, Vladimir I. 1952. *Materialism and Empirio-Criticism*. World Publishing House (orig. 1909).

Lockwood, Michael 1989. *Mind, Brain and The Quantum: The Compound 'I'*. Oxford University Press.

Mach, Ernst 1863. 'Vorträge über Psychophysik', *Österreichische Zeitschrift für praktische Heilkunde* 9: 146–148, 167–170, 202–204, 225–228, 242–245, 260–261, 277–279, 294–298, 316–318, 335–338, 352–354, 362–366.

    1871. 'Über die physikalische Bedeutung der Gesetze der Symmetrie', *Lotos* 21: 139–147.

    1872/1910. *History and Root of the Principle of the Conservation of Energy*. Open Court.

    1883/1960. *The Science of Mechanics: A Critical and Historical Account of its Development*. Open Court.

    1886/1959. *The Analysis of Sensations*. Dover.

    1895. 'On the Economical Nature of Physical Inquiry', in his *Popular Scientific Lectures*. Open Court, pp. 186–213.

    1905/1976 *Knowledge and Error*. D. Reidel.

Maxwell, Grover 1978. 'Rigid Designators and Mind–Body Identity', in C. Wade Savage (ed.), *Minnesota Studies in the Philosophy of Science*, vol. 9, pp. 365–403.

Nagel, Thomas 2000. 'The Psychophysical Nexus', in P. Boghossian and C. A. B. Peacocke (eds.), *New Essays on the A Priori*. Clarendon Press.

Perry, Ralph Barton 1925. *Present Philosophical Tendencies*. Longman's, Green and Co.

Popper, Karl R. 1953. 'A Note on Berkeley as a Precursor of Mach and Einstein', *British Journal for the Philosophy of Science* 4: 26–36.

Quine, Willard van O. 1966. 'Russell's Ontological Development', *Journal of Philosophy* 63: 657–667.

Russell, Bertrand 1984. *Collected Papers Vol. 7 Theory of Knowledge: The 1913 Manuscript*, eds. E. R. Eames and K. Blackwell. Allen & Unwin.

    1914. *Our Knowledge of the External World*. Allen & Unwin.

    1918. *Lectures on Logical Atomism*. Open Court.

    1919. 'On Propositions: What They Are and How They Mean', *Proceedings of the Aristotelian Society, Supplementary Volume* 2: 1–43.

    1921. *The Analysis of Mind*. Allen & Unwin.

    1927. *The Analysis of Matter*. Kegan Paul.

Stadler, Friedrich 2001. *The Vienna Circle – Studies in the Origins, Development, and Influence of Logical Empiricism*. Springer.

Strawson, Galen 2006. 'Realistic Monism: Why Physicalism entails Panpsychism', *Journal of Consciousness Studies* 13: 3–31.

Stubenberg, Leopold 2016. 'Neutral Monism', *The Stanford Online Encyclopedia of Philosophy*, updated November 2016. Available from http://plato.stanford .edu/archives/win2016/entries/neutral-monism/

Suppe, Fred (ed.) 1977. *The Structure of Scientific Theories*. University of Illinois Press.

Uebel, Thomas 2007. *Empiricism at the Crossroads: The Vienna Circle's Protocol Sentence Debate*. Open Court.

Wolters, Gereon 1987. *Mach I, Mach II und die Relativitätstheorie: Eine Fälschung und ihre Folgen*. Walter de Gruyter.

INDEX

Printed in the United States
by Baker & Taylor Publisher Services